EVOLUTIONARY APPROACHES IN THE BEHAVIORAL SCIENCES: TOWARD A BETTER UNDERSTANDING OF HUMAN NATURE

RESEARCH IN BIOPOLITICS

Series Editors: Steven A. Peterson and Albert Somit

Volume 1: Sexual Politics and Political Feminism, 1991

Volume 2: Biopolitics in the Mainstream, 1994

Volume 3: Human Nature and Politics, 1995

Volume 4: Research in Biopolitics, 1996

Volume 5: Recent Explorations Bio and Politics, 1997

Volume 6: Sociology and Politics, 1998

Volume 7: Ethnic Conflicts Explained by Ethnic Nepotism, 1999

RESEARCH IN BIOPOLITICS VOLUME 8

EVOLUTIONARY APPROACHES IN THE BEHAVIORAL SCIENCES: TOWARD A BETTER UNDERSTANDING OF HUMAN NATURE

EDITED BY

STEVEN A. PETERSON

Pennsylvania State University, Harrisburg, USA

ALBERT SOMIT

Southern Illinois University at Carbondale, USA

2001

JAI imprint

An imprint of Elsevier Science

Amsterdam – London – New York – Oxford – Paris – Shannon – Tokyo

ELSEVIER SCIENCE Ltd
The Boulevard, Langford Lane
Kidlington, Oxford OX5 1GB, UK

First edition 2001

Library of Congress Cataloging in Publication Data
A catalog record from the Library of Congress has been applied for.

British Library Cataloguing in Publication Data
A catalogue record from the British Library has been applied for.

ISBN: 0-7623-0769-2

∞The paper used in this publication meets the requirements of ANSI/NISO Z39.48-1992 (Permanence of Paper).
Printed in The Netherlands.

CONTENTS

FOREWORD

John C. Wahlke

This is a book about the state of social science, with a chapter on each discipline by an author or authors experienced in it. All share the viewpoint, often called "biobehavioral," which is a minority viewpoint in all of the disciplines discussed. It is not their purpose in this volume to argue the case for that viewpoint. It has been well stated and discussed within the respective disciplines many times elsewhere. Nor is it the aim here to convert anyone to share in applying that viewpoint in teaching and research, and on ways of dealing with them. Their ultimate goal, like that of all scientists, social and otherwise, is the advancement of knowledge. The following brief discussion and comments on the place of "the biobehavioral movement" in social science does little more than describe a few major landmarks in its development and offer a thumbnail sketch of its intellectual framework. But the authors will fill in more on that.

BACKGROUND

"Traditional" Social Science

Today's social sciences in the United States and Western Europe developed out of the structure of knowledge (curricular and organizational) of the mid-nineteenth century major European universities, which relied mainly on History, Philosophy, Jurisprudence, and later Economics. American universities adopted much the same institutional framework in the latter part of the nineteenth century, and separate disciplines of Economics, Political Science, Anthropology, Sociology, and others evolved essentially at that time.

The basic conception of knowledge in some disciplines – e.g. Political Science and Sociology – remained "traditional," i.e. within a framework based primarily on History, Philosophy, and Jurisprudence, the disciplines from which they had emerged. Others – e.g. Economics, Psychology, and Anthropology – were shaped more by intellectual developments of a more recent vintage.

In the more "traditional" disciplines, especially in the United States, there was from the beginning a significant criticism of formalistic and abstract conceptions of the subject, and a demand for attention to observable events as the object of disciplinary study. Woodrow Wilson (1885) took his own advice and analyzed roll call voting in the American Congress. By the 1920s, the perspective of most American social sciences had shifted clearly toward more concern with empirical analysis of contemporary as well as historical phenomena, and formal institutions and structures. From this time on, European universities were more often following American patterns rather than the other way around.

The events leading up to World War II and the war itself were influential in changing social science perspectives again. In most disciplines, there was a "behavioral revolution," which relied on "methodological individualism" – making the observed behavior of individual human actors the basic "unit of analysis," instead of abstract "collective actors" like political parties, interest groups, legislative bodies, nation-states, or the United Nations. The resulting perspective, combining the empirical premise with acceptance of the observed actions of individual human beings as the only observable evidence of a "group action" or "institutional action," is still the prevailing social science perspective today.

Events of the mid to late 1960s led in some disciplines to a "post-behavioral revolution." Post-behavioralist critics found mainline social science research and teaching trivial, irrelevant to basic concerns and feelings of people about events in their everyday world. Economics and political science were the most severely criticized for this. The authors in this volume represent the most recent criticism of social science, which emerged more or less contemporaneously with the "postbehavioral" revolution. All are established and respected scholars in their own fields and beyond. All of them share the conviction that knowledge in their disciplines is seriously flawed, most of all by scholars' failure to recognize and apply the increasing volume of established knowledge from the natural sciences. Each of the book's chapters presents an author's or co-author's view of their own discipline's dealing with this problem. A brief description and some brief comments on the "biobehavioral" movement may help set the stage for the reader.

THE BIOBEHAVIORAL MOVEMENT

The idea of drawing on biology for fuller understanding human behavior was first expressed in 1908 by Graham Wallas (1908), who was greatly influenced by Darwin (as were many other scholars of his time) and by the psychologists William James and William MacDougall. He argued against what he saw as

the " 'excessively anti-intellectualist notion' that man's main motives are consciously held goals, [and] emphasized, in addition to instincts, both habits and impulses" (Stone & Smith, 1983: 699). The British political scientist W. J. M. Mackenzie was the next to argue explicitly that political scientists should view their field as "biological politics" (Mackenzie, 1967).

Maurice Duverger (1966) included a section on "Biological Factors: Politics in Animal Kingdoms" in a chapter entitled "The Elements of Conflict" in his book, *The Idea of Politics*. But Albert Somit's article in the *Midwest Journal of Political Science* (Somit, 1968) is usually identified as the beginning of a biopolitics "movement" within political science.

Similar "movements" followed within a relatively short time in most other social science disciplines. In psychology, experimental psychology was already a field utilizing scientific concepts and methodology. A social psychology subfield early on introduced the theory and concepts of Freud and the other psychoanalysts, a version of science that has almost disappeared, replaced by the concepts described below. In sociology, a more important and lasting influence has been from sociobiology, not as a "field" within sociology but a more fundamentally scientific perspective, as first stated by E. O. Wilson in 1975 (Wilson, 1975). In his *Consilience: The Unity of Knowledge* (Wilson, 1994), he argued for even tighter integration of all knowledge about human behavior into a unified theoretical perspective. Evolutionary psychology was also among the early "biosocial sciences." Economics was among the latest. Within each discipline, the number of scholars taking the biosocial science approach has apparently increased slowly but fairly steadily. Informal gatherings now meet at professional conferences, but with relatively little formalization, except in political science, as described below in the chapter by Somit and Peterson.

CRITICISMS OF CONTEMPORARY SOCIAL SCIENCE

Obsolete Conception of the Mechanics of Human Behavior

The governing social science conception of the human actor is based analogically on the physics of moving objects. Its model, drawn from systems theory, is sometimes depicted as human-like being (called a *homunculus*) sitting inside a person's skull at the controls of a vehicle-like machine. The basic mechanics of the process is a mélange of concepts drawn from sociology and social psychology of the 1940s and 1950s. On the one hand, there are "interests," "affect," "beliefs," "ideas," and "cognitions" combined by Reason into decisions about steering, shifting, goal setting, etc. involved in taking a drive. On

the other hand, there are voting, participation, aggression, violence, and other behaviors.

All these words are used, however, without any consensual definition. They are, in effect, reified (i.e. they are operationalized by different users as if they were names of "things", although they refer only to some notion in the mind, and may mean something totally different to other users). That procedure is "science upside down"; science has to start with observable problems and looks for theories and concepts with which to explain them. But there can be no theory linking together "concepts" whose meaning varies from report to report and reader to reader. The inevitable result is what has been called the state of "total disarray of theory" in the social sciences (in the scientific, not the political theory, sense).

Biobehavioral critics see the root of the problem in the essentially 18th century rationalistic premises of contemporary social science conceptions of the mechanisms of human behavior (*vide* Graham Wallas' remarks just after the turn of the century). They find it even more untenable in light of: (a) the relevant knowledge being rapidly accumulated by natural science investigators, and (b) the unsatisfactory and chaotic results of present-day social science to produce meaningful results.

Professor Max Kaase has expressed this latter criticism more trenchantly and in more general terms in a prescient article analyzing the impact of the rapid development of the Internet and the World Wide Web on "the science system" in general and the social sciences and humanities in particular (Kaase, 2000: 272):

> Probably the most important point made in this chapter is that the social sciences are not cumulative and at the same time selective, as are other fields: they lack the generally accepted scholarly criteria by which research findings, such as in the natural sciences, can be eliminated if according to the present state of the discipline they are no longer valid . . . Thus in the social sciences not only is there an information overload, but there is also an element of arbitrariness immanent in scholarly debates: in principle, anything goes. Nor are there institutionalized penalties for ignoring previous evidence and research, and so the accumulation of knowledge is slow and cumbersome, if it takes place at all. . . .

IGNORANCE AND NEGLECT OF RELEVANT SOCIAL SCIENCE

Biobehavioralists attribute the social sciences' theoretical incoherence and the resulting triviality and irrelevance of much of their work to what they see as persistent, almost willful neglect of natural science knowledge. The task of biobehavioralist critics to cope with the problem has been most clearly and

succinctly stated by James Chowning Davies, an early pioneer in the movement (Davies, 1983: 731):

> To avoid trivializing or sensationalizing biopolitics, we have to specify how biology is relevant to which political phenomena . . . In addition to neurophysiology, ethology and now genetics eventually will help explain political behavior in which the organic component is of major significance.

As the following section suggests, the increasing breadth of scientific perspective and the growing communality of viewpoint among biobehavioral social scientists across disciplines indicates that Davies' advice is being followed. As the first biobehavioral work spanning the social sciences to address the problem of their theoretical disarray, this volume may itself be a landmark step in the biobehavioral movement.

BIOBEHAVIORAL SOCIAL SCIENCE: A THUMBNAIL SKETCH

This section attempts to provide a quick summary of the substantive framework of biobehavioral social scientists' analyses of human being's behavior. That framework is not a packaged body of knowledge, of explanations about events or behavior, but an instrument of identifying the problems confronting human beings in society and seeking where to look for explanations of their behavior. Like any science, it does not claim to have discovered the ultimate truth, only the best explanations yet attained. It is well described by Professor Dov Ronen as a new pair of spectacles, radically different from the traditional social science perspective (Ronen, 2000). But it requires more than just providing an inventory or seeking whatever particular "findings" might be used for the purposes at hand, and more than applying findings analogically to social science topics. The natural sciences are used as genuine knowledge, providing scientific explanations of behavior to be taken seriously as such.

Evolutionary Perspective

The focus of Darwin's work was the species, not the individual members of them. The unfortunate misrepresentation and misuse of the term "survival of the fittest" by "Social Darwinism" prejudiced generations of non-biological scholars against Darwin's work. Only recently are they coming to realize the significance of that distinction. Meanwhile, neo-Darwinians have extended the evolutionary focus to take in all living creatures, including the most microscopic forms. They have also stimulated the development of derivative and related "life sciences," like the neurosciences, and the theoretical linkage with

other fields of knowledge, including cognitive science and information processing.

Ethology: Pre-Programmed Behavior

The pioneering work of Konrad Lorenz (1966) and Iraenaus Eibl-Eibesfeldt (1970) was rapidly picked up and (in many cases popularized by) a number of social scientists and others, including Lionel Tiger (1971) and Desmond Morris (1969), and others. Generally speaking, ethology as a science accepts the viewpoint of evolutionary biologists about the origins and development of all living creatures, including microscopic, but concerns itself primarily with the behavior of many different species of animals (including *Homo sapiens*), insects, fish, and birds. Its concern is with the behavior of these, however, not their morphology. Although most people tend to think of "evolution" in terms of inherited physical characteristics, Darwin himself did demonstrate an ethological interest in *The Expression of Emotions in Animals and Man* (1872), published over a decade after his *Origin of Species* (1859).

Pioneers in what became the science of ethology observed and studied numerous species-specific fixed patterns of behavior of various animal and insect species, including *Homo sapiens*. These are behaviors that: (1) are displayed by all individuals in almost every species of creature observed over time, and (2) change very little or only slowly over time for the species. The focus of ethology is on those behaviors that are associated with the individual's coping with the environment and with other members of the species. Many of them could be described as "instinctive," that is, ingrained and not learned – such as recoiling from horrifying stimuli.

Ethologists have identified two additional types of "pre-programmed" ("hard-wired") behavior of direct concern to social scientists. One is species-distinctive behavior patterns relating to individuals' social relations, or ritualistic forms of behavior in higher-order primates or moose and elks, for example, competition among males for mates. Formation of human groups, communities, and societies likewise involves different persistent behaviors of this type.

The other type of pre-programmed behaviors of interest to social scientists consists of psychophysical and psychophysiological reactions to stimuli from the senses and certain emotions sensed or "felt" internally. Among such behaviors, exhibited by all individual human beings of every nationality, tribe, sex, etc. studied are measurable: bristling of hair on the head and nape of the neck when frightened or threatened; dilation of the pupils of the eye when sexually aroused or when viewing a desirable individual of the opposite sex. Measurable elevations in blood pressure, pulse rate, galvanic skin response, and many other

measurable reactions, indicate the degree of arousal induced by given stimuli. Many social scientists besides the biobehaviorally inclined have measured these responses in experimental research to anchor interpretations of respondents' answers in survey research.

It is difficult to deny the accuracy of the ethologists' description of many human behavioral phenomena. Higher primates were one of the species examined by early ethologists, who not only drew analogies of gorilla and chimpanzee behavior to human behavior but also began examining human behaviors directly. The alimentary functioning of all human organs from ingestion through digestion through elimination is clearly not something we have in any sense "learned." Nor are functions like walking and reproduction learned. But many people, including social scientists, refuse even to consider the possibility that human beings display such pre-programmed behaviors, much less the fact that many such patterns for human beings are virtual replicas of high-order primates' patterns. It is, of course, still more difficult for many laypersons and social scientists to accept the fact that genetic studies of DNA, RNA, and the human genome have found the human species to share 98% of the genetic structure of higher-order primates.

Many people, including some social and natural scientists, objected on the grounds that: (a) the human species is the only creature with a Soul, Mind, and Spirit, and, therefore, (b) human behavior is not properly subject to ethological or any other scientific analysis. This objection is essentially ontological and religious, and amounts to outright rejection of all principles of science in at least this particular realm. Biobehavioral social scientists cannot deal with this as with scientific objections. It is not surprising that, despite the fact that Human Ethology is now a subfield of Ethology, traditional social science has yet to recognize the application of ethology to the study of human behavior. Its conception of behavior still begins and ends with conscious decisions, with no way to account for preprogrammed or hard-wired behavior.

Other Relevant Natural Sciences: The Brain and Behavior

Work already being done demonstrates that knowledge from the neurosciences, cognitive science, and related biological sciences, along with ethology, can provide a much surer, empirically validated model of the "machinery" behind human behavior. The starting point is the nervous system, of which the brain is the central element (Plato was the first to locate the seat of "Mind" in the skull).

Different parts of the brain are now recognized to be associated with different kinds of mental activity (e.g. MacLean, 1972; and see Ronen, n.d.).

Neuroscientists today "map" "information" received by sensory organs into particular areas in the brain specialized for different functions. They are also able to identify sites within the brain at the level of individual cells. While much has been learned about this "machinery," there is still much more to be learned. The knowledge already available has greatly increased our understanding of the processes by which human beings (and many other creatures) process sensory input from the environment into short-term memory in the brain, store some of it in long-term memory, recall it back into short-term memory as needed, and send information based on that through the nervous system and thence to the muscles and other appropriate parts of the human *soma*.

The point of this brief description is not to supply details but to emphasize that all the facts about and principles of such knowledge about the "machinery" of the brain, established by natural scientists in biology, and the neurosciences are much better established than the conceptions of "mind" on which most social science depends. The new "biobehavioral lenses" do not provide an explanation of human beings' behavior. They are the viewpoint, the perspective from which to view the observed behavior to be explained.

BIOBEHAVIORALISM IN SOCIAL SCIENCE TODAY

In many ways, the social sciences have changed very little over the post-war years. The most obvious exception to the statement is the very great advance in the technology of research, which, except in the case of psychology, means primarily in the level of mathematical and statistical techniques used, and the development of survey research as a primary social instrument (in social psychology). Experimental psychology is still an important component in that field. Perhaps the most notable is development of rational-choice (or positive) theory by the economists, and its application to other disciplines, political science in particular. Even in economics, however, that approach is not always integrated into the rest of the discipline. On the whole, the outmoded rationalistic conception of the processes governing behavior in social science is still predominant.

The biobehavioral component in the social sciences has in some ways changed much and in some important ways not much. But it has grown fairly steadily, if slowly, and lost few defectors. Its scholars have produced an increasing volume of research, whose scientific and scholarly quality has notably improved steadily by all standards. On balance, though, it has made very little progress in terms of the legitimacy of its enterprise among its peers or its recognition by academic administrators from the department level up.

Research and Publications

It is impossible to summarize or evaluate the corpus of theory and research generated so far by biobehavioral scholars. The general impression, which is the least reliable evidence, is that, while the number of active, productive biobehavioral social scientists in each discipline is relatively small, the results of their efforts compares favorably with those of traditional social science. Their published works appear less frequently in journals of their own disciplines than do works of their disciplinary colleagues, but much more often in journals of other disciplines (including some in natural sciences) and multi-disciplinary journals. Unlike published articles in professional journals generally, their citations in biobehavioral works are not limited to works primarily within their own discipline, but include a much broader set (based on an unpublished study by this author).

Two articles from a recent issue of *Science* illustrate the theoretical coherence of some biobehavioral research and show how the biobehavioral perspective identifies and deals with "relevant" problems, important in terms of value to theory and in terms of touching genuine personal concerns of many human beings. One deals with the role of environmental and biological factors in drug addiction. It reports a laboratory study of mice that indicates that addiction of different strains of mice was reducible by environmental changes (food shortage). It concluded that, "Consequently, to identify biological determinants of behavior, an integrated approach considering the interaction between environmental and genetic factors needs to be used" (Cabib et al., 2000: 463).

The second article deals with the study of the parts of the human brain that process reasoning. It bears on the longstanding controversy among social scientists, educators in general, and much of the general public over so-called intelligence tests. In academic circles the dispute is between Jensenists (proponents) and anti-Jensenists (opponents) of tests for measuring "IQ" or "g" (for general intelligence) as they were named by Spearman (1927). The study (Duncan et al., 2000) found that IQ-type tests measuring the cognitive demands of three different types of skill (analytic, creative, etc.) are all "associated with selected recruitment of [processes located in] the [four lobes of] the lateral frontal cortex in one or both hemispheres of the brain" (Duncan et al., 2000: 457). Scientific understanding of such brain processes is the "pure science" goal of such research. The findings of such studies have implications not only for Jensenists and anti-Jensenists but for anyone claiming to know what human reason "is."

This particular case also illustrates the important principle that the truth-value of all scientific knowledge is always a probability less than 1.00, however close

it might come. The cited article tested particular elements of then generally established knowledge and justified changing it. An introductory article points out specific points of disagreement between the main article's authors' and the prevailing views, concluding that "The results of Duncan et al. provide a holey grail rather than the Holy Grail because as yet they have not provided the whole grail" (Sternberg, 2000: 399). Together the two articles provide an exemplary model of scientific thinking for social scientists.

Another important example of opening new ground is the work of Milton Lodge and his colleagues at the Laboratory of Behavioral Research at the State University of New York at Stony Brook on the "mental machinery" involved in voting and other decisions by individual human beings. Cognitive and neuroscientists have hypothesized that the key indicator on a "tag" is not substantive like, say, the alphabetized location of a word in a dictionary, but a "value," an elementary rating of the "goodness" or "badness" of the item for the individual. Lodge's work supports and works from this conception. The initial value tag of "Clinton," for instance, when first stored in memory would be the averaged value tags of any items linked to "Clinton" in the initial stimulus and to those readily recalled from long-term memory with which to compare it for interpretation. The key technology for research involved experiments (in milliseconds), the lapse of time between the administration of a stimulus (e.g. visual stimuli might include pictures, words, etc. on a card), and response to whatever question was asked or task was to be performed). Previous work has shown that the more "information" that must be recalled from long-term memory (the more cognitive activity that is required), the longer the time to response. It is easy to see how this might suggest research to test hypotheses about how voters choose among candidates.

Professional and Institutional Recognition

At the professional level, individual members have gained some professional reputation by organizing panels and presenting papers at annual professional association meetings. But in general they are rarely given roles in the planning and organization of those meetings themselves. But in the bulk of instances it is treated by program committees, as by colleges, divisions, and departments, as a special topic, not yet even a subfield. In general, the (traditional) social science establishment tolerates individual biobehavioral rebels and affords them accommodations as social scientists, but not legitimacy as biobehavioral scientists.

There are no data available, but many biobehavioral scientists feel discriminated against, although not excluded in the allotment of funds and facilities

for research from their own institutions. Fortunately, their field is more often recognized as a legitimate scientific enterprise by many governmental and private foundation sources of support. They are thus supported more or less in proportion to their merits in a broader field of competition. A much more serious handicap is failure of departmental, college, and university support for their most important educational function: teaching their subject. With only the rarest exceptions, their perspective is not listed in departmental and college catalogs and course lists. Most of them teach standard introductory level and "regular" courses. Those few who get the rare opportunity to apply their perspective in their teaching usually do so on an ad hoc basis as an irregular element in the curriculum.

In Sum

Despite over thirty years of effort within their own disciplines, and despite demonstrating the value of their perspective in a "liberal" education, and the value of their research contributions, biobehavioral social scientists today are clearly still a fringe group in "main line" social science. Their situation in traditional social science departments can fairly be described as peaceful coexistence and mutual tolerance. They are given accommodations but not disciplinary legitimacy. Their relationships with their colleagues are still collegial and not fratricidal. But few of their colleagues are aware of what they are doing or saying, and have little genuine interests in hearing about it.

CONCLUSION

The question they leave, not for themselves alone to face, or for only social scientists, but for all who are concerned about education and about advancing knowledge, might seem to be, "What is to be done?" The obvious next question would be, "By whom?" But the prior question in either case is, "What is the problem?"

For individual biobehavioral scholars, the problem is how to cope persistently, day-by-day, with dispensing and advancing our knowledge of human behavior in a scholarly environment where academic problems are of little concern to the general public, and the scholarly environment is one in which they are essentially strangers. So far, they are successfully coping. The question for them is, "Will their efforts interest and attract sufficient students to follow in their footsteps to carry on their work?"

For the traditional social sciences, the question is, "Can they confront realistically, and soon, the magnitude of the intellectual challenges of modern science

to their academic way of life?" The real problem is not the biobehavioral gadflies, but science itself. And that problem is one confronting education at all levels and in all societies. Biobehavioral scientists may be more sensitive to the dimensions of the problem, but it is still a challenge that they, too, must face. Is their grasp of science, as individual scholars, adequate, and is it keeping up with science? Beyond that, are they prepared and able to face the twin challenge of reconciling their perception of the world with the non-scientific perceptions of the world presented in the Arts and Humanities? The term "Liberal Arts" suggests that the worlds of science and the Arts and Humanities are two separate, conflicting worlds. The term "liberal education" suggests that a well-rounded education provides scholars with a grasp of what both worlds are about.

Beyond these intellectual questions, whatever the answers may be, there are still important questions of strategy and tactics to be faced. The most important one is that faced in almost all areas of life in modern industrialized societies – the institutionalization and bureaucratization of basic social functions, of which education and the production of knowledge are certainly two important ones. The natural tendency of most human beings operating within that context is to resist changes of routinized behavior, and the usual result is inertia. Academics might do well to look closely at the forces that shape and homogenize the textbooks published, the influence of professional associations in structuring of knowledge, of teachers unions in the design of curricula and the delivery of knowledge.

The only conclusion seems to be that, just as our scientific knowledge is never better than highly probable, we should not expect to "answer" the questions or "solve" the problems outlined above. In both cases, we must live with the ancient Greek paradox that, if progress is made at the rate of 50% of the distance to a goal every day, that is no doubt "remarkable progress," but it also means that the goal will never be reached, however close one might get. Depending on how far away you start toward the goal, that should "satisfice" biobehavioral scholars for some time.

REFERENCES

Cabib, S. et al. (2000). Abolition and reversal of strain differences in behavioral responses to drugs of abuse after a brief experience. *Science*, *29* (21 July), 463–465.

Darwin, C. (1859). *The Origin of Species by Means of Natural Selection*. London: Murray.

Darwin, C. (1872). T*he Expression of Emotions in Animals and Man.*

Davies, J. C. (1983). The proper biological study of politics. *Political Psychology*, *4*, 731–743.

Duncan, J. et al. (2000). A neural basis for general intelligence. *Science*, *290* (21 July), 457–460.

Duverger, M. (1966). *The Idea of Politics*. Indianapolis: Bobbs-Merrill (transl. by R. North and R. Murphy).

Eibl-Eibesfeldt, I. (1970). *Ethology*. New York: Holt, Rinehart & Winston.

Kaase, M. (2000). Political science and the Internet. *International Political Science Review*, *21*, 265–282.

Lorenz, K. (1952). *King Solomon's Mines: New Light on Animal Ways*. New York: Crowell.

Lorenz, K. (1966). *On Aggression*. New York: Harcourt Brace.

Mackenzie, W. J. M. (1967). *Politics and Social Science*. Baltimore: Penguin.

MacLean, P. (1972). *A Triune Concept of the Brain and Behavior*. Toronto: University of Toronto Press.

Morris, D. (1969). *The Human Zoo*. New York: McGraw-Hill.

Ronen, D. (2000). The Human Revolution and the Emerging World Order. (manuscript).

Somit, A. (1968). Toward a more biologically-oriented political science *Midwest Journal of Politics*, *12*, 550–567.

Spearman, C. (1927). *The Abilities of Man*. New York: MacMillan.

Sternberg, R. J. (2000). The Holy Grail of general intelligence. *Science*, 290 (21 July), 399–401.

Stone, W. F., & Smith, D. C. (1983). Human nature in politics. *Political Psychology*, *4*, 731–743.

Tiger, L. (1971). *The Imperial Animal*. New York: Holt, Rinehart & Winston.

Wallas, G. (1908). *Human Nature in Politics*. London: Constable & Co.

Wilson, E. O. (1975). *Sociobiology: The New Synthesis*. Cambridge: Harvard University Press.

Wilson, E. O. (1994). *Consilience: The Unity of Knowledge*. New York: Knopf.

Wilson, W. (1885). *Congressional Government*. Boston: Houghton, Mifflin and Company.

ANTHROPOLOGY AND THE EVOLUTIONARY STUDY OF HUMAN BEHAVIOR

Lee Cronk

INTRODUCTION

Anthropology is a discipline with an unusually broad subject matter, spanning virtually the entire human experience, from the origins of our species through human prehistory to modern cultural diversity. A diverse subject matter implies a diversity of approaches, and anthropologists run the gamut from the most technical genetic research to personal, poetic accounts of other cultures. This diversity makes for an interesting and lively discipline, but also one with a history of sharp divisions over issues as fundamental as the definition of the field and its subject matter. Traditionally, anthropologists have dealt with these internal divisions by following the old adage that good fences make good neighbors, i.e. by dividing the discipline into largely autonomous subdisciplines. These days American[1] anthropologists usually conceive of four such subdisciplines: social, cultural, or sociocultural anthropology; biological or physical anthropology, which includes human paleontology and primatology; archaeology; and linguistic anthropology. Sometimes applied anthropology is included as a fifth subdiscipline.

Evolutionary Approaches in the Behavioral Sciences, Volume 8, pages 1–30.

ISBN: 0-7623-0769-2

EARLY DEVELOPMENT OF EVOLUTION AND BEHAVIOR STUDIES IN ANTHROPOLOGY

Evolution and culture have long been the two core concerns of anthropology. In the nineteenth century, major anthropological thinkers such as Lewis Henry Morgan, E. B. Tylor, and Herbert Spencer considered both of these issues at the same time, using archaeological and cross-cultural information to develop ambitiously broad accounts of human social and cultural evolution. This changed dramatically around the turn of the twentieth century when Franz Boas, the founder of anthropology as an academic discipline in the United States, pointed out that a great deal of theorizing was being done without very much information at hand (Boas, 1896). As a remedy, Boas eschewed theory and recommended long term immersion by ethnographers in different societies around the world. He also argued that the method used by the evolutionists, which relied heavily on analogies between contemporary societies and prehistoric ones, was on shaky logical ground. Contemporary societies, he argued, far from being living fossils of past societies, had their own particular histories that must be understood, and to do otherwise is to run the risk of being ethnocentric. Thanks to these sorts of arguments by Boas and his followers, not only the culture concept but also the closely related doctrines of cultural determinism and cultural relativism became central to the discipline of anthropology in general and to the subdiscipline of cultural anthropology in particular.

Outside the cultural subdiscipline, anthropologists retained evolution as a core concern. Archaeologists continued to ponder the long sequences of societal development they learned about through their excavations, and physical anthropologists continued to focus on hominid evolution and the causes of physical variations among contemporary humans.[2] An equilibrium of sorts developed between those anthropologists interested in evolution and those interested in culture. Evolutionary anthropologists were free to apply their principles to the study of hominids, prehistory, and even physical variation among contemporary humans, but for the most part they refrained from saying much about the behavior of modern humans. Behavior was considered to be within the realm of culture, and that subject was largely left to the cultural anthropologists.

This equilibrium began to be upset in the 1960s and early 1970s as a result of breakthroughs in the study of animal behavior. Some of these breakthroughs were empirical and methodological and are well represented by the work of people like Niko Tinbergen and Konrad Lorenz. Theoretical breakthroughs soon followed, with W. D. Hamilton's (1964) inclusive fitness theory leading the way. Hamilton pointed out that because related individuals share some of the same genes by virtue of the fact that they have common ancestors, selection

may sometimes favor genes that lead their bearers to do things that harm their own reproductive interests if in so doing they increase the reproductive success of relatives. This became the centerpiece of what is sometimes called "selfish gene theory" (Dawkins, 1976), the idea that we can gain great insight into behavior by focusing on the differential reproduction not of groups or individuals, but of genes. Williams (1966) contributed a key argument for why selection is likely in most circumstances to act most strongly through the differential survival and reproduction of individual organisms and the genes they carry rather than through groups of organisms. That same year, Robert H. MacArthur and Eric Pianka (1966) laid the foundations for the development of optimal foraging theory and the now widespread use of economic models in animal behavior studies. In a series of seminal articles in the early 1970s, Trivers (Trivers, 1971, 1972, 1974; Trivers & Hare, 1976; Trivers & Willard, 1973) extended selfish gene thinking to topics such as reciprocity, parental behavior, and sexual selection.

This new, evolutionary approach to behavior was introduced to anthropology by the iconoclastic work of such scholars as Lionel Tiger and Robin Fox (e.g. Tiger, 1969; Tiger & Fox, 1971), who suggested that such an approach might be very revealing if applied to humans, as well. A number of other anthropologists also took an early interest in this approach, including Earl Count, Irven DeVore, Derek Freeman, and Robert Hinde. While most of the early work was rather speculative, that began to change in the mid-1970s following several key publications by biologists. First among these was University of Michigan biologist Richard Alexander's 1974 article on evolution and social behavior, in which he gave detailed suggestions for how hypotheses derived from evolutionary theory might be tested with data on behavior from different human societies. Soon thereafter, E. O. Wilson's *Sociobiology: The New Synthesis* (1975) and Richard Dawkins' *The Selfish Gene* (1976) attracted more attention, and more people became interested in testing the new models.

Armed with these new ideas, a small group of anthropologists began to reanalyze cross-cultural data and data from their own fieldwork. Some of this early work was showcased in two sets of sessions at the 1976 meetings of the American Anthropological Association, one organized by Napoleon Chagnon and William Irons and the other by Irven DeVore. These sessions formed the basis for the landmark 1979 edited volume *Evolutionary Biology and Human Social Behavior: An Anthropological Perspective* (Chagnon & Irons, 1979). Other important early work included John Hartung's (1976) cross-cultural study of inheritance practices in light of the theory of natural selection and Bruce Winterhalder's (1977) study of hunting and gathering among the Cree of Canada, the first use of optimal foraging theory on data from humans. These

studies addressed many traditional concerns in anthropology, including kinship, marriage, and stratification. The methods and data were also largely of traditional anthropological varieties. Their key novel element was hypotheses derived from evolutionary biological theory: Are humans generally nepotistic? Does a tendency towards nepotism influence family organization and residential patterns? Do parents direct their investment in their offspring in ways that increase the number of their grandchildren? Do efforts to achieve high status enhance reproductive success? Are the choices people make about food acquisition, such as where to hunt, how long to hunt, and which species to hunt, efficient in terms of the amount of food acquired in the time available? Do patterns of food exchange among individuals and households reflect a concern with reciprocity?

As is the case with many new fields, it took time for a consensus to develop on what to call this approach. In the 1970s, many followed E. O. Wilson's lead and called it sociobiology, while a few used anthropologist Lionel Tiger's older label "biosociology," instead. These neologisms reflected a felt need to distinguish the new approach based on the ideas of Hamilton, Williams, and Trivers from such earlier approaches as cultural ecology, human ecology, and ecological anthropology (see Orlove, 1980 for a review). When Wilson's application of evolutionary theory to human behavior attracted a great deal of criticism (e.g. Sahlins, 1976; Lewontin et al., 1984; Kitcher, 1985), sociobiology became less popular as a label, even though the approach itself was flourishing. Various other labels have been tried by both proponents and opponents of the new paradigm, including socioecology, evolutionary ecology, Darwinian ecological anthropology, evolution and behavior studies, and ethology. Most of the differences among these labels are cosmetic, but to some extent they do reflect slight differences between various groups of researchers in theoretical or substantive foci. In the 1980s and 1990s the label "behavioral ecology" came to be used more often than any other, and so that is the one chosen for use in the remainder of this article.

THE INITIAL RECEPTION BY ANTHROPOLOGISTS

The initial reaction to these ideas in anthropology was quite hostile. At the 1976 meeting of the American Anthropological Association, a motion was put before the business meeting to condemn sociobiology. The motion was defeated by a narrow margin following testimony against it by, paradoxically, some prominent anthropologists who were on record as opposing the new approach. Marshall Sahlins, for example, argued that passage of the motion would tend

to strengthen sociobiology by making its supporters look like martyrs, and Margaret Mead argued that the motion could be misconstrued as condemning any use of Darwinian theory. At about the same time, the journal of the American Anthropological Association, *American Anthropologist*, published negative reviews of some of the early books on the approach. For example, Elliot Chapple (1976) wrote that E. O. Wilson's *Sociobiology* "clearly wins the prize as the worst book in biology for 1975." John Buettner-Janusch (1976) described Richard Dawkins' *The Selfish Gene* as "an example of the silliness that can result from an Oxbridge education" based on his misbelief that Dawkins attributed conscious intentions to "selfish genes." University of Chicago anthropologist Marshall Sahlins even went so far as to publish his own short book entitled *The Use and Abuse of Biology* (Sahlins, 1976) in an effort to nip the weed of sociobiology in the bud.

The hostility and suspiciousness that greeted the new paradigm had a number of causes, some justifiable but most the products of misunderstandings, ignorance, and flawed logic on the part of the critics. In the first place, it must be admitted that some of the most widely read early sociobiologists had a tendency to make pronouncements about the evolutionary origins of particular human behaviors without first subjecting them to rigorous hypothesis testing. But certainly the same could be said of much cultural anthropology, a sub-discipline without a good track record of adherence to the hypothetico-deductive method. Furthermore, since the mid-1970s human behavioral ecologists have gone to great lengths to emulate the rigor and precision of their colleagues in nonhuman behavioral ecology, and it is now not an idle boast to say that their studies include some of the more rigorously and thoroughly tested hypotheses in the history of anthropology.

Overall, the negative reaction of most anthropologists to human behavioral ecology seems to have had much less to do with the quality of the research than with the way it challenged existing and cherished ideas in the discipline. Most cultural anthropologists consider the discipline's main contribution to have been a demonstration of the overwhelming influence of culture on human affairs. Sometimes this emphasis on culture takes on the air of a religious conviction, as, for example, when two leading cultural anthropologists write that the purpose of ethnography is "to persuade the reader that culture matters more than he might have thought" (Marcus & Fischer, 1986: 43). Many cultural anthropologists are concerned that any study of possible biological influences on behavior will undermine this accomplishment. While it is true that the culture concept is absent from most human behavioral ecological writings, this is rapidly changing as more people in the field find ways to incorporate the culture concept into their research (see below and Cronk, 1995, 1999).

Cultural anthropologists also fear that human behavioral ecologists will commit a logical error known as the "naturalistic fallacy," the idea that what is natural is somehow good (Moore, 1903). In particular they worry that if behaviors (e.g. rape: Thornhill & Palmer, 2000) are described as products of natural selection and therefore "natural," this will be used to justify them morally. Fortunately, the naturalistic fallacy is something that human behavioral ecologists are well aware of, and most are scrupulous to avoid it and to encourage their students and readers to avoid it, as well. It is routine for articles by human behavioral ecologists to include disclaimers about the impossibility of deriving moral conclusions from arguments about the evolutionary origins or adaptiveness of particular behaviors in their publications. Perhaps cultural anthropologists are nervous about the naturalistic fallacy because they routinely commit virtually the same fallacy in another guise. Most cultural anthropologists believe in some form of cultural relativism. One implication of cultural relativism is that it is difficult, and perhaps impossible, to make valid value judgements across cultural boundaries because everyone's values are valid in their own particular societies, and there is no moral ground outside specific cultural traditions on which we can stand. So when we see people in other societies doing things we find objectionable, we are not allowed to criticize them because their behaviors are determined by their cultures. In other words, what is "cultural" is "good," or at least immune to criticism. Because this *reductio ad absurdum* of cultural relativism makes no more sense than the notion that what is "natural" is "good," it should be known as the "culturalistic fallacy" (Cronk, 1999: 114). Both the naturalistic and culturalistic fallacies are members of a much broader type of fallacy, first pointed out by David Hume ([1739/40] 1966), in which one derives a moral precept from a description of the way things are. It is no more acceptable to violate Hume's separation of "is" and "ought" statements when explaining behavior using the culture concept than when explaining it using evolutionary theory. Human behavioral ecologists are abundantly aware of the potential of their research to lead to the naturalistic fallacy; regrettably, most cultural anthropologists seem oblivious to the potential of their approach to lead to the culturalistic fallacy.[3]

Many of the arguments against human behavioral ecology have a clear political dimension. As Sahlins has argued, "What is inscribed in the theory of sociobiology is the entrenched ideology of Western society: the assurance of its naturalness and the claim of its inevitability" (Sahlins, 1976: 101). In other words, the evolutionary approach to human behavior, far from being an objective account of the origins of human behavior, is merely a manifestation of the Western, capitalist, bourgeois worldview. As such, it is designed not to improve understanding but rather to further the interests of the capitalist ruling class.

Other critics have been equally clear about their political motivations: "We share a commitment to the prospect of the creation of a more socially just – a socialist – society. And we recognize that a critical science is an integral part of the struggle to create that society, just as we also believe that the social function of much of today's science is to hinder the creation of that society by acting to preserve the interests of the dominant class, gender and race" (Lewontin et al., 1984: ix). The idea that they are little more than capitalist stooges comes as a bit of a shock and an affront to the many human behavioral ecologists who are committed to political and economic change and enhanced social justice. It also betrays the critics' ignorance of the fact that the basic theoretical ideas of people like Hamilton and Trivers have as much or more to do with co-operation as with competition among organisms. Evolutionary biologist David Queller (1995: 488) notes that his own work on social insects "emphasizes altruistic behavior, focuses on females rather than the males, and suggests that collective worker interests are crucial determinants of advanced insect societies." He asks, "Does this make me a good guy, a nurturing feminist, and a stalwart of the working class?"

ACCOMPLISHMENTS

The project of early human behavioral ecology was conceived in very broad terms: To test the idea that human behavior across cultures tends to conform in broad outline with expectations derived from current evolutionary biological theory. The obvious alternative hypothesis was that because culture operates according to its own internal logic, human behavior should conform to predictions derived from evolutionary theory no more often than one would expect by chance (e.g. Sahlins, 1976). The conflict between these views of culture and its role in shaping human behavior was brought to the fore when William Irons proposed that, if evolutionary theory is relevant to the study of human behavior, there should be a correlation between individuals' achievement of the goals that are promoted by the cultures of the societies in which they live and their achievement of the goal hypothesized by evolutionary theory, enhanced inclusive fitness. In Irons' own terms, "in most human societies cultural success consists in accomplishing those things which make biological success (that is, high inclusive fitness) probable" (Irons, 1979: 258). It should be noted that this hypothesis is in no way contingent on anyone being aware of the connection between cultural and reproductive success. If culture is serving to guide people to behave in ways that in their particular local circumstances enhance reproduction, it should do so even if people are unaware not only of the goal of reproduction but even of how biological reproduction works.

One of the first tests of this hypothesis was Irons' own study of Yomut Turkmen pastoralists of northern Iran. He found that wealth correlated well with men's lifetime reproductive success (Irons, 1979). Similarly, Monique Borgerhoff Mulder (1987) found the same correlation among Kipsigis agro-pastoralists in western Kenya, as did I among Mukogodo pastoralists in north central Kenya (Cronk, 1991b). In all of these cases, the circumstances are somewhat similar: local definitions of male status and definitions of success are mainly determined by wealth, livestock are a large part of what constitutes wealth, substantial bridewealths must be paid by grooms to their in-laws in order to marry, and polygyny is allowed. Given these conditions, it may not seem surprising that men's wealth would correlate with both their numbers of wives and their reproductive success. But the situation is not as obvious as it might seem at first glance. If the local definitions of "success" were arbitrary cultural constructions and nothing more, then it would be surprising indeed to find so many societies in which success is defined largely in terms of wealth accumulation. The fact that wealth is a major determinant of status and prestige in so many societies in so many different parts of the world suggests that local definitions of success, far from being arbitrary, may be designed for a purpose.

But what about societies in which wealth is either not accumulated at all or not the sole or even a major correlate of status or prestige? In such cases, a variety of other locally valued characteristics have been found to correlate with reproductive success, including political status (Betzig, 1986; Chagnon, 1979), hunting skills (Kaplan & Hill, 1985), involvement in violence (Chagnon, 1988a), peace-keeping (Moore, 1990), and certain personality traits (Barkow, 1977). The pattern emerging seems to provide strong support for Irons' hypothesis in natural fertility societies (i.e. those without access to modern contraceptives): achievement of success as it is locally defined is a good predictor both of an individual's status and of his reproductive success, regardless of the details of the local environment, economy, or social system. This suggests that the highly variable definitions of high status across human societies may indeed be cultural constructions, but not randomly variable ones. Rather, local opinions about what constitutes success may be the outcomes of ongoing negotiations among fundamentally similar actors in highly variable environmental, political, social, and historical situations.

The widespread confirmation of Irons' prediction of a match between cultural and reproductive success helps justify a continuation of the human behavioral ecological project. Accordingly, over the past twenty-five years human behavioral ecologists have tested a wide variety of more specific hypotheses derived from evolutionary theory on a wide variety of topics. Because human behavioral ecologists are guided by the paradigm of non-human behavioral ecology, many

of the topics covered are the same as those one would find covered in animal behavior textbooks and journals: foraging behavior (e.g. Hawkes et al., 1982), conservation (e.g. Alvard, 1998), sex biased parental investment (e.g. Cronk, 2000; Irons, 2000), male parental behavior (e.g. Hewlett 1991; Lancaster & Kaplan 2000), menstruation (e.g. Strassmann, 1996), nepotism (e.g. Chagnon & Bugos, 1979), reciprocity (e.g. Hames, 2000; Patton, 2000; Sosis, 2000), conflict (e.g. Chagnon & Bugos, 1979; Chagnon, 1988a; Patton, 2000), termination of pair bonds (e.g. Betzig, 1989; Fisher, 1989; Blurton Jones et al., 2000), polygyny (e.g. Borgerhoff Mulder, 1990; Strassman, 2000), and polyandry (e.g. Haddix, in press).

The basic method of human behavioral ecology can be demonstrated by looking at a study of human behavior closely modelled on similar studies of animal behavior. Polygyny, a mating system in which one male mates with several females, is common among both humans and non-humans, and occurs for a variety of reasons. *Resource defense polygyny* occurs when males defend important resources such as nesting sites and food-bearing territories in order to mate with the females who need those resources (see Emlen & Oring, 1977). A question raised by resource defense polygyny is why a female would mate with an already mated male if an unmated male, who would provide the female with access to all of the resources he has monopolized, is also available. In order to explain such behavior, ornithologist Gordon Orians (1969) proposed that if the quantity or quality of the resources monopolized by the males varies a great deal from one male to another, then sometimes a female might gain access to more or better resources if she mates with a male who is already mated, forcing her to share the resources with at least one other breeding female, rather than by mating with a solitary male who has few or poor resources. This *polygyny threshold* model has been successfully applied to a variety of bird species (e.g. the indigo bunting: Carey & Nolan, 1975, 1979).

Resource defense polygyny is common in human societies, as well. Among the Kipsigis, an agropastoralist group in western Kenya, for example, men own land and livestock and can have more than one wife at a time. Women often marry men who are already married, and they usually move from their parents' households to their new husband's land. Monique Borgerhoff Mulder, an anthropologist and human behavioral ecologist at the University of California at Davis, wondered whether Kipsigis' decisions about marriage fit with the predictions of Orians' polygyny threshold model. Specifically, she wondered whether women and their families selected husbands with an eye toward their control of resources, taking into account the fact that if a woman marries an already-married man, she will have to share his land with his earlier wives. The data support this prediction: Kipsigis women have a statististically significant

tendency to choose men according to the amount of land they offered the woman to farm, regardless of the number of cowives they would have as a result (Borgerhoff Mulder, 1990).

Because most human behavioral ecologists are, for the most part, also anthropologists, their studies are guided not only by the concerns of animal behavior studies but also by the traditions of the field of anthropology. As a result, they also study a wide variety of topics that are unique to humans and for which appropriate models do not already exist in the literature on animal behavior. These include kinship terminologies (e.g. Chagnon, 1988b, 2000; Hawkes, 1983), inheritance practices (e.g. Hartung, 1976; Borgerhoff Mulder, 1989), menopause and long post-reproductive lifespans in women (e.g. Hawkes et al., 2000), systems of economic redistribution (e.g. Betzig, 1988), bridewealth and dowry (e.g. Borgerhoff Mulder, 1988; Gaulin & Boster, 1990), and morality (e.g. Alexander, 1987; Cronk, 1994; Irons, 1991). One way in which our species is unique is its extreme behavioral plasticity, which makes it possible for us to live in a wide variety of circumstances and swiftly to change basic elements of our lives, such as how we obtain food. My own behavioral ecological analysis of a quick transition in subsistence among the Mukogodo of Kenya explores this distinctively human phenomenon. Before 1900, the Mukogodo were foragers and beekeepers who lived in rockshelters in a rugged, forested range of hills in north central Kenya. They spoke their own distinct language, called Yaaku, and had their own religion and other customs, such as the payment of beehives as bridewealth. Within a few years early in the twentieth century, they dropped their old language, subsistence patterns, and religion in favor of those of the Maasai. This swift transition presents an opportunity to test an extension of Irons' hypothesis about cultural and reproductive success: when a society's environment changes, culture may help people track changing environments in a way that would enhance their fitness (Irons, 1979: 258). The reasons for the change lie in a combination of changes in the social environment of the Mukogodo brought on by the British colonial presence and the way the Mukogodo responded to those changes. Specifically, British policies led a variety of Maasai-speaking pastoralists groups into steady contact with the Mukogodo and also gave them strong incentives to marry the Mukogodo. Marrying daughters and sisters to pastoralists was attractive to Mukogodo because the bridewealths they received from the pastoralists in the form of livestock were worth many times the value of a traditional Mukogodo beehive bridewealth. Quickly, no Mukogodo fathers would accept mere beehives as bridewealth and remaining a hunter-gatherer was no longer a viable option: Mukogodo men had to obtain livestock and become pastoralists or never marry. In behavioral ecological terms, the environment had changed so drastically that

the cultural values and practices that previously had been good guides to enhanced inclusive fitness – hunting, gathering, beekeeping, Yaaku-speaking, and so on – no longer were, and the Mukogodo responded by shifting to the cultural values and practices of their new Maasai-speaking neighbors, such as pastoralism and the Maasai language (Cronk, 1989).

INSTITUTIONAL DEVELOPMENTS

Human behavioral ecology has sparked the development of several new professional associations, journals, and graduate programs. The most prominent of the new associations is the Human Behavior and Evolution Society (HBES). HBES has about a thousand members, mostly from anthropology and psychology but also from related fields such as biology, sociology, and even many humanities disciplines. HBES has held annual meetings since 1989, usually during the summer on a college campus, and it has adopted a previously existing journal, *Ethology and Sociobiology*, and renamed it *Evolution and Human Behavior*. Other groups to which many human behavioral ecologists belong include the International Society for Human Ethology, the International Society for Behavioral Ecology, and the European Sociobiological Society. In addition to *Evolution and Human Behavior*, human behavioral ecological research is routinely published in the journals *Human Nature: An Interdisciplinary Biosocial Perspective, Behavioral Ecology and Sociobiology*, and *Behavioral Ecology*, as well as in a variety of more general anthropology journals, particularly *Current Anthropology*.

Graduate study in human behavioral ecology is now possible in many of the world's top anthropology programs. The two most prominent programs in recent years have been at the Universities of Utah and New Mexico. Other institutions with human behavioral ecologists teaching in graduate anthropology programs include Emory, Harvard, Northwestern, Pennsylvania State, Rutgers, Texas A&M, and Washington State Universities; the Universities of Arkansas, Connecticut, Michigan, Missouri, Nebraska, North Carolina, Oregon, Pittsburgh, and Washington; and the University of California at Davis, Los Angeles, and Santa Barbara. Outside the United States, human behavioral ecology is taught at the graduate level at University College London and at Dalhousie University in Nova Scotia.

Anyone interested in learning more about this approach would do well to start with one of many good review articles available (e.g. Borgerhoff Mulder, 1991; Cronk, 1991a; Smith, 1992a, b; Voland, 1998; Winterhalder & Smith, 2000). The development of human behavioral ecology has also been marked by the publication of a series of excellent edited volumes: *Evolutionary Biology*

and Human Social Behavior (Chagnon & Irons, 1979), *Hunter-Gatherer Foraging Strategies* (Winterhalder & Smith, 1981), *Human Reproductive Behaviour* (Betzig, Borgerhoff Mulder & Turke, 1988), *Evolutionary Ecology and Human Behavior* (Smith & Winterhalder, 1992), and *Adaptation and Human Behavior* (Cronk, Chagnon & Irons, 2000). Some collections that consist primarily of research on nonhumans also contain important work in human behavioral ecology (e.g. Alexander & Tinkle, 1981; Rubenstein & Wrangham, 1986; Standen & Foley, 1989). Betzig's *Human Nature: A Critical Reader* (1997) is a useful collection of some landmarks in both human behavioral ecology and evolutionary psychology, together with review essays and retrospective comments by the original authors. Few textbooks have been produced that focus on human behavioral ecology, although Boyd and Silk's *How Humans Evolved* (1997), Daly and Wilson's *Sex, Evolution, and Behavior* (1983), Trivers' *Social Evolution* (1985), Badcock's *Evolution and Individual Behavior* (1991), Buss's *Evolutionary Psychology* (1998), and Alcock's *Animal Behavior* (1997) cover some of the same subject matter. Several good popular introductions to this and related approaches have also been published over the past decade, including Ridley's *The Red Queen* (1993) and *The Origins of Virtue* (1996), Low's *Why Sex Matters* (2000), Small's *What's Love Got To Do With It?* (1995) and *Our Babies, Ourselves* (1998), Hrdy's *Mother Nature* (1999), Allman's *The Stone Age Present* (1994), and Wright's *The Moral Animal* (1994).

MORE RECENT REACTIONS AMONG ANTHROPOLOGISTS

Since the mid-1980s, the situation for human behavioral ecologists in anthropology has definitely improved. This improvement has a lot to do with changes within anthropology, such as how the subdisciplines are defined and how they divide up anthropology's traditional subject matter. In the 1970s, anthropologists who studied the behavior of living humans were classified as cultural anthropologists, and those who studied human biology in any way were classified as physical anthropologists. Since that time, both subdisciplines have changed significantly. Cultural anthropology, which used to have a place for a variety of scientifically-oriented, materialistic approaches, has moved strongly away from science and towards the humanities. Whereas cultural anthropologists at one time gained inspiration from such sciences as ecology and from the other social science disciplines, more often these days they draw upon ideas from such fields as literary criticism and semiotics. Cultural anthropology's movement towards the humanities is mainly a result of a long-term trend that goes at least as far back as Boas, but it may have been accelerated by the challenge presented

by the new, evolutionary approach to behavior. At the same time, physical anthropology has increasingly redefined itself as biological anthropology, a name change that reflects a broadening of focus beyond physiology and morphology to include the biological study of behavior. In the 1970s, most of the anthropologists who became interested in the new, evolutionary approach to behavior, including such notables as Lionel Tiger, Robin Fox, Napoleon Chagnon, William Irons, Derek Freeman, Jerome Barkow, and Raymond Hames, were members of the cultural subdiscipline. Now it is hard to find cultural anthropologists who identify themselves as human behavioral ecologists, but easy to find biological anthropologists who identify themselves as such.

The differences in how members of the cultural and biological subfields react to evolutionary analyses of behavior was made very clear by a survey conducted in 1984 by Leonard Lieberman (1989). Animal behaviorists, biological anthropologists, and cultural anthropologists were asked whether they considered a series of things identified as "sociobiological concepts" to be useful for future research on humans. The list included "kin selection," "reciprocal altruism," "genetic basis of altruism," "male-female reproduction strategy," "gene-culture coevolution," and "interaction of genes and environment." In every case, a majority of animal behaviorists agreed that the concept was likely to be useful in the study of humans, with agreement reaching as high as 82% in the cases of "kin selection" and "interaction of genes and environment." Biological anthropologists were only slightly more skeptical than animal behaviorists, with a majority seeing some usefulness in all of the concepts except "genetic basis of altruism." Cultural anthropologists were by far the most skeptical, with a majority seeing some usefulness only in the idea of an "interaction of genes and environment." Interestingly, the responses of all three groups were strongly affected by certain emotive words. The word "genetic" in the phrase "genetic basis of altruism" appears to have produced strong negative emotions in all three groups, despite the fact that the theoretically extremely similar phrases "kin selection" and "reciprocal altruism" received relatively high marks. Similarly, the word "culture" in the phrase "gene-culture coevolution" appears to have slightly warmed the hearts of the cultural anthropologists, who gave that phrase a relatively high rating.

Another recent development is a nascent reordering of anthropology's traditional subdisciplines. Whereas the divisions among the four old subdisciplines were ostensibly about subject matter, not ideology, the new division is about fundamental epistemological and ideological issues. In brief, those anthropologists who value a scientific, materialistic, and broadly evolutionary approach find themselves at odds with the majority of cultural anthropologists, many of whom

not only value a more humanistic and relativistic approach but who are openly hostile to science in general. What is emerging is a new, amalgam subdiscipline called "evolutionary anthropology," united by the use of scientific method and an interest in materialistic and evolutionary approaches, that includes archaeologists, biological anthropologists, and some former cultural anthropologists, along with human behavioral ecologists. This change is reflected in the internal organization of some departments (e.g. Rutgers), in the division of other formerly unified departments into two separate ones (e.g. Duke & Stanford), and in the success of the new journal *Evolutionary Anthropology*. This rearrangement of anthropology's intradisciplinary fences appears to be a good thing for human behavioral ecology in that it keeps them in the company of relatively sympathetic scholars. However there is a danger that it could impoverish human behavioral ecology by distancing it from cultural anthropology's traditional expertise in field methods and from the concept of culture itself (more about this below).

As the rift within anthropology has widened, cultural anthropologists' criticisms have expanded from a narrow focus on human behavioral ecology and related evolutionary approaches to behavior to a much broader critique of the possibility of a science of human behavior and culture in general. Relatively insightful and helpful critiques of human behavioral ecology itself have come from sources closer to home, i.e. from researchers who share with human behavioral ecology a scientific approach and an interest in the role of evolution in human affairs. For example, some evolutionary psychologists have argued that because any organism's current condition is a product of natural selection in ancestral environments, not current ones, the reproductive costs and benefits of particular behaviors are largely irrelevant to the identification of behavioral adaptations (e.g. Symons, 1989). This criticism reflects the fact that most evolutionary psychologists follow John Bowlby (1969) in locating the natural selection that led to our species in a long-past "Environment of Evolutionary Adaptedness," or EEA, often equated with Pleistocene hunting and gathering societies. To evolutionary psychologists who adhere closely to this view, contemporary societies, even those in which anthropologists usually work, are so radically different from the EEA that it is useless to measure the fitness consequences of particular behaviors in them. In response, human behavioral ecologists have argued that the evolutionary psychologists' understanding of the human EEA is simplistic, naive, and misleading, and that it should at least be made more sophisticated (e.g. Hrdy, 1999) or perhaps replaced with something else (e.g. Adaptively Relevant Environments: Irons, 1998). If humans are thought to have evolved not in a chronologically isolable EEA but rather over hundreds of millions of years in a series of quite different environments, then we ought to be composites of traits acquired in all of those environments. Some traits we will

share with a wide range of organisms (e.g. all vertebrates), others with a smaller range (e.g. all primates), and others might be distinctively human. To the extent to which current environments resemble past ones, current reproductive rates should indeed be useful indicators of specific adaptations. For example, one interpretation of the research on status striving and reproductive success discussed above is that, despite the tremendous changes that took place in human societies with the domestication of plants and animals, the reproductive rewards to those with high status continued as before.

Instead of examining behaviors' current fitness benefits, evolutionary psychologists usually examine the proximate psychological mechanisms behind particular behaviors and try to discern how they may have been produced by natural selection. For example, research into the psychological mechanisms involved in mate attraction suggest that there are similarities in the ways in which individuals evaluate members of the opposite sex that are consistent both within and among societies and that fit predictions derived from evolutionary theory (e.g. Buss, 1994). Human behavioral ecologists, in contrast, typically treat the specific psychological mechanisms involved in particular behaviors as unavailable for study, and they focus instead on the measurement of the costs and benefits of various behaviors. The failure of human behavioral ecologists to address proximate psychological mechanisms is one of the bases for another recent critique from a relatively sympathetic source, ecological anthropologist Andrew P. Vayda (1995). Vayda is well known in anthropology for his early work that applied evolutionary biological but group selectionist models to human behavior and cultural institutions, though he later broke with that approach in favor of one that focuses on more on individuals (Vayda & McKay, 1975). Vayda argues that true explanations of phenomena are accounts of causal chains, and because human behavioral ecologists usually fail to demonstrate all of the links in their causal chains, their explanations are inadequate. Human behavioral ecologists respond that: (a) it is normal procedure in many sciences to provide explanations without demonstrating all of the causal links involved, (b) they are most often ill-equipped by training to examine proximate psychological mechanisms, (c) their field situations and the cultural differences of the people they study often make such studies very difficult, and (d) further exploration of proximate mechanisms and other details of complete causal chains, which is a worthy if distant goal for all science, will come with time.

CURRENT FRONTIERS AND FUTURE DIRECTIONS

Two broad trends will shape the immediate future of human behavioral ecology. First, human behavioral ecologists will continue to adopt and adapt new theories,

models, and methods developed by nonhuman behavioral ecologists. Second, human behavioral ecology will benefit from contacts with two rapidly developing and closely related approaches, evolutionary psychology and memetics. The first trend is already a well established habit. Human behavioral ecology exists not in isolation but rather as a part not only of anthropology but also of behavioral ecology in general. Unique as humans may be, one of the basic ideas of human behavioral ecology is that because, like all other living things, we are products of a history of natural selection, it would make little more sense to have a separate sort of behavioral ecology for humans than it would for any other individual species. Fortunately for human behavioral ecology, nonhuman behavioral ecology is a lively field percolating with new ideas. Here I will briefly discuss three recent development in nonhuman behavioral ecology that are already beginning to influence work in human behavioral ecology: life history theory, dynamic modeling methods, and the study of animal communication.

Life history theory developed in biology as a way to understand the evolution of the timing of events during an organism's lifetime, including the age at which the organism matures sexually, the age of its first reproduction, and its overall life span (Hill & Kaplan, 1999). Because organisms vary so widely in terms of life history traits, the body of theory involved is enormous and is still undergoing rapid development. What unifies this complex body of theory is the idea that, throughout their lifetimes, organisms must allocate a limited supply of energy. Life history theory predicts that natural selection will shape the way organisms allocate the energy they have available over the course of a lifetime in a way that will, on average, optimize the benefits to their inclusive fitness, taking into consideration all relevant constraints on their allocation patterns. For example, an organism may have a choice between investment in its own maintenance and development (referred to as "somatic effort"), investing in its own reproductive success either by trying to obtain a mate ("mating effort") or by helping its offspring ("parental effort"), or investing in relatives in order to enhance their reproductive success ("indirect reproduction," "nepotism," or "kin selection"). The costs and benefits to different patterns of allocation among these options might change considerably through an organism's lifetime. For example, when an organism is young an emphasis is often placed on somatic effort, with the emphasis shifting to reproductive effort when it reaches sexual maturity. Kristen Hawkes and her colleagues (Hawkes et al., 2000) have suggested that might help explain the evolution of menopause and long post-reproductive lifespans in human females. In brief, the idea is that women may reach a point where they can enhance their own inclusive fitness more effectively by avoiding the risks of pregnancy and childbirth and helping their daughters

instead. Such aid may be particularly beneficial if it comes at a time in the daughters' lives when there is a conflict between the demands of subsistence and the desire to reproduce, e.g. if their ability to nurse an infant were diminished by a need to forage. If the costs of continued direct reproduction are high enough and the benefits of shifting to a nepotistic strategy are great enough, then natural selection may have favored the evolution of menopause and a long postreproductive lifespan for human females. Data from some contemporary hunter-gatherers supports this. Older women among the Hadza of Tanzania, for example, are more productive foragers than younger women (Hawkes et al., 1989).

Life history theory bears many similarities to demography because both fields examine the scheduling of major life events, particularly fertility and mortality. Accordingly, some human behavioral ecologists have begun to use life history theory to organize and analyze demographic data from traditional societies (e.g. Hill & Hurtado, 1996). Life history theory may also help shed light on current demographic patterns in industrial and post-industrial societies, which are seen by some as presenting a major challenge to human behavioral ecology (e.g. Vining, 1986). Such societies have undergone a drastic shift, known as the demographic transition, from high rates of fertility and mortality to low rates of fertility and mortality. The choice made by the wealthiest people on earth to reduce their fertility rather than to use their tremendous resources to have more offspring appears to run counter to the expectations of evolutionary theory. Kaplan and his colleagues (e.g. Kaplan & Lancaster, 2000) have examined this issue in light of life history theory, suggesting that fertility reductions in modern societies reflect a response on the part of parents to an environment in which the wages workers receive, and hence their wealth, is directly linked to the costs of obtaining the skills necessary to be competitive in today's skills-based competitive labor markets. In such an economic environment, having fewer but better endowed children may be a better long-run strategy than having many children (for other behavioral ecological work on the demographic transition, see Barkow & Burley, 1980; Bereczkei, 1998; Borgerhoff Mulder, 1998; Low, 2000a, b; Luttbeg et al., 2000; Rogers, 1992; Turke, 1989).

Dynamic modeling (Mangel & Clark, 1988) is one aspect of an increasing methodological sophistication in behavioral ecology. Traditional optimization models in behavioral ecology posit a decision maker faced with a set of possible choices, each one of which is associated with certain costs and benefits, and a set of constraints. Dynamic models extend the decision-making process over time and give it a capacity for self-adjustment. They do this by including variables that describe the state of the decision maker and his "state dynamics," i.e. some rules that specify how the decision maker's state will change over

time as he makes various choices (Mace, 1993: 364). Human behavioral ecologists have applied dynamic modeling methods to such phenomena as subsistence change (Mace, 1993), family size (Anderies, 1996; Beauchamp, 1994; Mace, 1996, 2000), and marriage (Luttbeg et al., 2000), and the potential exists for this technique to be applied to a great many other phenomena. Like life history theory, dynamic modeling may even be helpful in cracking that tough old nut, the demographic transition. Toward that end, Mace (1996, 2000) developed a dynamic model of reproductive decisions among the Gabbra, a pastoralist group in northern Kenya. Wealth among the Gabbra, as among many peoples, is inherited, and inherited wealth contributes greatly to both future wealth and reproductive success. Mace modeled two decisions made by parents: when to have another baby and how to distribute the family's resources among the children when a parent dies. She defines the state variables as the parent's wealth and number of living sons, and an optimal decision rule is defined as one that produces the most grandchildren. Her main finding is that increased costs of children is the main thing likely to lead to a decision to limit fertility, which supports conclusions reached independently by demographers (e.g. Caldwell, 1982) and other human behavioral ecologists (e.g. Kaplan & Lancaster, 2000).

Animal communication has always been a central concern of behavioral ecologists. While communication might at first glance appear to be a simple matter of information transfer, in some circumstances signallers may be trying not so much to inform receivers as to manipulate them. A key variable is whether there is a conflict or a confluence of interests between signallers and receivers. When there is a clear conflict of interests between the two parties, i.e. where the signaller would benefit if the receiver were to do something that is not in the receiver's best interests, behavioral ecologists predict that an evolutionary arms race will develop. The signaller should send increasingly elaborate and conspicuous signals in order to influence the behavior of a receiver that is increasingly resistant to manipulation (Dawkins & Krebs, 1978; Krebs & Dawkins, 1984). This might explain such very refined signalling systems as an anglerfish's lure, which has evolved to attract smaller fish towards the anglerfish's waiting mouth, which is clearly not in the best interests of the prey. This line of thought has inspired a great deal of research on deception in animal communication systems (e.g. Mitchell & Thompson, Eds, 1986).

When there is a confluence of interests between signallers and receivers, in contrast, natural selection should favor the evolution of signals that are only as elaborate, conspicuous, and costly as they need to be in order to get the information across. When both parties are aware of such a confluence of interests, for example in a mated pair of organisms cooperatively rearing offspring, then

signals should be muted and economical (Dawkins & Krebs, 1978). However, this apparently simple situation becomes quite complicated by the fact that sometimes only the signaller is aware of the existence of a confluence of interests. In such circumstances, the signaller may need to send a very elaborate, conspicuous and costly signal in order to convince the receiver that they do indeed share some interests (Zahavi et al., 1997). Such confluences of interest may arise even in situations we normally think of as being essentially conflictual, such as an encounter between a predator and its prey. A Thompson's gazelle, for instance, actually has a confluence of interests with a lion trying to kill and eat it: the gazelle wants to avoid being eaten, and the lion wants to avoid wasting time and energy on a failed hunt. This helps explain why such gazelles "stot," a peculiar sort of theatrical hopping display, when pursued by lions: Stotting sends a signal to the lion that alerts her to their underlying confluence of interests (FitzGibbon & Fanshawe, 1988). By stotting, a gazelle tells a lion that it is aware of her presence and that it is clearly too fit to be easily caught. The implication is that both parties would benefit if the lion were to look elsewhere for a meal. Similarly, a high quality male and a female he is pursuing as a mate have a confluence of interests: he wants a mate, and she wants a high quality male. Their confluence of interests rests on the fact that he is indeed a high quality male, but, at the outset of the interaction, only the male can be aware of his own high quality and hence of their confluence of interests, and he must put on an elaborate and costly display to convince the female of it. In both of these cases, there is a broad conflict of interest between two classes of individuals (prey and predators; males and females) but a confluence of interests between the receivers (predators and females) and *certain members* of the signalling class (*fit, alert* prey and *high quality* males), allowing for a cooperative type of signal to evolve despite the wider conflict. Thus, surprisingly, some very elaborate and expensive signals may actually fit the predictions of cooperative signalling theory in that they are as economical as they can be and still get their message across.[4]

These ideas about animal communication have begun to have an impact on human behavioral ecology. Harpending, Draper and Rogers (1987) have suggested that the idea that situations of conflict will lead to loud, conspicuous, and manipulative signals might help make sense of propaganda and other public pronouncements typically seen in situations in which there is a marked difference in power between two groups, such as elites and nonelites or, in some societies, between males and females. Boone (1998) has used animal signalling theory to shed light on conspicuous consumption and magnanimity. Thus, the lavish displays of generosity, sometimes even including the destruction of wealth, at potlatches among Northwest Coast Indians (Drucker & Heizer, 1967)

or in the *moka* gift-giving ceremonies in highland New Guinea (Strathern, 1971) may be analogous to the way that a small bird called the Arabian babbler competes for status by magnanimously feeding others (Carlisle & Zahavi, 1986; Zahavi, 1990). Even big game hunting may be a way for men to show off, i.e. to send a signal to others about their value as a provider of food to the group and as a potential mate (Hawkes, 1991).

Evolutionary psychology and *memetics* have developed alongside human behavioral ecology as ways of applying selectionist thinking to human affairs. While there has been some competition between these three approaches in the past, a consensus is emerging that they are actually quite complementary (e.g. Smith, 2000), and there is sure to be more interaction between them in the future. Evolutionary psychology is the more developed of the two fields and has lately received widespread attention both among scholars and in the popular press. Some of the most impressive work in evolutionary psychology has been Leda Cosmides and John Tooby's (1992) studies of social cognition, which suggests that humans are especially good at dealing with logical problems when there is a social rule involved. They have shown that success rates on the Wason selection task, which involves questions with a logical structure of the "If p then q" variety, vary tremendously depending on how the test questions are phrased. If the question is simply descriptive ("If a card has a D on one side, then it must have a 3 on the other side") then success rates in all groups tested are much lower than if the question involves the application of a social rule ("If a person is drinking alcohol, then he or she must be at least 21 years of age"). This result is not contingent on the subject being familiar with the social rule in question. Even when the question is posed in terms of an exotic rule in an exotic setting ("If a person is eating taro, then he or she must have a tattoo"), success rates are still much higher than when the rule is phrased in a more abstract way. These results support the idea that the human mind is not a general purpose problem solver but rather a bundle of special purpose modules, one of which has been designed by natural selection to be especially adept at monitoring others' adherence to social rules.

Although these results on the Wason selection task are consistent across samples taken from American society, the case for a special mental module for monitoring adherence to social rules would be strengthened if the study were replicated in other societies, as well, particularly ones isolated from Western society and its particular forms of logic. This was recently done by Larry Sugiyama, an anthropologist and human behavioral ecologist trained by Cosmides and Tooby, among the Shiwiar of the Ecuadorian Amazon (Sugiyama, 1996). The Wason selection task is usually administered in written form, but, because few Shiwiar are literate, Sugiyama had to redesign the test in order to administer

it using only pictures and oral instructions. On the descriptive question, Shiwiar subjects performed much like American subjects: badly. That is, the descriptive question elicited a logically correct response about as often as would be expected by chance. The social contract questions, in contrast, elicited correct responses significantly more often than would be expected by chance. Furthermore, the errors committed by Shiwiar subjects were mostly errors of commission (turning over too many cards) not omission (turning over too few cards). Hence, most of those giving incorrect responses to the social contract questions would still have succeeded in enforcing the social rule. Their only error was in collecting more information than they needed in order to accomplish this task.

Memetics is the newest and least tested of the three selectionist approaches to human affairs. It also goes by such names as cultural transmission theory, dual inheritance theory, and evolutionary culture theory. Prominent figures in the field include Boyd and Richerson (1985), Cavalli-Sforza and Feldman (1981), Durham (1991), and Lumsden and Wilson (1981; for recent popular accounts, see Brodie, 1995; Dennett, 1995; Lynch, 1996). The name "memetics" is based on the word "meme," which was coined by Richard Dawkins (1976) to describe a culture trait. Just as the name implies, the central metaphor is between the meme and the gene. Both memes and genes are replicators and, as such, both may undergo an evolutionary process, with memes that are successful in being copied from mind to mind spreading just as successful genes spread in a gene pool. Because genes are passed from parents to offspring, in general they are beneficial for their bearers, and a broad expectation in memetics is that most memes will also be helpful to their bearers in solving life's problems and guiding behavior toward adaptive ends. The finding described above that there is a widespread correlation between culturally defined success and enhanced inclusive fitness fits this expectation neatly, but it also means that this prediction of memetics is neither unique nor very interesting. However, memes can also be spread between nonrelatives and even strangers, much like viruses, and just as viruses can be successful at spreading at the expense of those who become infected by them, so, in theory, might memes spread from mind to mind even if they have detrimental consequences for their bearers. Because the prediction of conflict between cultural traits and genes is unique and serves to distinguish memetics from human behavioral ecology, much of the research in memetics has focused on just such traits. So far the quest for maladaptive memes has been interesting but not very fruitful. Durham (1991) offers one possible example from New Guinea: cannibalism among a group called the Fore. The Fore adopted cannibalism because they found that human flesh was a surprisingly tasty type of food. Unfortunately, cannibalism also provided an avenue for the spread of an infectious agent that produced a lethal

form of spongiform encephalopathy called *kuru*, which is very similar to mad cow disease, scrapie, and Creutzfeld-Jakob disease. But the link between cannibalism and the spread of *kuru* was so difficult to discern, not only for the Fore themselves but also for Western medical researchers, that the practice of cannibalism continued for some time even though it was spreading a devastating disease. It is not just food consumed that can cause health problems: food not consumed due to taboos can also deprive people of needed nutrients. Robert Aunger examined the potential conflict between memes for food taboos and peoples' fitness interests among four different ethnic groups in the Ituri forest in eastern Zaire (now Congo), where individuals typically maintain a variety of complex food taboos. He found a discernable reduction in nutritional status attributable to food taboos among only pregnant women in one of the four groups (Aunger, 1992).

Whether maladaptive memes turn out to be common or rare, memetics has provided a service to human behavioral ecology by redirecting attention to the issue of culture. Most behavioral ecological theory is devoid of culture simply because most nonhuman species have little or no discernable culture. One way other than memetics to make greater use of the culture concept in human behavioral ecology would be through more studies of communication because, among humans, culture is the raw materials out of which signals are fashioned (Cronk, 1995, 1999). A nice example is Napoleon Chagnon's work on how Yanomamö men manipulate their kin terminology (Chagnon, 1988b, 2000). The Yanomamö kin terminology system includes terms that specify which women a man can marry and which he cannot. Chagnon has shown that men systematically fudge the rules of the system by reclassifying young women as marriageable kin. In other words, they use the raw material of a cultural system for defining marriageable kin to construct manipulative, fitness-enhancing signals. Work of this kind also dovetails nicely with a rapidly growing body of research on animal cognition (e.g. Ristau, Ed., 1991) and animal culture (e.g. Bonner, 1980), particularly among nonhuman primates (e.g. McGrew, 1992; Wrangham et al., Eds, 1994).

CONCLUSION

Because evolution has always been a central theme in anthropology, human behavioral ecology represents a sort of return to the discipline's roots. Over the past twenty five or thirty years, human behavioral ecology has developed from a largely speculative field to a theoretically and methodologically rigorous one with a good foundation in empirical research. In addition to confirming the validity and usefulness of an evolutionary approach to human behavior, human

behavioral ecology's findings also necessitate a rethinking of many of anthropology's central theoretical concepts and traditional topics. Although it has accomplished much, human behavioral ecology's work has barely begun. It is a field in which new and exciting possibilities are always opening up thanks to a rapidly developing body of theory and endless opportunities for new field research.

NOTES

1. The idea that anthropology is a discipline consisting of these four subdisciplines is distinctively American. In other countries with strong traditions of anthropological research, such as Britain and France, these are often treated as entirely separate disciplines. Nevertheless, because of the much larger number of anthropologists at work in the United States compared to any other country, theoretical concerns that arise out of the American four-field conception of the field tend to influence theoretical developments even in countries where that conception is not widely shared.

2. It is important not to put too fine a point on this. Even in the heyday of Boasian cultural relativism, some cultural anthropologists maintained an interest in evolution. Some, such as Julian Steward (1955) and his followers in cultural ecology and ecological anthropology, were influenced by the new science of ecology. Others, including Leslie White (1959) and Marvin Harris (1979), were inspired by Marxist theory, which was itself a form of nineteenth century evolutionism.

3. There are exceptions, including Scheper-Hughes (1995), a cultural anthropologist who rejects relativism for fear that it will place objectionable behaviors such as oppression and exploitation beyond the reach of moral judgement, and Bagish (1982).

4. I have presented models of animal communication developed by Dawkins and Krebs and by Zahavi as complementary, with Zahavi's account of how costly, honest signals can evolve in apparently conflictual situations as a special case of Dawkins and Krebs' more general model. This is not how these two models are usually presented in the literature on animal signalling. Rather, they are usually seen as being in conflict because they appear to make different predictions about the reasons for costly signals and about their honesty. In this section I have attempted to show that the conflict between them is more apparent than real. Dawkins and Krebs' prediction that conspicuous, perhaps costly, and even deceptive signals will evolve in situations of clear conflict between signallers and receivers still holds despite Zahavi's evidence that conspicuous, costly, but honest signals can also evolve in circumstances where, even within a situation that may be rife with conflict, there is a confluence of interests between signaller and receiver of which only the signaller is initially aware.

REFERENCES

Alcock, J. (1997). *Animal Behavior: An Evolutionary Approach.* Sixth edition. Sunderland, MA: Sinauer.

Alexander, R. D. (1974). The evolution of social behavior. *Annual Review of Ecology and Systematics, 5,* 325–383.

Alexander, R. D. (1987). *The Biology of Moral Systems.* Hawthorne, NY: Aldine de Gruyter.

Alexander, R. D., & Tinkle, D. W. (Eds) (1981). *Natural Selection and Social Behavior: Recent Research and New Theory*. New York: Chiron Press.

Allman, W. F. (1994). *The Stone Age Present*. New York: Simon and Schuster.

Alvard, M. (1998). Evolutionary ecology and resource conservation. *Evolutionary Anthropology*, *7*, 62–74.

Anderies, J. M. (1996). An adaptive model for predicting !Kung reproductive performance: A stochastic dynamic programming approach. *Ethology and Sociobiology*, *17*, 221–246.

Aunger, R. (1992). The nutritional consequences of rejecting food in the Ituri Forest of Zaire. *Human Ecology*, *20*, 263–291.

Badcock, C. (1991). *Evolution and Individual Behavior*. Oxford: Basil Blackwell.

Bagish, H. H. (1982). Confessions of a former cultural relativist. In: E. Angeloni (Ed.), *Annual Editions: Anthropology 82/83* (pp. 16–23). Guilford, CT: Dushkin.

Barkow, J. H. (1977). Conformity to ethos and reproductive success in two Hausa communities: An empirical evaluation, *Ethos*, *5*, 409–425.

Barkow, J. H., & Burley, N. (1980). Human fertility, evolutionary biology, and the demographic transition. *Ethology and Sociobiology*, *1*, 163–180.

Beauchamp, G. (1994). The functional analysis of human fertility decisions. *Ethology and Sociobiology*, *15*, 31–53.

Bereczkei, T. (1998). Kinship network, direct childcare, and fertility among Hungarians and Gypsies. *Evolution and Human Behavior*, *19*, 283–298.

Betzig, L. L. (1986). *Despotism and Differential Reproduction: A Darwinian View of History*. New York: Aldine.

Betzig, L. (1988). Redistribution: equity or exploitation? In: L. Betzig, M. Borgerhoff Mulder & P. Turke (Eds), *Human Reproductive Behavior: A Darwinian Perspective* (pp. 49–63). Cambridge: Cambridge University Press.

Betzig, L. (1989). Causes of conjugal dissolution: A cross-cultural study. *Current Anthropology*, *30*(5), 654–676.

Betzig, L. (Ed.). (1997). *Human Nature: A Critical Reader*. Oxford: Oxford University Press.

Betzig, L., Borgerhoff Mulder, M., & Turke, P. (Eds). (1988). *Human Reproductive Behavior: A Darwinian Perspective*. Cambridge: Cambridge University Press.

Blurton, J., Nicholas, G., Marlowe, F. W., Hawkes, K., & O'Connell, J. F. (2000). Hunter-gatherer divorce rates and the paternal investment theory of human pair-bonding. In: L. Cronk, N. Chagnon & W. Irons (Eds), *Adaptation and Human Behavior: An Anthropological Perspective*. Hawthorne, NY: Aldine de Gruyter.

Boas, F. (1896). The limitations of the comparative method of anthropology. *Science*, *4*, 901–908. Reprinted in *Race, Language, and Culture* (pp. 270–280). New York: MacMillan, 1940.

Bonner, J. T. (1980). *The Evolution of Culture in Animals*. Princeton: Princeton University Press.

Boone, J. L. (1998). The evolution of magnanimity: When is it better to give than to receive? *Human Nature*, *9*(1), 1–21.

Borgerhoff Mulder, M. (1987). On cultural and reproductive success. *American Anthropologist*, *89*, 617–634.

Borgerhoff Mulder, M. (1988). Kipsigis bridewealth payments. In: L. Betzig, M. Borgerhoff Mulder & P. Turke (Eds), *Human Reproductive Behaviour: A Darwinian Perspective*. Cambridge: Cambridge University Press.

Borgerhoff Mulder, M. (1989). Reproductive consequences of sex-biased inheritance. In: V. Standen, & R. A. Foley (Eds), *Comparative Socioecology of Mammals and Man* (pp. 405–427). London: Blackwell.

Borgerhoff Mulder, M. (1990). Kipsigis women's preferences for wealthy men: Evidence for female choice in mammals? *Behavioral Ecology and Sociobiology, 27*, 255–264.

Borgerhoff Mulder, M. (1991). Human behavioral ecology. In: J. R. Krebs & N. B. Davies (Eds), *Behavioral Ecology: An Evolutionary Approach*, (3rd ed.) (pp. 69–98). Oxford: Blackwell Scientific.

Borgerhoff Mulder, M. (1998). The demographic transition: Are we any closer to an evolutionary explanation? *Trends in Ecology and Evolution, 13*, 266–270.

Bowlby, J. (1969). *Attachment and Loss*. New York: Basic.

Boyd, R., & Richerson, P. J. (1985). *Culture and the Evolutionary Process*. Chicago: University of Chicago Press.

Boyd, R., & Silk, J. (1997). *How Humans Evolved*. New York: Norton.

Brodie, R. (1995). *Virus of the Mind*. Seattle: Integral Press.

Beutttner-Janusch, J. (1978). Review of Dawkins (1976). *American Anthropologist, 80*(2), 463–464.

Buss, D. M. (1994). *The Evolution of Desire: Strategies of Human Mating*. New York: Basic.

Buss, D. M. (1998). *Evolutionary Psychology: The New Science of the Mind*. Needham Heights, MA: Allyn & Bacon.

Caldwell, J. C. (1982). *Theory of Fertility Decline*. New York: Academic.

Carey, M., & Nolan Jr., V. (1975). Polygyny in indigo buntings: A hypothesis tested. *Science, 190*, 1296–1297.

Carey, M., & Nolan Jr., V. (1979). Population dynamics of indigo buntings and the evolution of avian polygyny. *Evolution, 33*, 1180–1192.

Carlisle, T. R., & Zahavi, A. (1986). Helping at the nest, allofeeding and social status in immature Arabian babblers. *Behavioral Ecology and Sociobiology, 18*, 339–351.

Cavalli-Sforza, L. L., & Feldman, M. W. (1982). *Cultural Transmission and Evolution*. Princeton, NJ: Princeton University Press.

Chagnon, N. A. (1979). Is reproductive success equal in egalitarian societies? In: N. A. Chagnon & W. Irons (Eds), *Evolutionary Biology and Human Social Behavior: An Anthropological Perspective* (pp. 374–401). North Scituate, MA: Duxbury.

Chagnon, N. A. (1988a). Life histories, blood revenge, and warfare in a tribal population. *Science, 238*, 985–992.

Chagnon, N. A. (1988b). Male Yanomamö manipulations of kinship classifications of female kin for reproductive advantage. In: L. Betzig, M. Borgerhoff Mulder & P. Turke (Eds), *Human Reproductive Behaviour: A Darwinian Perspective* (pp. 23–48). Cambridge: Cambridge University Press.

Chagnon, N. A. (2000). Adult Yanomamö classification accuracy for first ascending generation kin. In: L. Cronk, N. A. Chagnon & W. Irons (Eds), *Adaptation and Human Behavior: An Anthropological Perspective*. Hawthorne, NY: Aldine de Gruyter.

Chagnon, N. A., & Bugos Jr., P. E. (1979). Kin selection and conflict: An analysis of a Yanomamö ax fight. In: N. A. Chagnon & W. Irons (Eds), *Evolutionary Biology and Human Social Behavior: An Anthropological Perspective* (pp. 213–238). North Scituate, MA: Duxbury.

Chagnon, N. A., & Irons, W. (Eds) (1979). *Evolutionary Biology and Human Social Behavior: An Anthropological Perspective*. North Scituate, Mass.: Duxbury.

Chapple, E. D. (1976). Ethology without biology. *American Anthropologist, 78*(3), 590–593.

Cosmides, L., & Tooby, J. (1992). Cognitive adaptations for social exchange. In: J. H. Barkow, L. Cosmides & J. Tooby (Eds), *The Adapted Mind: Evolutionary Psychology and the Generation of Culture* (pp. 163–228). Oxford: Oxford University Press.

Cronk, L. (1989). From hunters to herders: Subsistence change as a reproductive strategy among the Mukogodo. *Current Anthropology, 30*(2), 224–234.

Cronk, L. (1991a). Human behavioral ecology. *Annual Review of Anthropology*, *20*, 25–53.

Cronk, L. (1991b). Wealth, status, and reproductive success among the Mukogodo of Kenya. *American Anthropologist*, *93*, 345–360.

Cronk, L. (1994). Evolutionary theories of morality and the manipulative use of signals. *Zygon: Journal of Religion and Science*, *29*(1), 81–101.

Cronk, L. (1995). Is there a role for culture in human behavioral ecology? *Evolution and Human Behavior*, *16*(3), 181–205.

Cronk, L. (1999). *That Complex Whole: Culture and the Evolution of Human Behavior*. Boulder, CO: Westview.

Cronk, L. (2000). Female-biased parental investment and growth performance among Mukogodo children. In: L. Cronk, N. Chagnon & W. Irons (Eds), *Adaptation and Human Behavior: An Anthropological Perspective*. Hawthorne, NY: Aldine de Gruyter.

Cronk, L., Chagnon, N. A., & Irons, W. (Eds), (2000). *Adaptation and Human Behavior: An Anthropological Perspective*. Hawthorne, NY: Aldine de Gruyter.

Daly, M., & Wilson, M. (1983). *Sex, Evolution, and Behavior*. Second edition. Belmont, CA: Wadsworth.

Dawkins, R. (1976). *The Selfish Gene*. Oxford: Oxford University Press.

Dawkins, R., & Krebs, J. (1978). Animal signals: Information or manipulation? In: J. Krebs & N. Davies (Eds), *Behavioural Ecology: An Evolutionary Approach* (pp. 282–309). Sunderland, MA: Sinauer.

Dennett, D. (1995). *Darwin's Dangerous Idea*. New York: Simon and Schuster.

Drucker, P., & Heizer, R. F. (1967). *To Make My Name Good: A Reexamination of the Southern Kwakiutl Potlatch*. Berkeley: University of California Press.

Durham, W. H. (1991). *Coevolution: Genes, Culture, and Human Diversity*. Stanford: Stanford University Press.

Emlen, S. T., & Oring, L. W. (1977). Ecology, sexual selection and the evolution of mating systems. *Science*, *197*, 215–223.

Fisher, H. E. (1989). Evolution of human serial pairbonding. *American Journal of Physical Anthropology*, *78*(3), 331–354.

FitzGibbon, C. D., & Fanshawe, J. H. (1988). Stotting in Thompson's gazelle: An honest signal of condition. *Behavioral Ecology and Sociobiology*, *23*, 69–74.

Gaulin, S. J. C., & Boster, J. S. (1990). Dowry as female competition. *American Anthropologist*, *92*(4), 994–1005.

Haddix, K. (In press). When polyandry falls apart: Leaving your wife and your brothers. *Evolution and Human Behavior*.

Hames, R. (2000). Reciprocal altruism in Yanomamö food exchange. In: L. Cronk, N. A. Chagnon & W. Irons (Eds), *Adaptation and Human Behavior: An Anthropological Perspective*. Hawthorne, NY: Aldine de Gruyter.

Hamilton, W. D. (1964). The evolution of social behavior. *Journal of Theoretical Biology*, *7*, 1–52.

Harpending, H., Rogers, A., & Draper, P. (1987). Human sociobiology. *Yearbook of Physical Anthropology*, *30*, 127–150.

Harris, M. (1979). *Cultural Materialism: The Struggle for a Science of Culture*. New York: Random House.

Hartung, J. (1976). On natural selection and the inheritance of wealth. *Current Anthropology*, *17*, 607–622.

Hawkes, K. (1983). Kin selection and culture. *American Ethnologist*, *10*, 345–363.

Hawkes, K. (1991). Showing off: Tests of another hypothesis about men's foraging goals. *Ethology and Sociobiology*, *11*, 29–54.

Hawkes, K., Hill, K., & O'Connell, J. (1982). Why hunters gather: Optimal foraging and the Ache of eastern Paraguay. *American Ethnologist*, *9*, 379–398.

Hawkes, K., O'Connell, J. F., & Blurton Jones, N. (1989). Hardworking Hadza grandmothers. In: V. Standen & R. Foley (Eds). *Comparative Socioecology: The Behavioural Ecology of Mammals and Man* (pp. 341–366). London: Blackwell.

Hawkes, K., O'Connell, J F., Blurton Jones, N. G., Alvarez, H., & Charnov, E. L. (2000). The grandmother hypothesis and human evolution. In: L. Cronk, N. A. Chagnon & W. Irons (Eds), *Adaptation and Human Behavior: An Anthropological Perspective*. Hawthorne, NY: Aldine de Gruyter.

Hewlett, B. (1991). *Intimate Fathers: The Nature and Context of Aka Pygmy Paternal Infant Care*. Ann Arbor: University of Michigan Press.

Hill, K., & Hurtado, A. M. (1996). *Ache Life History: The Ecology and Demography of a Foraging People*. Hawthorne, NY: Aldine de Gruyter.

Hill, K., & Kaplan, H. (1999). Life history traits in humans: Theory and empirical studies. *Annual Review of Anthropology*, *28*, 397–430.

Hrdy, S. B. (1999). *Mother Nature: A History of Mothers, Infants, and Natural Selection*. New York: Pantheon.

Hume, D. (1739/40). *A Treatise of Human Nature, Volume II, Book III: Of Morals*. Reprinted by Oxford University Press in 1966.

Irons, W. (1979). Cultural and biological success. In: N. A. Chagnon & W. Irons (Eds), *Evolutionary Biology and Human Social Behavior: An Anthropological Perspective*. (pp. 257–272). North Scituate, MA: Duxbury.

Irons, W. (1991). How did morality evolve? *Zygon: Journal of Religion and Science*, *26*, 49–89.

Irons, W. (1998). Adaptively relevant environments versus the environment of evolutionary adaptedness. *Evolutionary Anthropology*, *6*(6), 194–204.

Irons, W. (2000). Why do the Yomut raise more sons than daughters? In: L. Cronk, N. A. Chagnon & W. Irons (Eds), *Adaptation and Human Behavior: An Anthropological Perspective*. Hawthorne, NY: Aldine de Gruyter.

Kaplan, H., & Hill K. (1985). Hunting ability and reproductive success among male Ache foragers. *Current Anthropology*, *26*(1), 131–133.

Kaplan, H. S., & Lancaster, J. B. (2000). Skills-based competitive labor markets, the demographic transition, and the interaction of fertility and parental human capital in the determination of child outcomes. In: L. Cronk, N. A. Chagnon & W. Irons (Eds), *Adaptation and Human Behavior: An Anthropological Perspective*. Hawthorne, NY: Aldine de Gruyter.

Kitcher, P. (1985). *Vaulting Ambition: Sociobiology and the Quest for Human Nature*. Cambridge, MA: MIT Press.

Krebs, J. R., & Dawkins, R. (1984). Animal signals: Mind-reading and manipulation. In: *Behavioural Ecology: An Evolutionary Approach*, (2nd ed.) (pp. 380–402). Oxford: Blackwell Scientific.

Lancaster, J. B., & Kaplan, H. S. (2000). Parenting other men's children: Costs, benefits, and consequences. In: L. Cronk, N. A. Chagnon & W. Irons (Eds), *Adaptation and Human Behavior: An Anthropological Perspective*. Hawthorne, NY: Aldine de Gruyter.

Lewontin, R., Rose, S., & Kamin, L. (1984). *Not in Our Genes: Biology, Ideology and Human Nature*. New York: Pantheon.

Lieberman, L. (1989). A discipline divided: Acceptance of sociobiological concepts in anthropology. *Current Anthropology*, *30*(5), 676–682.

Low, B. (2000a). Sex, wealth, and fertility: Old rules, new environments. In: L. Cronk, N. A. Chagnon & W. Irons (Eds), *Adaptation and Human Behavior: An Anthropological Perspective*. Hawthorne, NY: Aldine de Gruyter.

Low, B. (2000b). *Why Sex Matters: A Darwinian Look at Human Behavior*. Princeton: Princeton University Press.

Lumsden, C., & Wilson, E. O. (1981). *Genes, Mind, and Culture*. Cambridge: Harvard University Press.

Luttbeg, B., Borgerhoff Mulder, M., & Mangel, M. (2000). To marry again or not: A dynamic model for demographic transition. In: L. Cronk, N. A. Chagnon & W. Irons (Eds), *Adaptation and Human Behavior: An Anthropological Perspective*. Hawthorne, NY: Aldine de Gruyter.

Lynch, A. (1996). *Thought Contagion*. New York: Basic.

MacArthur, R. H., & Pianka, E. R. (1966). On optimal use of a patchy environment. *American Naturalist*, *100*; 603–609.

Mace, R. (1993). Transitions between cultivation and pastoralism in sub-Saharan Africa. *Current Anthropology*, *34*(4), 363–382.

Mace, R. (1996). When to have another baby: A dynamic model of reproductive decision-making and evidence from Gabbra pastoralists, *Ethology and Sociobiology*, *17*, 263–274.

Mace, R. (2000). An adaptive model of human reproductive rate where wealth is inherited: Why people have small families. In: L. Cronk, N. A. Chagnon & W. Irons (Eds), *Adaptation and Human Behavior: An Anthropological Perspective*. Hawthorne, NY: Aldine de Gruyter.

Mangel, M., & Clark, C. W. (1988). *Dynamic Modeling in Behavioral Ecology*. Princeton: Princeton University Press.

Marcus, G. E., & Fischer, M. M. J. (1986). *Anthropology as Cultural Critique: An Experimental Moment in the Human Sciences*. Chicago: University of Chicago Press.

McGrew, W. C. (1992). *Chimpanzee Material Culture*. Cambridge: Cambridge University Press.

Mitchell, R. W., & Thompson, N. S. (Eds) (1986). *Deception: Perspectives on Human and Nonhuman Deceit*. Albany: State University of New York Press.

Moore, G. E. (1903). *Principia Ethica*. Cambridge: Cambridge University Press.

Moore, J. H. (1990). The reproductive success of Cheyenne war chiefs: A contrary case to Chagnon's Yanomamö. *Current Anthropology*, *31*(3), 322–330.

Orians, G. (1969). On the evolution of mating systems in birds and mammals. *American Naturalist*, *103*, 589–603.

Orlove, B. S. (1980). Ecological anthropology. *Annual Review of Anthropology*, *9*, 235–273.

Patton, J. Q. (2000). Reciprocal altruism and warfare: A case from the Ecuadorian Amazon. In: L. Cronk, N. A. Chagnon & W. Irons (Eds), *Adaptation and Human Behavior: An Anthropological Perspective*. Hawthorne, NY: Aldine de Gruyter.

Queller, D. C. (1995). The spaniels of St. Marx and the Panglossian paradox: A critique of a rhetorical programme. *The Quarterly Review of Biology*, *70*(4), 485–489.

Ridley, M. (1993). *The Red Queen: Sex and the Evolution of Human Nature*. New York: Macmillan.

Ridley, M. (1996). *The Origins of Virtue: Human Instincts and the Evolution of Cooperation*. New York: Viking.

Ristau, C. A. (Ed.). (1991). *Cognitive Ethology: The Minds of Other Animals*. Hillsdale, NJ: Lawrence Erlbaum.

Rogers, A. (1992). Resources and population dynamics. In: E. A. Smith & B. Winterhalder (Eds), *Evolutionary Ecology and Human Behavior* (pp. 375–402). Hawthorne, NY: Aldine de Gruyter.

Rubenstein, D. I., & Wrangham, R. W. (Eds) (1986). *Ecological Aspects of Social Evolution: Birds and Mammals*. Princeton: Princeton University Press.

Sahlins, M. (1976). *The Use and Abuse of Biology*. Ann Arbor: University of Michigan Press.

Scheper-Hughes, N. (1995). The primacy of the ethical: Propositions for a militant anthropology. *Current Anthropology*, *36*(3), 409–440.

Small, M. (1995). *What's Love Got to Do With It? The Evolution of Human Mating*. New York: Anchor.

Small, M. (1998). *Our Babies, Ourselves: How Biology and Culture Shape the Way We Parent*. New York: Anchor.

Smith, E. A., & Winterhalder, B. (Eds) (1992). *Evolutionary Ecology and Human Behavior*. Hawthorne, NY: Aldine de Gruyter.

Smith, E. A. (1992a). Human behavioral ecology: I. *Evolutionary Anthropology*, *1*(1), 20–25.

Smith, E. A. (1992b). Human behavioral ecology: II. *Evolutionary Anthropology*, *1*(2), 50–55.

Smith, E. A. (2000). Three styles in the evolutionary analysis of human behavior. In: L. Cronk, N. A. Chagnon & W. Irons (Eds), *Adaptation and Human Behavior: An Anthropological Perspective*. Hawthorne, NY: Aldine de Gruyter.

Sosis, R. (2000). The emergence and stability of cooperative fishing on Ifaluk atoll. In: L. Cronk, N. A. Chagnon & W. Irons (Eds), *Adaptation and Human Behavior: An Anthropological Perspective*. Hawthorne, NY: Aldine de Gruyter.

Standen, V., & Foley, R. (Eds) (1989). *Comparative Socioecology: The Behavioural Ecology of Mammals and Man*. London: Blackwell.

Steward, J. H. (1955). *Theory of Culture Change: The Methodology of Multilinear Evolution*. Urbana: University of Illinois Press.

Strassmann, B. I. (1996). Menstrual hut visits by Dogon women: a hormonal test distinguishes deceit from honest signaling. *Behavioral Ecology*, *7*, 304–315.

Strassmann, B. I. (2000). Polygyny, family structure, and child mortality: a prospective study among the Dogon of Mali. In: L. Cronk, N. A. Chagnon & W. Irons (Eds), *Adaptation and Human Behavior: An Anthropological Perspective*. New York: Aldine de Gruyter.

Strathern, A. (1971). *The Rope of Moka*. Cambridge: Cambridge University Press.

Sugiyama, L. S. (1996). In Search of the Adapted Mind: A Study of Human Cognitive Adaptations Among the Shiwiar of Ecuador and the Yora of Peru. Ph.D. dissertation, University of California at Santa Barbara.

Symons, D. (1989). A critique of Darwinian anthropology. *Ethology and Sociobiology*, *10*, 131–144.

Thornhill, R., & Palmer, C. T. (2000). *A Natural History of Rape: Biological Bases of Sexual Coercion*. Cambridge. MIT Press.

Tiger, L. (1969). *Men in Groups*. New York: Random House.

Tiger, L., & Fox, R. (1971). *The Imperial Animal*. New York: Holt, Rinehart and Winston.

Trivers, R. (1971). The evolution of reciprocal altruism. *Quarterly Review of Biology*, *46*, 35–57.

Trivers, R. (1972). Parental investment and sexual selection. In: B. Campbell (Ed.), *Sexual Selection and the Descent of Man* (pp. 136–179). Hawthorne, NY: Aldine de Gruyter.

Trivers, R. (1974). Parent-offspring conflict. *American Zoologist*, *14*, 249–264.

Trivers, R., & Hare, H. (1976). Haplodiploidy and the evolution of social insects. *Science*, *191*, 249–263.

Trivers, R., & Willard, D. E. (1973). Natural selection of parental ability to vary the sex ratio of offspring. *Science*, *179*, 90–92.

Trivers, R. (1985). *Social Evolution*. Menlo Park, CA: Benjamin/Cummings.

Turke, P. (1989). Evolution and the demand for children. *Population and Development Review*, *15*, 61–90.

Vayda, A. P. (1995). Failures of explanation in Darwinian ecological anthropology, parts I and II. *Philosophy of the Social Sciences*, *25*, 219–249 and 360–375.

Vayda, A. P., & McKay, B. (1975). New directions in ecology and ecological anthropology. *Annual Review of Anthropology*, *4*, 293–306.

Vining, D. R. (1986). Social versus reproductive success: The central theoretical problem of human sociobiology. *Behavioral and Brain Sciences*, *9*, 167–216.

Voland, E. (1998). Evolutionary ecology of human reproduction. *Annual Review of Anthropology*, *27*, 347–374.

White, L. A. (1959). *The Evolution of Culture: The Development of Civilization to the Fall of Rome*. New York: McGraw-Hill.

Williams, G. C. (1966). *Adaptation and Natural Selection*. Princeton, NJ: Princeton University Press.

Wilson, E. O. (1975). *Sociobiology: The New Synthesis*. Cambridge: Harvard University Press.

Winterhalder, B., & Smith, E. A. (Eds). (1981). *Hunter-Gatherer Foraging Strategies: Ethnographic and Archaeological Analyses*. Chicago: University of Chicago Press.

Winterhalder, B., & Smith, E. A. (2000). Analyzing adaptive strategies: Human behavioral ecology at twenty-five. *Evolutionary Anthropology*.

Winterhalder, B. (1977). Foraging Strategy Adaptations of the Boreal Forest Cree: An Evaluation of Theory and Models from Evolutionary Ecology. Ph.D. Dissertation, Cornell University.

Wrangham, R. W., McGrew, W. C., & De Waal, F. B. M. (Eds) (1994). *Chimpanzee Cultures*. Cambridge: Harvard University Press.

Wright, R. (1994). *The Moral Animal: The New Science of Evolutionary Psychology*. New York: Pantheon.

Zahavi, A. (1990). Arabian babblers: The quest for social status in a cooperative breeder. In: P. B. Stacey & W. D. Koenig (Eds). *Cooperative Breeding in Birds: Long-Term Studies of Ecology and Behavior*, (pp. 103–130). Cambridge: Cambridge University Press.

Zahavi, A., Zahavi, A., Zahavi-Ely, N., & Ely, M. P. (1997). *The Handicap Principle*. Oxford: Oxford University Press.

ECONOMICS IS AN EVOLUTIONARY SCIENCE

Paul J. Zak and Arthur T. Denzau

INTRODUCTION

In a widely read 1898 essay published in the *Quarterly Journal of Economics*, Thorstein Veblen asked why economics had strayed from its foundation in the natural sciences. The title of Veblen's paper is "Why is economics not an evolutionary science?" In it, Veblen identifies the many advances of classical and the emerging neoclassical economics, but critiques their use of hypothetical constructs of human decision processes. The paper is an appeal to return the focus of economic analysis to human beings. With one hundred years perspective, Veblen's essay is amazingly prescient. Nevertheless, as this chapter demonstrates, this call has been largely ignored by professional economists. To the contrary, the predominant approach to economic analysis in the twentieth century was not the biology of human behavior, but a systems approach mimicking the analysis of classical physics.

This chapter examines the mutual evolution of the biological and economic sciences, tracing their intimate relationship in the previous two centuries, their nearly complete divorce in the twentieth century, and their current gradual rapprochement. We do this by examining the individuals, many with familiar names, who forded the currents between disciplines, bringing to light the bi-directional contributions of one upon the other. By doing this, we reach the fundamental conclusion which is this chapter's title: Economics *is* an evolutionary science. This chapter does *not* argue that evolutionary biology is

Evolutionary Approaches in the Behavioral Sciences, Volume 8, pages 31–65.

ISBN: 0-7623-0769-2

the proper metaphor for economic analysis, but that methods and findings in the biological sciences need to be incorporated *directly* into economics if the discipline is to continue to produce relevant insights into human behavior.

A HISTORICAL CHRONOLOGY OF THE FEEDBACK BETWEEN ECONOMICS AND BIOLOGY

German medical doctor Karl Burdach (1776–1847) coined the term "biology" to refer to the study of human beings. His masterwork *Physiology as a Science of Experience* advocated the deduction of the bases of human behavior by observation. The purview of biology was expanded to include botany, zoology, geology, chemistry and meteorology by Jean Baptiste Pierre Antoine de Monet Lamarck in 1812 who was convinced that the scientific methods of observation and experiment applied uniformly to the study of humans, animals, cells, and molecules. A near-contemporary of Burdach, Adam Smith, wrote repeatedly in 1759's *The Theory of Moral Sentiments* of the "oeconomy of nature," when describing human behavior, borrowing the phrase from a 1751 book with this title by the father of biological taxonomy Carolus Linnaeaus. By his 1776 *The Wealth of Nations*, Adam Smith had dispensed with biology as his overarching metaphor, and adopted Newtonian mechanics. Henceforth the schism between the evolutionary basis in biology and economics, for the most part, grew.

The lacuna between biology and economics deepened in the 19th century because of a fundamental misunderstanding of the relationship between the two by the polymath Herbert Spencer (1820–1903). Spencer published a simple version of biological evolution before Charles Darwin, and was the first to use the terms "evolution" and "survival of the fittest" to describe biological dynamics. Unfortunately, Spencer did not clearly identify the mechanisms through which evolution occurs as did Darwin, and the latter is rightfully identified as the father of modern evolutionary theory, though for many years Spencer outshined Darwin and was considered one of the world's leading biologists. Spencer went on to write important works in sociology and psychology (Spencer, 1851, 1855, 1880), and to work as an editor at the *Economist* magazine. Nevertheless, Spencer is often now overlooked because he was unwilling to accept the non-directional nature of Darwinian evolution; instead his take on evolution continued to be both Lamarckian and teleological. Spencer (1851) wrote that "Progress . . . is a part of nature" (p. 65). As a result, Spencer's later work advocated laissez-faire economics in order to promote adaptation and social progress via Lamarckian learning. Nevertheless, Spencer was the first to integrate biological evolution with social theory, and his influence – for better or worse – is still present, if only tacitly.

In this section, we examine the historical co-evolution of biology, especially evolutionary theory, and economics, by examining the major influences during the past two centuries. Note that by "evolutionary theory" we mean the modern synthetic or neo-Darwinian theory that adjoins genetics, biochemistry, paleobiology, and natural history as supporting pillars to the basic theory of natural selection developed by Darwin.

Adam Smith

Several features of classical economic analysis have strong ties to biology. Natural law was a fundamental notion of the Enlightenment, focusing on the innate basis of human behavior. A key notion of the Scottish Enlightenment is the idea of *unanticipated consequences*, which relates closely to the Darwinian approach to evolution. Also related is the classical economic model of population growth. These three elements are all expressed in the work of Adam Smith (1723–1790).

Smith was fully aware of one of the earliest unanticipated consequences arguments, that of Bernard de Mandeville. Mandeville, in *The Fable of the Bees* (1714), argued that the selfish motives of humans did not necessarily cause the results of their actions to be bad for society as a whole. We would now term this a discovery of an emergent phenomenon, and this is a very sensible interpretation of the unanticipated consequences idea in Scottish Enlightenment philosophy.

Smith's contemporary David Hume specifically designed the specie flow model to reference economic phenomena in his 1752 work *On the Balance of Trade*. Once again, an emergent phenomenon, in this case the stability of equilibrium was invoked to explain the implications of a mercantilist surplus in the balance of trade between two countries in the context of the use of specie, gold, for the settlement of international payments.[1] This example of a negative feedback mechanism supporting an equilibrium naturally leads to similar uses both of equilibrium and of stability in Smith's work.

Smith's 1759 book, *The Theory of Moral Sentiments*, is explicit in the ties with biology. In various parts, Smith attempts to argue for the implications of natural sentiments assumptions, based on his conception of human nature. These were tied to the natural law approach promulgated by Enlightenment scholars,

---

1. It is interesting to note that the anachronistic economic use of "specie" and the biological "species" have the same Indo-European root indicating a particular form or type.

both in Scotland, and on the continent. But these quite explicit ties seem both quaint and far less convincing to the modern reader.

Much more convincing, and tied more to the emergent properties idea are many arguments in *The Wealth of Nations*. For example, a version of the Iron Law of Wages is developed in which a model of population growth is made explicit. The optimistic choices of couples to procreate, and the environmental support of high wages enable more than the usual number of children to survive to adulthood, driving wages back toward the natural equilibrium. This negative feedback mechanism works in a similar manner to that in Hume's specie flow, but involves more time, and illustrates the strong power of market analysis presented in Smith.

The most famous emergent property argument of Smith is his notion of the Invisible Hand:

> . . . he intends only his own gain, and he is in this, as in many other cases, led by an invisible hand to promote an end which was no part of his intention. Nor is it always the worse for the society that it was no part of it. By pursuing his own interest he frequently promotes that of the society more effectually than when he really intends to promote it. I have never known much good done by those who affected to trade for the public good. It is an affectation, indeed, not very common among merchants, and very few words need be employed in dissuading them from it (Smith, 1776, p. 423).

Here the unanticipated consequences are quite explicit, and the analysis is a more rigorous study of the basic idea in Mandeville. By isolating the crucial elements, Smith generates the type of argument style that is followed for another two centuries. The emergent property feature is an element that becomes the center of many important economic works that follow.

Thomas Malthus

The Reverend Thomas Malthus (1766–1834) produced one of the first among a substantial series of studies that followed in the path of Adam Smith, in his 1798 *Essay on Population*. In this work, Malthus develops in greater detail and extent the population model already stated briefly in Smith. Malthus was responding to the radical institutional critics of the traditional order who arose during the American and French Revolutions. Much of the work is a statement of and argument for claims of unanticipated consequences of existing institutions, and thus the emergent properties in higher-level equilibria.

The extended deductive argumentation and relatively rigorous style of much of it was a departure from Smith, who was seldom far from an example to illustrate his theoretical analysis. The greater level of abstraction followed by Malthus' friend, David Ricardo, allowed the development of arguments that

could be used by fields other than economics or moral philosophy. One of the fields which benefited from these abstract arguments was biology, when Charles Darwin encountered the bewildering display of plants and animals on his round-the-world tour of duty as naturalist on the HMS *Beagle*.

Darwin had read Malthus' essay, and its deductive logic was the basis for his non-theistic explanation for the fit between organism and environment via differential survival of organisms with diverse traits. Indeed, Darwin identified Malthus' notion of "crowding and struggle" among human populations in his *Essay* as the germ from which he formulated natural selection as a general principle. Once again, emergent properties became a crucial element that united economics and biology in the eighteenth and nineteenth centuries.

The Neoclassical Reaction to Darwin and Darwinism

In English-language countries, Darwin's writings produced reactions of both horror and emulation. For those following naturalistic religions and the Argument from Design (William Paley, 1802), Darwin was a challenge to their praecognita for the existence of God. If one did not need to explain the remarkable fit between organisms and their environment by divine means, then religious belief was forced to defend itself only on traditional grounds, such as faith.

The reaction in the social sciences was diffuse. Some social scientists responded by incorporating evolutionary ideas into their fields of study. For instance, anthropologists applied Social Darwinism to study human groups. Different fields incorporated different interpretations of Darwin and evolution, but these fields generally attempted to make their efforts at least consistent with this bold new approach.

Economic theorists in the English-language countries responded in a very different manner. Mainstream analysis in the mode of what came to be called neoclassical economics, can be viewed as a reaction against Darwin and the ensuing revolution in biology. Instead of attempting to admix evolutionary findings into economic analysis, neoclassical economists looked instead to Newtonian physics for methodological guidance.

The fundamental organizing principle of classical physics is *equilibrium* (though this is not the case for quantum physics). This construct appealed to neoclassical economists as it enabled the derivation of rigorous predictions ("theorems") from economic models. As a result, neoclassical economists nearly always built mathematical models that had well-defined equilibria. Neoclassical economic models, based on differential calculus invented by Newton (and

Leibnitz), focused on analyzing static equilibria, with little or no concern at all for dynamics. Equilibrium relationships are powerful as they generate testable hypotheses, and are the primary reason for the ascension of economics among the social sciences. Nevertheless, a number of economists in the vanguard understood that equilibrium models are not the be-all and end-all of economic analysis.

Alfred Marshall

Alfred Marshall (1842–1924), whose 1890 *The Principles of Economics* became the basis for much of twentieth century microeconomics, initiated the marginalist revolution by utilizing differential calculus in imitation of classical physics. Thirty-five years after this fundamental work, Marshall (1925) asked post-humanists whether the more advanced problems of economics can be solved by analogies from physical dynamics. His answer is no: biology is where these answers reside.

Foreshadowing this argument, Marshall writes in Book I of *Principles* that "In this matter economists have much to learn from the recent experiences of biology: and Darwin's profound discussion of the question throws a strong light on the difficulties before us." Further, Marshall understood that economics is the study of the forces of change, and is therefore "a branch of biology, broadly interpreted." In Marshall (1925), he extends this exhortation, "Since that time biology has more than repaid her debt; and economists have in their turn owed much to the many profound analogies which have been discovered between social and especially industrial organization on the one side and the physical organization of the higher animals on the other."

Marshall was convinced that the economics of the future would be based not on classical physics, but on biology. Outside of some notions in Marshall's work on the aging of firms, and the interaction of firms in an industry, he did not live long enough to pursue this shift himself. With each step of advance, the *ceteris paribus* crutch gets further from reality. Instead, he suggested that the study of fluctuations around an equilibrium should be understood as a dynamic, evolving process. This is a very different Marshall than is usually presented in historical surveys. Even in Blaug (1985), there is almost no mention of this methodological shift advocated by the architect of the neoclassical school.

Most of the theoretical focus of the next two generations of economists solidified and added precision to the assumptions and arguments of Marshallian neoclassical analysis. This program by his followers at Cambridge University was complemented in the United States by the work of Frank Knight (1921) and John Bates Clark, increasing the rigor and extending the implications of the static analysis he fostered in his 1890 treatise. After World War II, many

mathematically trained researchers, both from economics and from the physical sciences continued the effort to add precision to microeconomics. This effort expanded and formalized mathematical structures, but seemed to thrive on rigor alone. This formalization of economics has resulted in a theory that "is more explicitly mathematical than the work of theoretical physicists." Thus, the legacy of Marshall is marginal analysis using differential calculus and physical reasoning, and not on his stated hopes of basing economic theory on biological reasoning. Marshall's final statement in his 1925 article is that "The Mecca of the economist is economic biology rather than economic dynamics."

Carl Menger

A scholar with leanings similar to Marshall's is another of the founders of neoclassical economics, the Austrian Carl Menger (1840–1921). Menger combined differential calculus and the Newtonian approach to microeconomics with biological forms of analysis. His most important contribution in this area is his analysis of the origin of money. Menger posited that money arose organically as an unanticipated consequence of the inefficiencies of barter exchange. As a result, money is valued without human intervention, and therefore does not require, for example, a government to set its value. He writes that "the origin of money can truly be brought to our full understanding only by our learning to understand the social institution [money] discussed here as the unintended result, as the unplanned outcome of specifically individual efforts of members of society" (1883, p. 155). Menger's analysis showed that money evolved to act as a medium of exchange, a finding verified by modern monetary theorists (Kiyotaki & Wright, 1989, 1991; Wright, 1995).

Menger showed that coordinating on a single money allows substantial efficiencies to be achieved (1892). The approach he takes to illustrate this is explicitly evolutionary, albeit informal. But the ideas have readily been turned into simulations, as illustrated in the recent work of Luo (1999). For this methodological advance, Menger is rightfully viewed as an early evolutionary economist.

American Institutional Economics and Thorstein Veblen

In contrast to the neoclassical reaction to Darwinism is the work of American Institutional economists of the late nineteenth and early twentieth centuries. This approach to economics is less theoretical and deductive than neoclassical theory. Instead, it takes a sociological approach to economic behaviors, focuses on the institutions and organizations of the state and economy. Among the most

prominent early Institutionalists were John R. Commons, Richard Ely, and Thorstein Veblen.

On balance, Institutionalists had little effect on mainstream theoretical economics in the twentieth century. Part of the reason for this is the growing professionalization and development of academic economics in the United States, beginning in the 1880s. Increasingly, those interested in economics were working at colleges and universities, and a professional journal literature gradually developed. As a result, the writing of economics was aimed at other professional economists, which increased the homogeneity of economic training and thinking. A few academic centers retained an Institutionalist bent, but most made a full transition to neoclassical analysis. The standards of theory shifted, and the Institutionalists failed to develop alternatives to the Marshallian theory of choice. Institutionalists, without a means of explaining the fundamentals of neoclassical theory – prices and transaction quantities – were largely ignored by neoclassical economists, and Institutionalist writings became viewed as a variant of sociology rather than a complement to neoclassical analysis.

Certainly the most interesting American Institutionalist for our purposes is Thorstein Veblen (1857–1929), who was well-acquainted with the works of Herbert Spencer, Charles Darwin, and the social Darwinist William Graham Sumner, and was also among the first to unite natural selection via environmental influences with the laws of Mendelian inheritance. Veblen was a harsh critic of the neoclassical reaction against Darwinism, and used caustic language about its thin psychological foundation and abandonment of social influences. For example, he characterizes neoclassical economics as using an outdated conception of man:

> The psychological and anthropological preconceptions of the economists have been those which were accepted by the psychological and social sciences some generations ago. The hedonistic conception of man is that of a lightning calculator of pleasures and pains who oscillates like a homogeneous globule of desire of happiness under the impulse of stimuli that shift him about the area, but leave him intact. He has neither antecedent nor consequent. He is an isolated definitive human datum, in stable equilibrium except for the buffets of the impinging forces that displace him in one direction or another. Self-imposed in elemental space, he spins symmetrically about his own spiritual axis until the parallelogram of forces bears down upon him, whereupon he follows the line of the resultant. When the force of the impact is spent, he comes to rest, a self-contained globule of desire as before (Veblen, 1898, p. 242).

Note the explicit comparison of neoclassical choice theory to static models in physics. This approach involves no learning, and thus cannot be evolutionary.

An evolutionary approach would involve both learning and a social context. He wrote that "The individual is but a single agent in each case," and that "an evolutionary economics must be the theory of a process of cultural growth as

determined by economic interest, a theory of cumulative sequence of economic institutions stated in terms of the process itself" (p. 243). These elements are still essential to building an evolutionary economics.

Clearly, Veblen wanted economics to follow psychology and anthropology along an evolutionary path, but the neoclassical approach at the dawn of the twentieth century took a tangential path. For Veblen, neoclassical economists are content with "repairing structure and doctrines and maxims resting on natural rights, utilitarianism, and administrative expediency." Instead, the modern sciences are evolutionary sciences, and he was convinced economics should join them.

Joseph Aloysius Schumpeter

While not himself a practitioner of evolutionary ideas in economics, Joseph Aloysius Schumpeter (1883–1950) developed concepts and made claims that have been investigated with evolutionary methods. In his most widely read work, *Capitalism, Socialism and Democracy* (1942), he introduced the idea of *creative destruction*, perhaps his most important insight into economic behavior. He argued that the most significant type of economic competition is not price competition. Instead, "The essential point to grasp is that in dealing with capitalism we are dealing with an evolutionary process" leading to competition between firms. Because "Capitalism, then, is by nature a form or method of economic change and not only never is but never can be stationary," a point also recognized by Marx, as change "alters the data of economic action." As a result, economic structures endogenously change, creating winners and losers via a selection process. Those who run firms, knowing this, will seek to anticipate structural changes in the struggle for economic survival. This occurs because "The fundamental impulse that sets and keeps the capitalist engine in motion comes from the new consumers' goods, the new methods of production or transportation, the new markets, the new forms of industrial organization that capitalist enterprise creates" (p. 82).

This fundamentally dynamic view of the economy clearly follows an evolutionary approach. Innovations, by their nature, are largely unpredictable, and an evolutionary model is the best way to incorporate them into economic theory. Instead of attempting to predict specific innovations, evolutionary models characterize the impact of innovation on the economy. For example, one can simulate the effects in an evolutionary system of a more diverse set of trials from which to select. The greater the diversity, the more rapid is the rate of change of the population of designs. Similarly, the more trials there are, of a given degree of diversity, the more rapid is evolutionary change. The

importance of changing mutation rates and noise in the filtering process can also be studied with ease in an evolutionary simulation.

The link between computer simulations and evolutionary economics is crucial. With computer capabilities improving 25% to 50% per year, the increasing ability to model evolutionary phenomena computationally certainly exceeds the ability of researchers to acquire new mathematical techniques. Further, computational experiments often guide and suggest analytical theorems. For example, see the feedback between these approaches in Mahaffy, Zak and Joiner (1995). It is becoming clear that better behavioral models in economics are possible as computational capabilities grow. Popper in *The Self and Its Brain* (1984) wrote that

> A computer is just a glorified pencil. Einstein once said "my pencil is cleverer than I." What he meant could perhaps be put thus: armed with a pencil, we can be more than twice as clever as we are without. Armed with a computer (a typical World 3 object), we can perhaps be more than a hundred times as clever as we are without; and with improving computers there need not be an upper limit to this (p. 208).

Evolutionary simulations, following the approach of Schumpeter, are an important approach utilizing this dynamic.

Friedrich August Von Hayek

One of the more explicitly evolutionary thinkers in economics is the Austrian and 1974 Nobel laureate in economics, Friedrich Hayek (1899–1992). Hayek viewed human evolution as having prepared us to live in small groups at the hunter-gatherer stage of development. He thought there had been too little time and too few generations for much further development of the human brain by evolutionary means. Thus, economic "rationality" was often Pleistocene rationality.

In his 1989 Presidential address to the American Economics Association, he summarized his understanding of markets and the role of theory. He argued that the market is a profoundly subtle information aggregation device. The market contains more information than any one individual could. Further, the market is "a more efficient mechanism for digesting dispersed information than any that man has deliberately designed" (Hayek, 1989, p. 7). As a result, the market efficiently coordinates economic activity, and does so without needing government intervention. Hayek's focus on the informational properties of markets is a natural outcome of viewing economic activity in an evolutionary manner, as evolution is an information parsing process.

For example, Hayek argues that one cannot determine a particular equilibrium price except through markets. He further contended that social sciences and natural sciences both deal with structures of essential complexity. Human

sciences like economics study phenomena of organized complexity as opposed to the unorganized complexity in physics. As a result, we cannot replace models involving individual elements with statistical data. Instead, we must pay attention to fluctuations from equilibrium and to diversity, as these produce the evolutionary dynamics in human and non-human systems.

Hayek's thinking about economic theory is clearly based on evolutionary reasoning. He claims that while we cannot predict the outcome of set of interactions, we can make useful statements about the direction of change. In other words, while only some properties can be predicted by theory, one is still able to produce falsifiable propositions. To expect more than this results in charlatanism (p. 7): "To act on the belief that we possess the knowledge and the power which enable us to shape the processes of society entirely to our liking, knowledge which in fact we do not possess, is likely to make us do much harm." He also states that "in the social field the erroneous belief that the exercise of some power would have beneficial consequences is likely to lead to a new power to coerce other men being conferred on some authority." In an earlier paper, Hayek (1983) focused on the idea of unanticipated consequences. "We are led to do things by circumstances of which we are largely unaware" (p. 46). The mechanism he argued that produced this effect was cultural evolution at the group level instead of genetically-based Darwinian evolution. This evolutionary group selection argument concerned human institutions – groups which adopted the rules of private property, honesty and family prospered and multiplied, displacing other groups. Even though some political and religious leaders have advocated social organizations opposed to private property and the family, these institutions have survived, while those organizations opposed have seldom lasted more than a century. This is natural selection at work, albeit cultural Lamarckism. Kelm (1997) is a recent Darwinian analysis of instutitonal evolution.

Hayek used evolutionary principles throughout the body of knowledge he developed, with special focus on the information flowing through markets. Rather than simply thinking of the economy as being at a static equilibrium, he examined the evolutionary processes that move the economy between equilibria.

Paul Anthony Samuelson

Among the many scientific contributions of Paul Samuelson (1915–), perhaps his most important – at least in terms of the literature it generated – is the introduction in 1958 of the overlapping generations model. A demonstration of the impact of the overlapping generations model is the 504 page graduate

textbook by Azariadis (1993) which is entirely based on this approach. The overlapping generations model originally appeared in the ecological literature in an analysis of multiple species by Leslie (1945), though Samuelson was not influenced by Leslie's work and developed the model *sui generis*. Samuelson was, however, keenly aware of the analysis of population dynamics in Alfred Lotka's *Elements of Mathematical Biology* (1956) as well as in Lotka's earlier journal articles, and maintained a correspondence with Lotka (Samuelson, 2000).

In the overlapping generations model, Samuelson showed that a primary reason for loans between individuals is that young people desire to save, while the old seek to convert their savings into consumption during retirement. Through this arrangement, Samuelson finds "a biological rate of interest equal to the rate of population growth" (p. 474). This insight brought demographics, and more generally, biology, into macroeconomics after the model was extended to include production by Peter Diamond in 1965. The primary insight that Samuelson made was that heterogeneity is fundamental for economic exchange. The notion harkens back to Adam Smith's "double coincidence of wants," but uses a scientifically rigorous method to demonstrate this claim. The biological basis for interest rates has recently been derived by Gordon Getty (1999) from a much different point of departure, showing the wide applicability of this finding.

It is not surprising that Paul Samuelson integrated biology and economics. He has contributed to the theory of population biology (Samuelson, 1974, 1978, 1983; Samuelson & Yellin, 1977; Yellin & Samuelson, 1974), and has advocated the use of biological methods in economics (1993, 1985). In particular, Samuelson has sought to re-incorporate demography into economic models, writing that "Population analysis is at the intersection of social science and biology. Once upon a time, throughout the heyday of classical economics, demography belonged to political economy" (1985, p. 166). He laments that neoclassical economics has "removed population as a variable" subject to analysis, and that economists are far from a "satisfactory state of knowledge in these between-discipline domains" (p. 167–168). Samuelson fully understands that the development of the biological foundations in economics will be impeded by those wedded to traditional methods. Ultimately, Samuelson drolly observed, knowledge advances "funeral by funeral."

Influences from Economics to Biology

One of the fundamental principles in biology is scarcity. Michael Ghiselin (1974), following Linneaus (1751), calls this "the economy of nature." Because of scarcity, the natural economy is competitive. This is the essence of natural selection – the more efficient species in a given environment survive. Just as

recessions have a "cleansing" effect on the economy by causing inefficient firms to go extinct, environmental changes lead some species into extinction and others to ascendancy. It is in this sense that heterogeneity has an adaptive advantage. Just-in-time production processes and the wild boar are presumably both adapted to their particular niches. A change in the economic environment may cause a variant of a production "species" to have a differential survival advantage. Entrepreneurship is precisely niche-seeking.

Technically, both economics and biology are concerned with optimization (e.g. of fitness, or profits, or utility) under environmental constraints (e.g. predators, or crowding, or budgets, or a given market structure). In addition, biological entities from genes to species behave strategically, opening up biological modeling to game-theoretic methods. For example, Ridley (1995) documents male chimpanzees preferentially hunting for large game and sharing it with other members of their troop after the successful hunter first eats his fill. Subsequently, while the other males are eating, the successful hunter seeks to have sexual liaisons with the female partners of the other males. Female chimpanzees acquiesce to these liaisons since hunting prowess is a strong signal of desirable qualities in a mate, viz., the ability to provide resources, which makes the successful hunter a valued mate. This example also illustrates a division of labor in the natural economy. The division of labor occurs within organisms, among organisms, and within societies.

The chimpanzee behavior described above is explicitly strategic, and therefore can be described in game-theoretic terms. The first application of game theory to biology was Lewontin (1961), with subsequent important work by Maynard Smith (1972, 1982) who introduced the notion of an Evolutionarily Stable Strategy (ESS). An ESS is behavior among competing organisms that maximizes fitness in a given environment – even as other organisms adopt competing strategies. Robert Axelrod (1984) showed that a tit-for-tat strategy in a round-robin competition can support cooperative outcomes by purely selfish players. In a more general model, Hirshleifer and Martinez Coll (1988) show that tit-for-tat is less successful at maintaining cooperation in more general environments, though strategies that include punishment do sustain cooperation for a proportion of agents. This result was recently extended by Segal and Sobel (1999). The theory of kin selection developed by William Hamilton (1964) implicitly has organisms (or genes) comparing marginal benefits and costs when determining when to reciprocate cooperative behavior, just as human agents do in the cited game-theoretic models.

Robert Trivers (1972) and Trivers and Willard (1973) use explicit marginal cost-marginal benefit analysis to characterize, respectively, intrasex mate competition driven by differential parental investments in offspring, and variations in

organic sex ratios as environmental resources change. These fundamental papers in sexual selection apply "economic" reasoning to explain observed animal behaviors. Such reasoning quite nicely describes the effect of selection pressures on species, organisms, and genes. At each level of biological analysis, evolution induces a constrained optimization problem representing the grand compromise of nature. Trivers (1985) also develops this line of reasoning. Nevertheless, species are designed to be attracted to not the average attributes in a mate, but the extraordinary, in order to raise the reproductive advantage of one's offspring. One indicator of this behavior is exaggerated features that indicate a mate's reproductive value, for example, extra-bright or large plumage on birds, larger wings on butterflies, and surgically enhanced breasts in human females. When mating, species of all types seek to optimize, resulting in a "red queen" effect in mate preference: each sex must evolve just to keep from falling behind.

Paul Krugman (1996) notes the close methodological correspondence between evolutionary biology and economics, writing that

> To read the real thing in evolution – to read, say, John Maynard Smith's *Evolution and the Theory of Games*, or William Hamilton's new book of collected papers, *Narrow Roads in Gene Land*, is a startling experience to someone whose previous idea of evolution comes from magazine articles and popular books. The field does not look at all like the stories. What it does look like, to a remarkable degree, is – dare I say it? – neoclassical economics.

A partial list of open issues in biology that economic theory can fruitfully be used to examine include the analysis of multi-species communities, biodiversity and ecology, a fundamental theory of the gene as an informational unit, niche-seeking as entrepreneurship driven by competition for resources, social institution formation among non-human species, and better models of population genetics and dynamics. Much of this list comes from an editorial by economist Janet Tai Landa and zoologist Michael Ghiselin (1999), co-editors of the *Journal of Bioeconomics* which debuted in 1999. This journal seeks to promote the synthesis between economics and biology by providing a forum for the bi-directional transfer of ideas. Landa and Ghiselin's stated goal is "extending and integrating economics and biology" (p. 10). This is an important step toward the fusion that is now occurring between these two fields. An early example of this synthesis is Gordon Tullock's (1971) analysis of foraging by the coal tit. In the next two sections, the morphology of this emerging synthesis will become clear.

EPISTEMOLOGY AND METHODOLOGY

What is the methodology that we are advocating? Thomas Kuhn (1962) suggested an approach to understanding science using the concept of paradigms.

In the battle between conflicting paradigms, he argued that a judgment between them cannot be made on the basis of each paradigm's standards alone, as each competing paradigm will certainly be a better fit with the evaluative features of that paradigm. Judgment must be on some other basis. In postmodern times, we have some who question the existence of objective reality, and interpret Kuhn's ideas to imply that all scientific knowledge is a social construct, and therefore relativistic. Quite simply there is no difference in this view between scientific knowledge and knowledge in nonscientific areas, such as music or art.

Other approaches to understanding knowledge lead to similar reinterpretations of the status of scientific knowledge. In the hermeneutic approach to knowledge, understanding is embodied in language communities, and meaning is only shared within that community. The idea of external, objective referents is outside this view of knowledge. In the deconstruction movement, meaning is in the mind of the reader of a "text," and there is again no relevance for external, objective references of the sort required for a science. In these approaches to the meaning of truth and knowledge, the difference between science and non-science is not a sharp qualitative divide, as argued by Popper (1959), as constituting statements that are potentially falsifiable by observation of an external reality. The existence of external reality is in fact either denied, or at least a contentious idea in these newer philosophical schools.

This raises the issue of the basis of knowledge in economics. Evolutionary analysis provides an answer to this fundamental question. Epistemological logic is the foundation upon which one can assert that a claim is true. *Evolutionary epistemology* – based in substantial part on the work of Karl Popper – developed over the past forty years to provide serious answers to this question; answers that underpin economic analysis. Popper posited that every organism has inborn reactions or responses, which are often, though not always, appropriate to expected situations. For these reactions to have survived, they must have been advantageous more often than not. "One of the most important of these expectations is the expectation of finding a regularity" (1959). This argues for two claims, linking evolutionary biology and epistemology. First, surviving organisms will come to embody models of the external reality that reflect important regularities of the environment. There would be no basis for these models without such an external reality, and some degree of regularity in it. Second, there is an inborn propensity to find regularities in the environment, and it is likely that our approach to learning reflects this expectation.

On 15 November, 1960, Popper presented what is known as the three worlds approach, the three worlds being the external world of reality, the world of sensations, and the internal world of mental models. He linked this approach

to biology, providing support for Hegel's theory of the objective mind (Radnitzky, 1987, p. 18–20). Popper (1984) wrote that

> The main task of the theory of knowledge is to understand it as continuous with animal knowledge; and to understand also its discontinuity – if any – from animal knowledge.

This was a break from his earlier *Logic of Scientific Discovery* (1959), which defined scientific knowledge with the demarcation criterion: a statement is scientific if it is, at least in principle, falsifiable. All other statements are non-scientific.

Popper proceeded to argue that scientific knowledge and prehuman knowledge are the best means to underpin epistemology. Both scientific knowledge and the biologically based cognitive structures of animals can be viewed as products of evolution. He argued that both are produced by the same Darwinian mechanism: they are both the product of blind variation and selective retention, also known as trial and error. This same process governs the biological emergence of knowledge and the growth of knowledge in science. Instead of focusing on the subjective interior experience of the individuals – beliefs and perceptions – Popper focused on the objective products of the cognition. He contended that truth lies in the outcome of tests of conjectures. "We do not know: we can only guess" (1984) This process of discovering truth must be controlled by a method involving systematic tests of external reality.

This is the basis of evolutionary epistemology, an approach to the philosophical question of the source of knowledge that employs a scientifically based answer.

The Basic Evolutionary Model of Selective Retention

The basic model in evolutionary epistemology and evolutionary economics applies a three step learning algorithm. Consider an evolving population, whether of living things, or of ideas, in which:

(1) Variation exists – a large number of "experiments" or trials take place – this may be due to random variation, or a more directed process based on mental models;
(2) Selection by a filter occurs in which designs judged to be "better" survive at a higher rate than other designs; and
(3) Retention by reproduction or memory of those selected by the filter (e.g. written records, digital bits, or tasks repeated in a work system).

This is known as the model of *selective retention*, and is the basis of any evolutionary model. It is the heart of the neo-Darwinian approach to biological evolution, and was present in the original work published by Darwin in 1859.

Is there support for this simple evolutionary approach to understanding the creation of knowledge? If all three of the factors above are not included in the process, little knowledge is learned over time due what is known as a *combinatorial explosion* (Campbell, p. 105). We illustrate this using the *British Museum algorithm.* Kurt Lasswitz (1958) wrote a science fiction story, "The Universal Library," describing how all of knowledge was being produced and codified by a collection of monkeys punching keys on typewriters. The improbability of this story is due to the way that possibilities explode with each new choice of a letter. At each step, one of 26 keys could be typed, and in the course of typing ten letters, there are $26^{10} = 1.4 \times 10^{14}$ possible combinations that could be produced by random permutations. Clearly, this number rises as one types more than ten keys. Consider the standard English sentence used in typing classes: "Now is the time for all good men to come to the aid of their country." which has 71 characters, including spaces and punctuation. Typing at random among the 37 letter and punctuation keys gives us $37^{71} = 2.2 \times 10^{111}$ possible lines of 71 characters. A simple calculation demonstrates the near impossibility of arriving at the desired sentence this way. Suppose every atom in the universe operated as a computer capable of deciding if a particular set of 71 characters were the desired sentence. Even working with maximum speed, making decisions in the time it takes a light photon to cross the nucleus of a hydrogen atom, it would, on average, take more time to find this sentence than the current age of the universe. This is an obviously futile approach to producing a Shakespeare play!

This example demonstrates that random variation alone – that is, trial and error without memory and retention of successes – cannot be the basis of knowledge, or of biological life. The process by which knowledge grows requires cumulation and maintenance of successes, if not memory of failures. Popper (1969) writes that "our knowledge grows only through the correcting of our mistakes" (p. i).

Evolution solves the problem of combinatorial explosion by solving a long sequence of small steps, rather than a global, one-shot trial such as monkeys seeking to type a Shakespeare play. Each step of the evolutionary sequence is tested by a selection filter, and the more successful trials are retained. Further, the process takes advantage of parallelism. The number of trials can be enormous, and it may only take one successful trial to avoid a large number of failed pathways.

Once a small improvement occurs, it is selected for and retained in the population. Further improvements build on this foundation. Such a sequence of small steps gradually and continually improves the fit of the selection unit – whether ideas or organisms – to a particular environment. Instead of succeeding

in one large leap, the process continually climbs up the hill toward a maximum. The model of selective retention is even more important when the environment itself is changing. In this case there is no global optimum, only a constrained best approximation to the optimum. Isaac Newton (1675) understood this when he wrote in a letter to his rival Robert Hooke, "If I have seen further [than you and Descartes], it was only because I stood on the shoulders of giants."

Equilibrium or Kaizen?

One of the first explicit uses of an evolutionary approach to economic analysis is the remarkable paper by the Armen Alchian (1950). Alchian proposed a completely new class of economic models. He began by removing the unrealistic assumption of perfect competition typically used in standard economic analysis. Then, he included incomplete information, and uncertain forecasts of the future. Most radically, he dispensed with optimizing choices based on profit maximization, "where foresight is uncertain, 'profit maximization' is *meaningless* as a guide to specifiable action" (p. 211). Instead he modeled environmental adaptation through imitative and trial-and-error schemes by firms in pursuit of positive profits.

In this way, Alchian built a model of variation and (Schumpeterian) selection by the marketplace. The firms in his model seek maximum profits, but do not know the best way to achieve this. This model humbly admits that the ability to measure the implications of choices on costs and profits is imperfect, and in the face of true Knightian uncertainty (1921), models of choice under risk are vacuous. This follows because such models assume that decisions are based on a complete menu of possible actions and their related outcomes. Obviously, this is not true – innovation is the discovery of new choices not previously imagined, and the menu of alternatives is continually changing.

The standard economic model of firm choice is similar to Frederick Taylor's approach to industrial engineering. Taylor (1911) advocated engineering studies that include time and motion measurements in order to determine the unique best method of production. The job of the worker was merely to follow instructions in implementing the best method as specified by the industrial engineer. This approach to mass production delivered major benefits to American and European industry in the first half of the twentieth century, but then a better approach appeared – the Toyota Production System. This system, a set of adaptations to bank-required restructuring and inventory reductions (see Shingo, 1981, for a discussion of the system and its history) is in nearly complete opposition to Taylor's approach. Every production worker in the Toyota system has the power to change the way they work, in order to increase quality and

productivity. Instead of assuming there is a unique, identifiable maximum, the Toyota view is evolutionary: improvements in terms of the selection filter always exist, and can be found by allowing trials to occur. Each production worker trying a new approach is another trial, to be selected for its value. Instead of a few industrial engineers determining the static equilibrium production process, thousands of production workers with a wide variety of ideas are harnessed in the dynamic process of improving productivity and quality. This method is called *kaizen* in Japan. A basic result from the selective retention model is that the more diverse and numerous the trials in the first stage (the greater the variation in the initial population), the faster the process evolves.

This approach is also the basis of a method of maximizing functions without uniformly signed second-order conditions – the *genetic algorithm*. It is clear that wanting to maximize and knowing how to maximize are not the same thing, but neoclassical economics treats them identically. The difference between them is the basic difference between the usual equilibrium modeling of the firm, and an evolutionary model. Improvements must be discovered in an evolutionary model, and are not simply the result of satisfying specific marginal conditions.

In fact, if we take the problem of second-order conditions seriously, instead of assuming them away as is normally done, all the standard optimization method can guarantee is that a local optimum has been found. Optima with higher utility or profits may be attainable, but undiscoverable except through trial and error. The genetic algorithm accepts this challenge, and produces a sequence of local optima, though there is no guarantee that any of these is the global optimum even if one can prove that such a global optimum exists. Existence proofs are not necessarily constructive, while the genetic algorithm explicitly constructs local optima in a practical way. This method also applies to econometric optimization. Dorsey and Mayer (1995) show that several published nonlinear econometric models fail to find the global maxima of their respective likelihood functions. In most cases, a genetic algorithm produces quite different parameter estimates.

Alchian (1950, p. 213) uses a selection filter based on ex post realized profits, not ex ante expected profits. He assumed that firms with positive profits survive, and those with losses perish. Variations in production processes result both from chance and from explicit attempts to adapt to the environment. For either source of variation, the most important aspect of the model is the filtering and retention of successes. This occurs during interactions with the population of other firms, including competitors, suppliers and buyers. The environmental filter shapes the characteristics of the population of surviving firms. Put differently, the agents in this system attempt to solve a signal extraction problem. The environment is sending signals regarding production processes that improve profits, but the

signals are noisy. No individual firm is able to solve this problem due to its complexity, but, over time, the population of firms gradually improves the ability to do so.

Is there an equilibrium to such a system? Alchian (p. 215) argues that individuals choosing actions randomly may produce outcomes that are consistent with models of perfect foresight. That is, the results generated by a genetic algorithm are often observationally equivalent to choices by agents with complete knowledge of the environmental parameters. This suggests that evolutionary models may produce outcomes identical to those predicted by standard economic models, though this is by no means certain. Blaug (1980, p. 117) calls this the Alchian thesis, "the notion that all motivational assumptions in microeconomics may be construed as as-if statements." Friedman (1953, p. 22–23) follows this path and interprets the long-run characteristics of the surviving population of firms as exactly that described by an equilibrium model of microeconomics. This is fundamentally at odds with the notion of *kaizen* and the methodological approach of much of the work employing evolutionary ideas. Nevertheless, as the next section shows, evolutionary economics has value independent of the veracity of the Alchian thesis.

THE STATE OF THE ART AND PROSPECTS FOR THE FUTURE

Let us return to Thorstein Veblen to ask where economics is headed. In his 1898 essay he states that

> The process of change in the point of view, or in the terms of definitive formulation of knowledge, is a gradual one; and all the sciences have shared, though in an unequal degree, in the change that is going forward. Economics is not an exception to the rule, but it still shows too many reminiscences of the "natural" and the "normal," of "verities" and "tendencies," of "controlling principles" and "disturbing causes" to be classified as an evolutionary science . . . it is precisely in this use of figurative terms for the formulation of theory that the classical normality still lives its attenuated life in modern economics; and it is this facile recourse to inscrutable figures of speech as the ultimate terms of theory that has saved the economists from being dragooned into the ranks of modern science.

The error in this mode of analysis lies in viewing humans as simply hedonistic calculating machines, separate from their biology. Veblen also observed that

> In all the received formulations of economic theory, whether at the hands of English economists or those of the Continent, the human material with which the inquiry is concerned is conceived in hedonistic terms; that is to say, in terms of a passive and substantially inert and immutably given human nature.

He proposes a method through which economics will be an evolutionary science,

> The later psychology, re-enforced by modern anthropological research, gives a different conception of human nature . . . [I]n the view of the science, they are elements of the existing frame of mind of the agent, and are the outcome of his antecedents and his life up to the point at which he stands. They are the products of his hereditary traits and his past experience, cumulatively wrought out under a given body of traditions, conventionalities, and material circumstances; and they afford the point of departure for the next step in the process.

The significance of Veblen's tract is his exhortation that economists return to the primary method used in biology: methodical observation of the creature under analysis, both in the laboratory and in the field. This requires that economists read the literatures in evolutionary biology, psychology, behavioral genetics, and natural history, and integrate these findings and methods into their analysis. Economists have generally preferred to behave like philosophers, sitting in towers without windows and believing that their ruminations accurately reflect the world, rather than verifying that this is so. Popper (1985) called this approach to science "the world is my dream."

Let us be clear here, the method we are advocating is a difficult and time-consuming task, especially as the literature in the natural sciences is becoming more difficult to comprehend (Hayes, 1992). It is at the same time vital to the health and continued relevance of the field of economics. As E. O. Wilson eloquently argues in his 1998 book *Consilience: The Unity of Knowledge*, there is not an economic methodology and a biological methodology, but simply a scientific methodology which embraces findings in any field that improves a theory's ability to characterize and predict behavior. The disciplinary wall between evolutionary biology and economics is entirely false. As it crumbles, both fields benefit.

A typical criticism of biologically based theories of behavior, beginning with the reaction to E. O. Wilson's book *Sociobiology* (1975), is that they are deterministic – all behaviors are evolutionarily hardwired. Wilson was subject to intense criticism, including on several occasions being pelted with tomatoes when he spoke, for seeking to demonstrate that biological factors, along with environmental cues, affect behavior. The efforts to refute Wilson's findings were led by Wilson's Harvard colleagues Stephen Jay Gould and Richard Lewontin. Their views are perfectly captured by the title of a 1984 book by Lewontin (with Steven Rose and Leon J. Kamin), *Not in our genes*. This counter-assault was more an effort to support a philosophical viewpoint, than good science. Research in the last half-century provides copious support for the biological basis of behavior – clearly not in a deterministic manner, but in the sense of a built-in repertoire of possible actions, some of which are individually

idiosyncratic. If the study of the human genome has taught us anything, it is the vital importance of individual diversity. This essential Darwinian insight into human behavior has largely been ignored by economists.

A small but growing group of economists has begun to develop the modern biological basis for decision-making, including Gary Becker (1976), Jack Hirshleifer (1977), and Gordon Tullock (1979). Becker (1976) builds a model of genetic fitness and altruism in the vein of Hamilton's (1964) theory of kin selection to explain the evolutionary advantage of economic altruism. Hirshleifer (1977) use the four basic evolutionary operators, variation, inheritance, competition, and selection, to explain firm lifecycles, altruistic behavior, and the emergence of economic institutions. In introducing the basic analytical aspects of evolutionary biology, Hirshleifer advocated a return to Marshall's view of economics as a branch of biology. More generally, he states that "the social sciences can fruitfully be regarded as the sociobiology of the human species" (p. 51). All three of these trailblazers have continued since the 1970s to utilize evolutionary biology in their analysis, with Hirshleifer the most strongly committed to this approach, especially in models of conflict (e.g. Hirshleifer, 1998; Durham, Hirshleifer & Smith, 1998). Hirshleifer's objective is "to make bioeconomics a household word" (Hirshleifer, 2000). Becker continues to study human capital, fertility, and the evolution of preferences with "innate components" (Becker, 1993; Becker & Mulligan, 1997), while Tullock (1994) has contributed to evolutionary biology itself.

Another early pioneer who has used insights from evolutionary biology in economics is Vernon Smith who began the field of experimental economics when he performed his first laboratory experiments in 1956. Prominently, Smith has found significant deviations from noncooperative Nash equilibrium in experiments where participants play sequential prisoner's dilemma games with cash payoffs. Indeed, more than half of first-players in these experiments signal that they want to cooperate and share a larger payoff. After such a signal, nearly three-quarters of second-players cooperate, rather than defect and take a larger payoff but leaving the first-mover with nothing (Smith, 1997). Cooperation is not the predicted Nash equilibrium.

In related work, Smith has found deviations from theory in the "ultimatum" game where the first player has N dollars to split with player two, and player two either accepts or rejects. If the split is accepted, the proposed payment is made; if rejected, both players receive nothing. The Nash equilibrium is to offer player two \$1, since player two is better off with \$1 than with nothing. In the laboratory, offers less than 40% of N are almost always rejected. That is, individuals choose to punish the other player at the cost of hurting themselves (see also Fehr & Schmidt, 1999; and Fehr & Gachter, 1998). Smith and

colleagues attribute both the high levels of cooperation in sequential prisoner's dilemma games, and the desire to punish those who don't "play nice" in the ultimatum game as evidence of evolutionary imprinting (Landa, 1999). Humans are social animals because there is a survival advantage to living and cooperating in social groups. Maintenance of social order and cooperation were therefore essential to survival, and the genes that code for such activities were preferentially selected for, explaining these "irrational" behaviors. Evolutionary psychologists hypothesize that this type of "social accounting" is the reason humans evolved large frontal cortexes, including a specialized brain module to detect cheating (Cosmides & Tooby, 1995).

A genetically programmed bias to cooperate notwithstanding, the context ("environment") of exchange matters. Smith reports that in face-to-face exchanges, nearly 100% of players cooperate (Smith, 1999). Conversely, cooperation breaks down, though not completely, when payoffs increase from tens to hundreds of dollars. Inside and outside the laboratory, changing the social, institutional, and economic environments causes variations in measured levels of trust, showing that the evolutionary basis for cooperation is highly context-dependent (Zak & Knack, 2001). Adam Gifford Jr. (1999) identifies the role of evolution in forming cultural institutions that reduce cognitive costs by specifying rules for personal exchanges. Smith and Kevin McCabe are currently examining the evolutionary basis for decision-making by imaging subjects' brains while they form and execute strategies, a field they have dubbed "neuroeconomics." The goal of these functional magnetic resonance imaging studies is to develop a cognitive theory of exchange that integrates rational choice, specialized brain functions, and natural selection.

Herbert Simon, the 1978 Nobel laureate in Economics, has done related work. Simon began integrating findings from psychology to develop cognitive theories of bounded rationality based on the difficulty of making decisions in uncertain environments (see his collected works in Simon, 1997). One of the primary insights that Simon's work revealed is that individuals seldom maximize, rather they seek to "satisfice" (Simon, 1955, 1959). When satisficing, individuals seek to reach a point at which they are sufficiently happy with an outcome when using a rule of thumb to make choices. If rules of thumb are insufficient to reach a desired outcome, then Simon showed that people typically revise their expectations and therefore satisfice at an achievable outcome. Simon writes (1959, p. 263)

> Models of satisficing behavior are richer than models of maximizing behavior, because they treat not only equilibrium, but the method of reach it as well . . . (a) When performance falls short of the level of aspiration, search behavior . . . is induced. (b) At the same time, the level of aspiration begins to adjust itself downward until goals reach levels that are

> practically attainable. (c) If the two mechanisms just listed operate too slowly to adapt aspiration to performance, emotional behavior – apathy or aggression, for example – will replace rational adaptive behavior.

Having a repertoire of rules of thumb for a variety of circumstances is consistent with psychological studies of behavior. For example, Cosmides & Tooby (1995) present results regarding relatively simple mathematical puzzles that most subjects had trouble calculating in the abstract, but most had no trouble with when translated into a question about social behavior. This reveals the contextual nature of decision-making rules. Context- and heredity-dependent rules of thumb are called "epigenesis" by biologists, and permit rapid solutions to commonly arising problems, preserving scarce neuro-resources for other activities.

Psychologists Daniel Kahnemann and Amos Tversky also contributed influential work integrating psychology into economic choices. In a 1979 paper in *Econometrica*, Kahnemann and Tversky introduced "prospect theory." Based on extensive laboratory studies they had done, Kahnemann and Tversky developed a substantial extension to expected utility theory where individuals value and weight various outcomes subjectively. Specifically, they found that individuals systematically overweight small probabilities and underweight moderate to high probabilities. In addition, the value of alternative outcomes depends on a reference point such that for losses, value (their analog of utility) is convex and relatively steep, while for gains it is concave and less steep. This theory predicts that people are risk-loving over losses, and risk averse over gains as the reference point shifts – as suggested by their data. Reference points are shown to shift due to diminishing sensitivity in which marginal valuations close to one's reference point are larger than marginal values far from a reference point. Models of reference point dynamics predate Kahnemann and Tversky (e.g. work by economists Harl Ryder & Geoffrey Heal, 1973; and James Dusenberry, 1949), but a fully developed theory of transmutating preferences dates to Kahnemann and Tversky (1979).

While prospect theory generated great enthusiasm and follow-up research in the 1980s, by the 1990s, little of the results have entered standard economic analysis. Rabin (1998) writes

> Over the years, economists have proffered many reasons for downplaying the relevance of behavioral research challenging our habitual assumptions . . . As the aggressive uncuriosity shown in the past toward behavioral research continues to diminish, we can look forward to focusing entirely on its substance.

Associated theories of mind have been developed recently by Matthew Rabin and coauthors (O'Donoghue & Rabin 1999a, b; Bowman, Minehart & Rabin, 1999; Rabin & Schrag, 1999; Rabin, 1993, 1994), Arthur Denzau and Douglass

North (1994), Timur Kuran (1991), and Bentley MacLeod (1999). This work incorporates observed psychological findings into economic decisions, in particular, by modifying the model of the decision-making process. Specifically, Rabin uses psychological findings to include "reference levels" for consumption, loss aversion, and adaptation to new environments. Preferences such as these permit economic models to replicate experimental findings, including the endowment effect (once a good is owned, its value increases relative to the value when unowned), framing effects (when an option is presented relative to other outcomes, choices vary), and valued social goals (e.g. cooperation in public goods experiments). Denzau and North bring the psychological notion of "mind reading" or mental models (forecasting what others will do) into economics. They argue that decisions are based on perceived subjective values, not objective ones. These perceived values come from mental models built to understand the world. Such models may, if repeated feedback is available, be good approximations to objective reality, but need not be. In cases where feedback is unavailable, such mental models may have only a casual relation to reality. Kuran presents a theory of individual preference formation that depends on choices made by one's social group, with both individual and group choices jointly evolving. This construct permits errors to compound over time, producing lasting aggregate effects. MacLeod (1999) uses an evolutionarily based model of cognition as the foundation for a model of learning in noisy environments.

Psychological findings have had a significant influence on some game theorists, financial economists, and experimentalists. For example, loss aversion and reference points affect people's determination of what is "fair" in common contribution analyses (Richard Thaler, 1985; Kahnemann, Knetsch & Thaler, 1986; Rabin, 1993). Psychological factors also explain deviations from the standard theory's prediction of consumption smoothing by building in imperfect self-control (Thaler & Hersh Shefrin, 1981). Thaler, as well as financial economists such as Robert Shiller, have used biases in judgment to explain volatility in financial markets. Overconfidence, or "the law of small numbers" (Tversky & Kahnemann, 1971), indicates that individuals too heavily weight recent events, especially those from small samples, and overweight their own ability relative to others. This behavior can explain speculative booms and busts in financial markets (Thaler, 1993; Shiller, 2000). The results of evolutionary psychology have had little impact in equilibrium theory, and especially dynamic macroeconomic models upon which policy prescriptions are based (for a macroeconomic perspective, see Sargent, 1993; Day, 1987, 1989a, b).

The most promising area of strategic analysis that utilizes Darwinian principles is the field known as evolutionary game theory. Evolutionary games differ from standard pure strategy or mixed strategy approaches in that a player's

success at playing the game determines his or her fitness. Evolutionary games are explicitly dynamic, and include experimentation in strategies via replicator dynamics (i.e. reproduction). In these strategy choices, a Nash equilibrium is equivalent to a learned behavior, with players occasionally choosing strategies at random (Kreps, 1990; Weibull, 1995). Some evolutionary game theorists have used fitness strategies to develop a theory of preference formation as a foundation for utility theory (Güth & Yaari, 1992; Hanson & Stuart, 1990). Explicitly biological theories of preference formation have also been developed by Roger (1994) and Rubin and Paul (1979).

Related to theories of cognition is the way that humans commit to a particular choice among alternatives. An important advance in this area blends neuroscience, psychology and economics into a coherent theory of decision execution that identifies the emotions as, in Hirshleifer's (1987) memorable phrase, "the guarantors of threats and promises." Frank (1988) develops this notion into a general theory where emotions are directly tied to choices, especially in a social context which extends the study of the emotions by Darwin (1872). For example, Frank argues that blushing upon telling a lie may be adaptive in that it constrains short-run non-cooperative behavior by signaling intentions to others. Thus, what appears to be irrational behavior at a point in time can be understood as adaptive since life itself is a repeated game. Natural selection has placed survival within a social setting as a primary adaptive mechanism affecting economic exchange. Moreover, Frank's evolutionary explanation contextualizes Herbert Simon's (1959) claim that emotions are maladaptive when making choices, i.e. at times they may be, but they are also critical to choice execution as well, especially in repeated interactions. The role of emotions as motivators for decisions is supported by findings in evolutionary psychology (Cosmides & Tooby, 2000), and among brain-damaged individuals who lack emotion and as a result cannot execute choices (Bechara, Damasio, Tranel & Damasio, 1997; Damasio, 1994).

An emerging methodology that fruitfully integrates evolution, neuroscience, and economics is known as Alife, for Artificial Life, or ACE, for Agent-based Computational Economics. Alife models have their foundation in von Neumann machines. In the 1940s, mathematician extraordinaire John von Neumann conjectured that a machine could be built that could replicate itself, complete with a full instruction book. Just as DNA is a digital code (Ridley, 2000), von Neumann understood that information is the basis for life – actual or artificial. By the 1960s, John Horton Conway was experimenting with cellular automata that simulated life and death in the celebrated Game of Life (Berlekamp, Conway & Guy, 1982). This game was essentially a checkerboard grid in which under- or over-exposure to other automata lead to death. Conway's innovation was that von Neumann machines can be profitably designed as software agents. This is

currently a very active area of research that permits researchers to observe evolutionary selection and organization in environments with high degrees of heterogeneity, i.e. more heterogeneity than can be handled analytically. The primary mode of analysis utilizes the complex adaptive systems paradigm developed by John Holland (1992), using a bottom up approach (Epstein & Axtell, 1996). The method of analysis designs a computational laboratory where alternative agent-based and institutional structures can be examined.

There are several fundamental issues that have been discovered in Alife modeling. The first is the emergence of economies as self-organizing systems without top-down control – Adam Smith's "invisible hand," Hayek's "economic order," and Schelling's (1978) macrobehavior are shown to emerge spontaneously under very general conditions. Bak (1996) identifies evolution as the fundamental organizing principle leading to spontaneous organization. Second, Axelrod (1997, 1984), Hirshleifer (1999) and many others have shown that cooperation occurs in social exchange, supported by a variety of strategies, even when mutants who always defect enter the system, as long as their number is small. Third, institutional arrangements evolve with the environment as argued would occur by Schumpeter, Alchian, and as discussed in Nelson (1995). Fourth, learning via genetic algorithms (GA) (Holland, 1975; Dawid, 1996) exploits the creation of new approaches via recombination, linkage, mutation, and selection for fitness. GA has been used to explore areas ranging from auctions (Dawid, 1999), networks (Montesinos, Garcia-Guzman & Ayuso, 1999), trading rules (Allen & Karjalainen, 1999; Dorsey, Johnson & Van Boening, 1994), and economic stagnation and growth (Arifovic, Bullard & Duffy, 1997).

Fifth, herding behavior and path dependence is shown to occur in asset markets with costly information acquisition in the work of W. Brian Arthur and coauthors at the Santa Fe Institute, a hot-bed for complexity studies that intentionally cross disciplinary boundaries with much success (Arthur, Durlauf, & Lane, 1997; Arthur, 1995, 1994, 1993, 1991). Sixth, explicit models of the distribution of genetic material over generations characterizing human population genetics reveal the impact of institutions such as marriage and divorce, and technologies such as genetic engineering on output growth (Zak, 2000; Zak, Feng & Kugler, 2000; Zak & Park, 2000a, b). In these models, biology enters explicitly, for example, impacting mate choice by including the innate evolutionarily based evaluation of a potential spouse's physical attributes and economic resources, and constraining attainable human capital via the genetic endowment from one's parents and the nurturing parents give children. Related biologically based models of fertility have been developed by zoologist Alan Grafen (1998, 2000).

The Alife literature has shown that in the limit, agents behave like *homo economicus*, but in finite samples exhibit a richer set of behaviors (Vriend,

1996). A related area of research is artificial neural networks that seek to model the distributed processing of the human brain. This work is, to date, rather speculative, but very provocative nonetheless (Arbib, 1995; Fausett, 1994). Reviews of the Alife literature can be found in special issues of *Journal of Economic Dynamics and Control* and *Computational Economics*, with introductions by LeBaron (2000), and Leigh Tesfatsion (2000a, b). Also see the often updated ACE website maintained by Leigh Tesfatsion at http://www.econ.iastate.edu/tesfatsi/ace.htm. On the power of this method, Tesfatsion (2000b) concludes that "... agent based computational modeling should one day join analytical modeling, econometrics, field study, and human subject laboratory study as a standard tool in every economist's toolkit."

CONCLUSION

Economics is an evolutionary science, only most economists have not discovered this – to their detriment. Until they do, the economics of "what if" will predominate over the economics of "is." Darwinian evolution is not an analogy or metaphor for economic analysis as argued by Hodgson (1993) and Witt (1999). Rather, economics is fundamentally concerned with human decision-making that is at its core a process dictated by the interaction of evolution and environment. What this chapter has documented is that while the adoption of this approach has waxed, waned, and is again ascending (Day, 1995), there is no reason to separate biological evolution from economic analysis – they are one and the same. Hirshleifer (1977) writes "The ... social processes studied by economics ... are not mere analogs but are rather instances of sociobiological mechanisms" (p. 17). As findings from the natural sciences enter into the purview of economists, the grounding in economic analysis in the evolution of the human species will both deepen and broaden economics to include theories of mind that lead to theories of action. That is, to do economics properly requires, in many instances, that Darwinian evolution enter explicitly into the analysis. Metaphor is simply not enough.

This viewpoint is seen as radical by most "mainstream" economists. But let us appeal one last time to Veblen who ends his 1898 essay with a similar call to action. Veblen is convinced that economists who continue to ignore advances in the natural sciences risk working with speculations rather than facts. Indeed, over one hundred years ago Veblen was convinced that

> The social and political sciences must follow the drift, for they are already caught in it.

Because the quiddity of economic analysis is human decision-making, the nascent evolutionary approach to economics recognizes, to paraphrase Theodosius Dobzhansky, that nothing in economics makes sense except in light of evolution.

Postscript: Related Graduate Programs

The research program outlined in this chapter has yet to be fully realized. One of the primary stumbling blocks is the interdisciplinary (read: interdepartmental) nature of this endeavor. There are a number of programs worth noting that explicitly combine evolutionary biology and economics in one way or another. The leading program to date is the University of Michigan's Program on Complexity which combines social science, behavioral science and computation. The new graduate program in economic system design at the Interdisciplinary Center for Economic Science at the University of Arizona combines experimental economics, operations research, computer science, and neuroeconomics. Iowa State University is currently designing a graduate program with an emphasis in complex adaptive systems that looks promising. The Santa Fe Institute, while it does not grant degrees, holds summer courses for graduate students and post-doctoral fellows in complex systems and in computational economics. UCLA is spearheading a Southern California center for Social Informatics which commences in 2001 that unites the biological and social sciences under the banner of computation, though experimental methods and field work will also be encouraged (the authors of this chapter are contributing members). The Rand Graduate School (Santa Monica, CA) teaches courses on complex adaptive systems, but does not have a complete program in this area.

There are several good programs in bioinformatics, notably at Stanford University, though these are rather far from economics as it stands today. Carnegie Mellon University's Graduate School of Industrial Administration has a flexible, rigorous program that can be designed to include interdisciplinary coursework, especially from the program in Algorithms, Combinatorics, and Optimization, as well as biology and psychology. Similarly the Center for Complex Systems Research at The University of Illinois at Champaign, Urbana has operated since 1986 and has a solid reputation. We caution, though, that these programs are focused primarily on computation rather than on human decision processes.

ACKNOWLEGMENTS

We thank Margaret Gruter, Monika Gruter Morhenn, Jack Hirshleifer, Paul Samuelson, Vernon Smith and Lori Uber-Zak for helpful comments. Zak is grateful to Art Blakemore and the Department of Economics at Arizona State University for providing a congenial working environment while working on this essay.

REFERENCES

Alchian, A. A. (1950). Uncertainty, Evolution, and Economic Theory. *Journal of Political Economy, 58*, 211–222.

Allen, F., & Karjalainen, R. (1999). Using Genetic Algorithms to Find Technical Trading Rules. *Journal of Financial Economics, 51*(2), 245–271.

Arbib, M. (Ed.) (1995). *The Handbook of Brain Theory and Neural Networks*. MIT Press.

Arifovic, J., Bullard, J., & Duffy, J. (1997). The Transition from Stagnation to Growth: An Adaptive Learning Approach. *Journal of Economic Growth*, 2(2), 185–209.

Arthur, W. B., Durlauf, S. N., & Lane, D. A. (Eds) (1997). *The Economy As an Evolving Complex System II : Proceedings (Santa Fe Institute Studies in the Sciences of Complexity, Vol 27)*. Perseus Press.

Arthur, W. B. (1995). Complexity in Economic and Financial Markets, *Complexity, 1*(1).

Arthur, W. B. (1994). *Increasing Returns and Path Dependence in the Economy*. University of Michigan Press.

Arthur, W. B. (1993). On Designing Economic Agents that Behave Like Human Agents. *Journal of Evolutionary Economics, 3*, 1–22.

Arthur, W. B. (1991). Designing Economic Agents that Act Like Human Agents: A Behavioral Approach to Bounded Rationality. *American Economic Review Papers and Proceedings, 81*, 353–359.

Axelrod, R. (1997). *The Complexity of Cooperation*. Princeton.

Axelrod, R. (1984). *The Evolution of Cooperation*. Basic Books.

Azariadis, C. (1993). *Intertemporal Macroeconomics*. Blackwell.

Cosmides, L., & Tooby, J. (1995). Cognitive Adaptations for Social Exchange. In: J. H. Barkow, L. Cosmides, & J. Tooby (Eds), *The Adapted Mind: Evolutionary Psychology and the Generation of Culture*. Oxford.

Bak, P. (1996). *How Nature Works: The Science of Self Organizing Criticality*. Springer-Verlag.

Bechara, A., Damasio, H., Tranel, D., & Damasio, A. (1997). Deciding Advantageously before Knowing the Advantageous Strategy. *Science, 275*, 1293–1295.

Becker, G. S. (1993). *Human Capital: A Theoretical and Empirical Analysis with Special Reference to Education*, 3rd Edition. Chicago University Press.

Becker, G. S. (1976). Altruism, Egoism, and Genetic Fitness: Economics and Sociobiology. *Journal of Economic Literature, 14*(3), 817–826.

Becker, G. S., & Mulligan, C. B. (1997). The Endogenous Determination of Time Preference, *Quarterly Journal of Economics, 112*(3), 729–758.

Berlekamp, E. R., Conway, J. H., & Guy, R. K. (1982). *Winning Ways for your Mathematical Plays*, Volume 2. Academic Press.

Blaug, M. (1985). *Economic Theory in Retrospect*, 4th ed. Cambridge University Press.

Bowman, D., Minehart, D., & Rabin, M. (1999). Loss Aversion in a Consumption-Savings Model. *Journal of Economic Behavior & Organization*, *38*(2), 155–178.

Campbell, D. T. (1987). Blind Variation and Selective Retention in Creative Thought as in other Knowledge Processes, Ch. 3. In: G. Radnitzky & W. W. Bartley (Eds), *Evolutionary Epistemology, Rationality, and the Sociology of Knowledge*. Open Court.

Cosmides, L., & Tooby, J. (2000). Evolutionary Psychology and the Emotions. In: M. Lewis & J. M. Haviland-Jones (Eds), *Handbook of Emotions*, 2nd Edition. Guilford.

Damasio, A. (1994). *Descartes' Error: Emotion, Reason, and the Human Brain*. Avon Books.

Darwin, C. (1859/1982). *The Origin of Species by Means of Natural Selection*. Viking Press.

Darwin, C. (1872/1965). *The Expression of the Emotions in Man and Animals*. University of Chicago Press.

Dawid, H. (1999). On the Convergence of Genetic Learning in a Double Auction Market. *Journal of Economic Dynamics & Control*, *23*(9–10), 1545–1567.

Dawid, H. (1996). Adaptive Learning by Genetic Algorithms: Analytical Results and Applications to Economic Models. *Lecture Notes in Economics and Mathematical Systems*, No. 441. Springer.

Day, R. H. (1995). Review of Economics and Evolution: Bringing Life Back into Economics, by Geoffrey M. Hodgson. *Journal of Economic Literature*, *33*(3), 1329–1330.

Day, R. H. (1989a). Nonlinear Dynamics and Evolutionary Economics. In: R. H. Day & P. Chen (Eds), *Nonlinear Dynamics and Evolutionary Economics*. Oxford.

Day, R. H. (1989b). Dynamical Systems, Adaptation, and Economic Evolution. USC Modeling Research Group Working paper M8608.

Day, R. H. (1987). The Evolving Economy. *European Journal of Operational Research*, *30*(2), 251–257.

Denzau, A. T., & North, D. C. (1994). Shared Mental Models: Ideologies and Institutions. *Kyklos*, *47*(1), 3–31.

Diamond, P. (1965). National Debt in a Neoclassical Growth Model. *American Economic Review*, *55*, 1026–1055.

Dorsey, R. E., & Mayer, W. J. (1995). Genetic Algorithms for Estimation Problems with Multiple Optima, Non-Differentiability, and other Irregular Features. *Journal of Business and Economic Statistics*, *13*(1), 53–66.

Dorsey, R. E., Johnson, J. D., & Van Boening, M. V. (1994). The Use of Artificial Neural Networks for Estimation of Decision Surfaces in First Price Sealed Bid Auctions (pp. 19–40). In: W. W. Cooper & A. Whinston (Eds), *New Directions in Computational Economics*. Kluwer.

Durham, Y., Hirshleifer, J., & Smith, V. L. (1998). Do the Rich Get Richer and the Poor Poorer? Experimental Tests of a Model of Power. *American Economic Review*, *88*(4), 970–983.

Dusenberry, J. S. (1949). *Income, Savings and the Theory of Consumer Behavior*. Harvard University Press.

Epstein, J. M., & Axtell, R. (1996). *Growing Artificial Societies: Social Science from the Bottom Up*. MIT Press.

Fausett, L. (1994). *Fundamentals of Neural Networks*. Prentice-Hall.

Fehr, E., & Gachter, S. (1998). Reciprocity and Economics: The Economic Implications of Homo Reciprocans. *European Economic Review*, *42*(3–5), 845–859.

Fehr, E., & Schmidt, K. M. (1999). A Theory of Fairness, Competition, and Cooperation, *Quarterly Journal of Economics*, *114*(3), 817–868.

Frank, R. J. (1988). *Passions within Reason: The Strategic Role of the Emotions*. Norton.

Friedman, M., (1953). The Methodology of Positive Economics (pp. 3–43). In: *Essays in Positive Economics*. University of Chicago Press.

Getty, G. (1999). Total Return Economics. *mimeo.*

Ghiselin, M. T. (1974). *The Economy of Nature and the Evolution of Sex.* University of California Press.

Gifford Jr., A. (1999). Being and Time: On the Nature and the Evolution of Institutions. *Journal of Bioeconomics, 1*(2), 127–149.

Grafen, A. (2000). The Biological Approach to Economics through Fertility. *Economics Letters, 66*, 241–248.

Grafen, A. (1998). Fertility and Labour Supply in Femina Economica. *Journal of Theoretical Biology, 3*, 383–407.

Guth, W., & Yaari, M. (1992). Explaining Reciprocal Behavior in Simple Strategic Games: An Evolutionary Approach (pp. 24–34). In: U. Witt (Ed.), *Explaining Process and Change: Approaches to Evolutionary Economics.* University of Michigan Press.

Hamilton, W. H., (1964). The Genetical Evolution of Social Behavior. *Journal of Theoretical Biology, 7*, 1–52.

Hansson, I., & Stuart, C. (1990). Malthusian Selection of Preferences. *American Economic Review, 80*, 529–544.

Hayes, D. P. (1992). The Growing Inaccessibility of Science. *Nature, 356*, 739–740.

Hayek, F. A. V. (1989). The Pretence of Knowledge. *American Economic Review, 79*(6), 3–7.

Hayek, F. A. V. (1983). *Knowledge, Evolution, and Society.* Adam Smith Institute.

Hirshleifer, J. (1977). Economics from a Biological Viewpoint. *Journal of Law & Economics, 20*(1), 1–52.

Hirshleifer, J. (1987). On the Emotions as Guarantors of Threats and Promises. In: J. Dupre (Ed.), *The Latest on the Best: Essays on Evolution and Optimality.* MIT Press.

Hirshleifer, J. (1998). The Bioeconomic Causes of War. *Managerial & Decision Economics, 19*(7–8) 457–466.

Hirshleifer, J. (1999). There are Many Evolutionary Pathways to Cooperation. *Journal of Bioeconomics, 1*, 73–93.

Hirshleifer, J. (2000). Personal communication, June 11, Squaw Valley, CA.

Hodgson, G. M. (1993). *Economics and Evolution: Bringing Life back into Economics.* University of Michigan Press.

Holland, J. (1992). *Adaptation in Natural and Artificial Systems: An Introductory Analysis with Applications to Biology, Control, and Artificial Intelligence.* University of Chicago Press.

Holland, J. (1975). *Adaptation in Natural and Artificial Systems.* University of Michigan Press.

Kahnenman, D., & Tversky, A. (1979). Prospect Theory: An Analysis of Decision under Risk. *Econometrica, 47*(2), 263–251.

Kahnenman, D., Knetsch J. L., & Thaler, R. H. (1986a). Fairness as a Constraint on Profit Seeking: Entitlements in the Market. *American Economic Review, 76*(4), 728–741.

Kelm, M. (1997). Schumpeter's Theory of Economic Evolution: A Darwinian Interpretation. *Journal of Evolutionary Economics, 7*(2), 97–130.

Kiyotaki, N., & Wright, R. (1991). A Contribution to the Pure Theory of Money. *Journal of Economic Theory, 53*(2), 215–235.

Kiyotaki, N., & Wright, R. (1989). On Money as a Medium of Exchange, *Journal of Political Economy, 97*(4), 927–954.

Knight, F. H. (1921). *Risk, Uncertainty and Profit.* Houghton Mifflin.

Kreps, D. M. (1990). *A Course in Microeconomic Theory.* Princeton University Press.

Kuhn, T. S. (1962). *The Structure of Scientific Revolutions*, Rev. 2nd ed. University of Chicago Press.

Kuran, T. (1991). Cognitive Limitations and Preference Evolution. *Journal of Institutional & Theoretical Economics, 147*(2), 241–273.

Krugman, P. (1996). What Economists Can Learn from Evolutionary Theorists. Presentation to the European Association for Evolutionary Political Economy.

Landa, J. T. (1999). Bioeconomics of Some Nonhuman and Human Societies: New Institutional Economics Approach. *Journal of Bioeconomics*, *1*(1), 95–113.

Landa, J. T., & Ghiselin, M. T. (1999). The Emerging Discipline of Bioeconomics: Aims and Scope of the Journal of Bioeconomics. *Journal of Bioeconomics*, *1*(1), 5–12.

Lasswitz, K. (1958). The Universal Library. In: C. Fadiman (Ed.), *Fantasia Mathematica*. Simon & Schuster.

Lebaron, B. (2000). Agent-based computational finance: Suggested readings and early research, *Journal of Economic Dynamics and Control*, *24*(5–7), 679–702.

Leslie, P. H., (1945). On the Use of Matrices in Certain Population Mathematics. *Biometrika*, *33*, 183–212.

Lewontin, R. C., Rose S., & Kamin, L. J. (1984). *Not in Our Genes*. Pantheon Books.

Lewontin, R. C. Evolution and the Theory of Games. *Journal of Theoretical Biology*, *1*, 382–403.

Lotka, A. J. (1956). *Elements of Mathematical Biology*. Dover.

Luo, G. Y. (1999). The Evolution of Money as a Medium of Exchange. *Journal of Economic Dynamics and Control*, *23*(3), 415–458.

Macleod, W. B. (1999). Cognition and the Theory of Learning by Doing. Working paper, USC Economics Department.

Mahaffy, J. M., Zak, P. J., & Joiner, K. (1995). A Geometric Analysis of Stability Regions for a Linear Differential Equation with Two Delays. *International Journal of Bifurcation and Chaos*, *5*(3), 779–796.

Marshall, A. (1925). Mechanical and Biological Analogies in Economics. In: A. C. Pigou (Ed.), *Memorials of Alfred Marshall*. MacMillan.

Maynard Smith, J. (1982). *Evolution and the Theory of Games*. Cambridge University Press.

Maynard Smith, J. (1972). *On Evolution*. Edinburgh University Press.

Menger, C. (1883/1963). *Problems of Economics and Sociology*, F.J. Nock, trans., Illinois University Press.

Menger, C. (1892). On the Origin of Money. *Economic Journal*, *2*(2), 239–255.

Montesinos, P., Garcia-Guzman, A., & Ayuso, J. L. (1999). Water Distribution Network Optimization Using a Modified Genetic Algorithm. *Water Resources Research*, *35*(11), 3467–3473.

Nelson, R. (1995). Recent Evolutionary Theorizing about Economic Change. *Journal of Economic Literature*, *33*, 48–90.

O'Donogue, T., & Rabin, M. (1999a). Incentives for Procrastinators. *The Quarterly Journal of Economics*, *114*(3), 769–816.

O'Donogue, T., & Rabin, M. (1999b). Doing It Now or Later. *American Economic Review*, *89*(1), 103–124.

Paley, W. (1802/1986). *Natural Theology; Evidences of the Existence and Attributes of the Deity, Collected from the Appearances of Nature*. Classworks.

Popper, K. (1985). Realism and the Aim of Science, Volume I. In: Postscript to *The Logic of Scientific Discovery*, Volumes I, II & III. Routledge.

Popper, K. (1984). *The Self and Its Brain*. Routledge.

Popper, K. (1969). *Conjectures and Refutations*. Routledge.

Popper, K. (1959). *The Logic of Scientific Discovery*. Hutchinson.

Rabin, M., & Schrag, J. L. (1999). First Impressions Matter: A Model of Confirmatory Bias. *The Quarterly Journal of Economics*, *114*(1), 37–82.

Rabin, M. (1998). Psychology and Economics, Journal of Economic Literature, 36(2), 11–46.

Rabin, M. (1994). Cognitive Dissonance and Social Change. *Journal of Economic Behavior & Organization*, *23*(2), 177–194.

Rabin, M. (1993). Incorporating Fairness into Game Theory and Economics *American Economic Review*, *83*(5), 1281–1302.

Radnitzky, G., & Bartley, W. W. III. (Eds) (1987). *Evolutionary Epistemology, Rationality, and the Sociology of Knowledge*. Open Court.

Radnitzky, G., & Bartley, W. W. III. (1987). Introduction, Ch. 1. In: G. Radnitzky & W. W. Bartley (Eds), *Evolutionary Epistemology, Rationality, and the Sociology of Knowledge*. Open Court.

Ridley, M. (2000). *Genome*. Harpercollins.

Ridley, M. (1995). *The Red Queen: Sex and the Evolution of Human Nature*. Penguin.

Rodger, A. R. (1994). Evolution of Time Preference by Natural Selection. *American Economic Review*, *84*, 460–481.

Rubin, P., & Paul, C. W. (1979). An Evolutionary Model of the Taste for Risk. *Economic Inquiry*, *17*, 585–596.

Ryder Jr., H. E., & Heal, G. M. (1973). Optimal Growth with Intertemporally Dependent Preferences. *Review of Economic Studies*, *40*(1), 1–33.

Samuelson, P. A. (2000). Personal communication, May 30.

Samuelson, P. A. (1993). Altruism as a Problem Involving Group versus Individual Selection in Economics and Biology. *American Economic Review*, *83*, 143–148.

Samuelson, P. A. (1985). Modes of Thought in Economics and Biology. *American Economic Review*, *75*, 166–172.

Samuelson, P. A. (1983). Complete Genetic Models for Altruism, Kin Selection, and Like-Gene Selection. *Journal of Social and Biological Structures*, *6*, 3–15.

Samuelson, P. A. (1978). Generalizing Fischer's 'Reproductive Value': Overlapping and Nonoverlapping Generations with Competing Genotypes. *Proceedings of the National Academy of Sciences*, *75*(8), 4062–4066.

Samuelson, P. A., & Yellin, J. (1977). Comparison of Linear and Nonlinear Models for Human Population Dynamics. *Theoretical Population Biology*, *11*(1), 105–126.

Samuelson, P. A. 1974, Biological Least Action Principle for Ecological Model of Volterra-Lotka. *Proceedings of the National Academy of Sciences*, *71*(8), 3041–3044.

Samuelson, P. A. 1958, An Exact Consumption-Loan Model of Interest with or without the Social Contrivance of Money. *Journal of Political Economy*, *66*, 467–482.

Sargent, T. (1993). *Bounded Rationality in Macroeconomics*. Oxford University Press.

Schelling, T. C. (1978). *Micromotives and Macrobehavior*. W. W. Norton.

Schumpeter, J. A. (1942). *Capitalism, Socialism, and Democracy*. Harper and Row.

Segal, U., & Sobel, J. (1999). Tit for Tat: Foundations of Preferences for Reciprocity in Strategic Settings. Working paper, University of Western Ontario.

Shiller R. J. (2000). *Irrational Exuberance*. Princeton.

Shingo, S. (1981). Study of 'Toyota' Production System from Industrial Engineering Viewpoint. Japan Management Assn.

Simon, H. A. (1997). *Models of Bounded Rationality: Empirically Grounded Economic Reason*. MIT Press.

Simon, H. A. (1955). A Behavioral Model of Rational Choice. *Quarterly Journal of Economics*, *69*, 99–118.

Simon, H. A. (1959). Theories of Decision-Making in Economics and Behavioral Science. *American Economic Review*, *49*, 253–283.

Smith, A. (1776/1937). *The Wealth of Nations*. Modern Library.

Smith, V. L. (1997). The Two Faces of Adam Smith. Distinguished Guest Lecture, Southern Economic Association.

Smith, V. L. (1999). Personal communication, March 11th. University of Arizona.

Spencer, H. (1851). *Social Statics*. Chapman.

Spencer, H. (1855). *The Principles of Psychology*. Williams and Norgate.

Spencer, H. (1880). *The Study of Sociology*. Williams and Norgate.

Taylor, F. W. (1911). *The Principles of Scientific Management*. Harper & Row.

Tesfatsion, L. (2000a). Introduction to the CE Special Issue on Agent-Based Computational Economics. *Computational Economics*, forthcoming.

Tesfatsion, L. (2000b). Introduction to the JEDC Special Issue on Agent-Based Computational Economics, *Journal of Economic Dynamics and Control*, forthcoming.

Thaler, R. H. (Ed.) (1993). *Advances in Behavioral Finance*. Russell Sage Foundation.

Thaler, R. H. (1985). Mental Accounting and Consumer Choice. *Marketing Science*, *4*(3), 199–214.

Thaler, R. H., & Shefrin, H. M. (1981). An Economic Theory of Self-Control. *Journal of Political Economy*, *89*(2), 392–406.

Trivers, R. L. (1985). *Social Evolution*. Benjamin/Cummings.

Trivers, R. L. (1972). Parental Investment and Sexual Selection (pp. 136–179). In: B. Campbell (Ed.), *Sexual Selection and the Descent of Man*. Aldine.

Trivers, R. L., & Willard, D. E. (1973). Natural Selection of Parental Ability to Vary the Sex Ratio of Offspring. *Science*, *179*, 90–92.

Tullock, G. (1994). *The Economics of Nonhuman Societies*. Pallas Press.

Tullock, G. (1979). Sociobiology and Economics. *Atlantic Economic Journal*.

Tullock, G. (1971). The Coal Tit as a Careful Shopper. *The American Naturalist*, *105*, 77–80.

Tversky, A., & Kahnman, D. (1971). Belief in the Law of Small Numbers. *Psychological Bulletin*, *76*, 105–110.

Veblen, T. (1898). Why is Economics Not an Evolutionary Science?. *Quarterly Journal of Economics*, *12*.

Vriend, N. (1996). Rational Behavior and Economic Theory. *Journal of Economic Behavior and Organization*, *29*, 263–285.

Wilson, E. O. (1998). *Consilience: The Unity of Knowledge*. Knopf.

Wilson, E. O. (1978). *Sociobiology: The New Synthesis*. Harvard University Press.

Witt, U. (1999). Bioeconomics as Economics from a Darwinian Perspective. *Journal of Bioeconomics*, *1*, 19–34.

Weibull, J. (1995). *Evolutionary Game Theory*. MIT Press.

Wright, R. (1995). Search, Evolution, and Money. *Journal of Economic Dynamics & Control*, *19*(1–2), 181–206.

Yellin, J., & Samuelson, P. A. (1974). Dynamical Model for Human Population. *Proceedings of the National Academy of Sciences*, *71*(7), 2813–2817.

Zak, P. J. (2000). Genetics, Family Structure, and Growth. Working paper, Claremont Graduate University.

Zak, P. J., with Feng, Y., & Kugler, J. (2000). Immigration, Fertility, and Growth. *Journal of Economic Dynamics & Control*, forthcoming.

Zak, P. J., & Knack, S. (2001). Trust and Growth. *Economic Journal*, April, 2001.

Zak, P. J., & Park, K. (2000a). Population Genetics and Economic Growth. Working paper, Claremont Graduate University.

Zak, P. J. & Park, K. (2000b). Population Genetics, Unstable Households, and Economic Growth. Working paper, Claremont Graduate University.

HISTORY AND EVOLUTIONARY THEORY

Carl N. Degler

It is dangerous to make man see too clearly his equality with the brutes without showing him his greatness. It is also dangerous to make him see his greatness too clearly, apart from his vileness. It is still more dangerous to leave him in ignorance of both. But it is very advantageous to show him both. Man must not think he is on a level either with the brutes or with the angels, nor must he be ignorant of both sides of his nature; but he must know both.

Blaise Pascal, 1670.

When I was invited to participate in this volume, the editors asked me to do for history what all of the authors in this volume were doing for their social disciplines. Unfortunately, I was not able to agree to the editors' request simply because, so far as I could tell, few if any historians saw much connection between their work and evolutionary theory. It is true that I had written a book in 1991 about the subject and had even received a prize for it. Nonetheless there is still very little interest among historians in evolutionary theory, despite my continued personal and printed efforts to persuade historians to recognize, as so many social scientists have been doing over the last thirty or forty years, how evolutionary theory might help historians in their work.

Given the close connection between history and most of the social sciences this lack of interest among historians requires some explanation if we are to understand both history and evolutionary theory. After all, most of the social sciences derived from history. Moreover, Darwinian evolution, after all, is a historical science and thus ought to be a natural associate of traditional history. It is true, too, that a hundred and fifty years ago science made a serious impact

Evolutionary Approaches in the Behavioral Sciences, Volume 8, pages 67–90.
2001 by Elsevier Science Ltd.
ISBN: 0-7623-0769-2

on historical investigation, but the science drawn upon, like so many of the early social sciences, was physics, not biology. Indeed, through a good part of the late nineteenth and early twentieth centuries in the United States as well as in western Europe, historians looked upon science, meaning physics, as a model historians hoped to use in writing history. One president of the American Historical Association in the early years of the twentieth century, for example, wrote about the "laws of history". Perhaps the most prominent historian – also a president of the American Historical Association – who looked to science in accounting for the meaning and future of history was Henry Adams, whose 1909 essay "The Rule of Phase Applied to History" has long been forgotten.[1]

Among historians, in short, physics never caught on in thinking about a relation to science, but neither has biology, despite the remarkable success of Darwinian evolutionary theory across the human as well as the physical sciences. So why have historians shied away from evolutionary theory? Part of the reluctance stems from the way in which historians work. They often deal with narrow or limited aspects of a subject field, seeking out archival materials, correspondences of individuals, newspapers, and other, often minutely particularistic sources. Darwinian evolution with its emphasis upon species behavior in accounting for the way in which human beings act seems too broad, too "global" to fit into the details of human life that are often the mainstay of the historian's work. Even statistical analyses are not the way in which most historians work, despite their recognition that numbers can count in finding trends or explaining behavior patterns.

Perhaps more significant in accounting for historians' lack of interest in evolutionary theory is that in past efforts to draw upon evolutionary ideas the history have been found deeply and sadly flawed. Long before Darwin wrote the *Origin of Species*, Herbert Spencer explained how biological evolution accounted for differences in the behavior of human groups, an explanation we now rightly and pejoratively denominate as Social Darwinism. Today, of course, social scientists as well as historians repudiate that misuse of evolutionary theory. That misuse is still employed today as when recently I asked a prominent historian about possible gains that historians might obtain in drawing upon evolutionary theory. Many historians also remember eugenics as a misuse of biological evolution in the work of Francis Galton, and the deleterious impact eugenics has had upon public policy in the United States. As if that were insufficient to raise doubts about the use of biology in writing about human history, there is the monstrous misuse of evolution in Nazi Germany. Historians of the United States are especially sensitized to think negatively about biology. For today racial and sexual differences have been intensely studied historically and historians generally perceive those differences as shaped by culture and not by evolution.

Historians' attention to race and sex differences brings us to a further obstacle for historians in seeking to draw upon evolutionary theory. It is that historians' primary subject of investigation are differences and conflicts among human groups, such as classes, races, nations, tribes, political parties, and cultures. Evolution, on the other hand, treats the species as a whole, a subject that appears to find no place for the day-to-day work of historians. For many historians, then, evolution seems irrelevant to their effort to explain the past.

We have not yet reached the deepest level in seeking to account for the lack of interest historians seem to have in relating their discipline to evolutionary theory. After all, most of the objections that I have mentioned already might well be reasons political scientists, sociologists, or economists would also lack an interest in drawing upon evolutionary theory. Each of those disciplines, after all, came out of history, yet each of them now draws upon evolutionary theory.

The underlying difference between these social sciences and history is that history and the social sciences (as their name implies) differ fundamentally in their outlook toward science. All of the social sciences, unlike history, try to use methods and assumptions of the sciences in following their disciplines.

In a sense, we have now returned to our starting point. It is true, that in the middle of the nineteenth century there were a few historians, as we have seen, who sought to draw upon physics or biology in seeking to explain history, but most historians then and now did not accept science as a discipline helpful or related to history. The measure of that difference is best identified in the work of the nineteenth century German philosopher Wilhelm Dilthey. His message was that science and history were different ways of dealing with both the past and the present. History in Dilthey's terms was not simply identifying the past, that is, finding the "facts" or sources of the past for that might well be called science. For Dilthey and for many historians the discipline as a whole sought what men and women thought and meant about the past. There was no science there, no experiments, no proofs, no mathematical certainties. Science was the study of what was measurable, material, and explicable. For many, if not most historians today, I suspect Dilthey's conception of history is theirs.

Since I do not accept Dilthey's absolute dichotomy between history and the social sciences in regard to evolutionary theory, let me turn now to an effort to explain to historians why they might find evolutionary theory useful. At the outset it is essential to say that the conception of evolutionary theory followed here says nothing about more familiar, yet irrelevant concepts like "social Darwinism" or racism or race. As mentioned earlier, historians treat many subjects and concepts such as class, race, economic changes, foreign affairs, conflicts between nations and so forth. Almost none of those subjects will be touched upon here in discussing evolutionary theory. The subjects that may be

analyzed or explained by evolutionary theory are limited in number though centrally important in dealing with the actions or behavior of human beings as a species, that is, cross culturally. Evolutionary theory helps to explain modes of reproduction, sexuality, female-male relations, relations between parents and children, violence, selfishness and altruism, and social cohesion. Evolutionary theory will not account for, or have much to explain in a whole range of historical subjects or concerns, but it will have relevance since all of history deals with human beings and it is an assumption of the theory that we all have a common history – that is, a biological past – that influences, (although does not necessarily determine) the behavior of human beings which historians deal with over the centuries. Thus it may help to account for some general human behavior patterns.

THE SCIENCE OF EVOLUTION

Evolutionary theory works because it is based on three scientific conceptions. The first is that human beings are animals and that we are descended from animals and that we have inherited from animals many of our muscles, bones, nerves, and other tissues. Thanks to our recent knowledge about DNA we now know of connections with plants as well as with animals. That we have aspects of human behavior that animals and plants do not exhibit is also true, as in the case of language, which is a uniquely human behavioral pattern. Nevertheless language is equally derived from Charles Darwin's general principle of natural selection, which explains how our bodies have developed as integral elements in all living organisms.

The second conception is that evolution or our descent from animals is not the only source of our behavior. Anthropologists speak of "culture", which is behavior that is passed on from one individual or group without any genetic or biological connection. It is now known that some animals, too, can pass on behavioral patterns to other members of their species without any genetic connection, and thereby exhibit cultural behavior. For example, chimpanzee mothers often teach their offspring to seek for food by sticking blades of grass into the holes of termite hills in order to bring up the insects for food. Certain monkeys have learned how to pass on to members of their species other ways of gaining food or water.[2] Examples of the acquisition of cultural traits by apes and monkeys suggest that culture and biology are not necessarily at odds, as is sometimes contended. The two processes may well work together, as in the just mentioned feeding habits of chimpanzees, and often in the example of human cultural behavior as well. The behavior patterns of most animals are generally programmed genetically, but the existence of the cultural behavior of

chimpanzees exemplifies a major principle of evolutionary theory. Some animals, like chimpanzees, who come very close biologically to human beings, (they have 98% of the DNA of human beings) have the ability to adjust to environmental opportunities so as to fulfill a biological need like food. Put another way, culture, unlike language, is not a uniquely human behavioral pattern.

A third scientific conception that underlies evolutionary theory is the field of ethology or the scientific study of animal behavior. Unlike the kind of observations that Darwin and his hundreds of corresponding naturalists reported, such as those made at zoos or by accident among wild animals, this new field, which was recognized by Nobel awards in 1973 for the work of three well-known ethologists, consists of systematic observations, recordings, experiments and statistical analyses.[3] Today, the field flourishes nationally and internationally through a wide range of publications in scientific journals. Many modern students of the social sciences now draw upon a growing body of ethological knowledge in attempting to understand better human behavior and emotions.

Just as we know that some kinds of animals can exhibit cultural behavior patterns, so the study of animal behavior makes clear that certain emotions or feelings can be and have been bred into animals. As dog owners know, certain breeds of dogs can be bred to be vicious fighters, such as bull terriers, or made to be peaceful and even helpful, as with certain sheep dogs. Or as recently reported, wild foxes can be bred through only forty generations to become peaceful pets.[4] In fact, Darwin himself, in his third book on evolution, *The Expression of Animals and Man* (1873), made the point that emotional expressions of animals bore a striking resemblance to those of human beings. The book was, perhaps, his strongest evidence that human beings were not only physically descended from animals, but that their emotions as depicted in bodily forms were often congruent with those of animals.

Darwin and some of his followers generally concluded at the time that human beings were a species like other species, displaying patterns of behavior across the members of the species. By the early twentieth century, however, when the new discipline of anthropology and the general study of human behavior had matured, the great diversity of human cultures around the world came to be perceived as the central subject of the emerging social sciences, as many historians had long recognized. Indeed, most students of human behavior were so aware of the diversity of peoples around the globe that the very concept of a human nature was often called into question. As anthropologist Clifford Geertz remarked, to accept the Enlightenment conception of man as uniform leaves one with only two possibilities. Man is all of the obvious diversities of cultures or "he becomes a conscripted soldier in a vast Tolstoian army, engulfed in one or another of the terrible historical determinisms, which we have been plagued

from Hegel forward". Geertz's view is that "to be human here is thus not to be Everyman; it is to be a particular kind of man and of course men differ".[5]

One of the consequences of the new study of ethology or animal behavior by the 1980s was that, despite Geertz's admonitions, some social scientists once again picked up the idea that the human species possessed cross-cultural traits, or what came to called "universals", that is, behavior patterns that transcended societies and cultures, past and present. The most obvious and most important universal was language itself. Many universals are culturally rooted, though often related to biological or alimentarily sources, such as fire, cooking and metal tools. Some stem from biology, such as language itself. Furthermore, all societies have words that distinguish "black" and "white" (presumably from light and dark) and exhibit signs for male and female, mother and father, that is, reproductive beings. The great majority of societies recognize the status and age of people just as they also make tools to a higher degree of skill than any tool-making animals are able to. Human beings always live in groups wherever they may be and recognize their locality as theirs. Genetically close kin are generally denied sexual access or mating, that is, observe some form of incest avoidance. Sexual activity is always secluded or private while eating is always social. Sweets are widely relished. Facially expressed emotions are recognizable across all cultures as Darwin pointed out years before.[6] Religion or belief in the supernatural seems to be present in all societies.

The distinguished historian of ancient religion, Walter Burkert at the University of Zürich, in his fascinating book *Creation of the Sacred* (1996), observes that language, art, and religion are among the most obvious of universals. As Burkert points out, if one accepts ritual as a representation of religion, then the beginning of religion surely precedes art and perhaps even the full development of language in as much as the fossil evidence from Neanderthal people suggests limits on the full expression of speech. Burkert's thought is that some of the practices which we now understand among animals sometimes appear to be reflected in fundamental rituals in ancient religions. He identifies such practices as survival techniques of animals such as hierarchy (submission), order, and sacrifice which Burkert contends emulate ancient religious practices. "Religion's seriousness, so manifest in our feelings, reflects the hard rock of the biological landscape, the dangers, the limits and drive for the preservation of life. Religion keeps to the tracks of biology", Burkert contends. "Some extremists may happen to get off the track [modern Shakers, perhaps?] but they will disappear unless they somehow make their way back".[7]

Donald Brown, in his recent study of universals, concludes that there is a human nature that encompasses people in all societies, however different they may seem. In fact, the behavioral and cultural similarities across cultures are

once again attracting anthropologists beyond Brown. They recognize that differences among cultures may have been important in the historical development of anthropology, but the discipline is still the nature of all human beings across all cultures, as the subject was conceived at the beginning and not simply, as Clifford Geertz has insisted, as just the great diversity of humankind. Even Alfred Kroeber, an early proponent of culture against the fell influence of biology, closed his career in the 1960s, asking for the study of the nature of human beings, not simply that of a particular people or culture. One way of doing so, Kroeber suggested, was to study the behavior of animals, or, as he put it, "the relation of cultural history to organic history". Thinking back upon Darwin's close study of animal behavior, Kroeber urged anthropologists to follow that example. The growing interest in the study of human nature, he continued, makes it "clear that we cannot permanently ignore the basic genetic part of our psychology".[8]

A consequence of recognizing universals is that, despite the great diversity of cultures, we appear to possess a common nature – or dare I say, a common heritage? Conditions, environment, events may develop or retard that nature over time but people of all societies can participate in each other's time and place. That people today in other cultures can appreciate the dramas of the Greeks, or relate to Indian philosophy, or understand primitive peoples in far away places, suggests that behind that acceptance stands a human nature that allows us to understand those differences, deep as they may seem to be, and they in turn can understand us. Or put another way, if human beings were so fundamentally different as opponents of the commonality of human nature contend, how could anthropologists, or historians for that matter, learn and study the culture of societies other than their own? All human beings have an emotional and intellectual nature that communicates transhistorically. Evolutionary theory then may offer clues for explaining the common nature that underlies diversity.

How does that common understanding, that communication among peoples, take place or develop? At this juncture we turn to recent animal studies that have given new meaning to Darwin's conclusion that human beings have descended from animals. Today we have a fresh awareness of how Darwin's central idea – natural selection – works. The principles and study of ethology today have begun to be employed in examining human behavior according to Darwinian principles. It is true that most educated people have long acccepted the basic Darwinian idea that human beings are descended from animals. Current knowledge of DNA makes it finally clear that we are chemically related to animals, and for most of us, natural selection is the primary explanation for the great variety of species in the world, among which are human beings.

One of the several reasons for discussing human universals is the inference that there may be a human nature that is biologically rooted just as other species have a biological nature. Human biological universals remind us that nature does not divide or separate human groups, however much we may be divided by culture. After all, we are a species, which means we can all reproduce, despite any behavioral or other apparent differences that may seem to separate us. Physical differences, such as skin color, hair texture, or other striking differences in physical form, so far as we know, are not genetically related to any behavioral or mental patterns. Race, in short, is not a scientific way of separating the members of our species.

Unlike anthropologists who may be quite content to talk about a human nature, historians have no such common conception of humanity except a vague belief that emotional and physical behavior has a common Standard. Its best known example of the idea of a common Standard in judging or evaluating historical causality or motives was Carl Hempel's interpretation half a century ago. Hempel's conception of human nature, like that of most historians was defined personally and eclectically. Most historians explain motives or behaviors on the basis of experience or from information derived from the social sciences, running from anthropology and law through psychology and economics.[9] Many of these social sciences, today, have introduced concepts from evolutionary theory.[10]

YET EVOLUTION CAN BE USEFUL

Let me turn now from general perceptions of evolutionary theory to look at some concrete instances of how aspects of human behavior could be understood – and better yet explained – by evolutionary theory. These same examples may provide some ground for understanding the connections between ethological evidence and actual collective (as opposed to individual) human behavior. One of the earliest gains from the study of animal behavior in regard to human behavior is a relatively recent understanding of the incest tabu.

An early issue that exercised anthropologists in the late nineteenth century was the recognition, cross-culturally, that marriage between close relatives was forbidden. Today modern students of the tabu, like Sigmund Freud and the well-known French anthropologist Claude Levi-Strauss saw the tabu as a marker of that moment in the early history of humankind that human awareness or consciousness first appeared. (It is worth remembering that Freud's and Levi-Strauss's explanations were suppositional metaphors, not historical fact.) Freud in *Totem and Tabu* designated the establishment of the tabu as "the primal crime," by which he meant that the killing of the father by his sons in order to possess the mother ultimately forced the sons to marry out of the family,

thereby creating the incest tabu. Levi-Strauss explained the tabu as the origin of connections between families through the exchange of women. In both metaphorical explanations the aim was to avoid incestuous reproduction.

The hidden assumption in each of these explanations was that animals mated without regard to biological relation, that is, there was no avoidance of incest. (For Freud and Levi-Strauss the achievement of the tabu was the defining moment that separated human self-awareness or consciousness from the life of animals.) One of the discoveries of modern ethology was that natural selection often separated the sexes in a species at the time of mating, usually by causing young males to leave the troop or nest. In that way mating between two close relatives would be avoided and thereby escape the deficiencies of inbreeding. Of course, the animals knew nothing about genetics, but presumably natural selection had gradually weeded out those animals which mated too closely genetically. The primary engine behind natural selection is reproduction and its goal is reproductive success, that is, for all living things to reproduce as protectively as possible. In particular, then, incest avoidance enhanced reproductive success because it reduced genetic deficiencies. The new ethology provided an historico-biological answer that nicely trumped the myths and metaphors of earlier anthropologists and psychologists.

A striking sign of a possible validation of the ethologically based explanation of the tabu has been provided in the work of two social scientists, who explored the workings of the tabu in modern societies. The Israeli sociologist, the late Joseph Shepher, studied children who had grown up in a Children's House in a kibbutz. As was the practice in such cooperative settlements in Israel, the children lived together without making traditional distinctions between the sexes, living together more or less as young siblings.[11] But that was exactly where the issue of the incest tabu entered. Although the males and females were not related, none of them became sexually interested, even though some parents thought it was natural that a boy and girl who had been closely matched as children would be fine candidates for matrimony later on. The children's response to such suggestions was invariably" "How could I marry her, she (he) is my sister (or brother)". In short, the kibbutz children had learned what mates among animals also had "learned" – through natural selection – that nest mates leave before mating time or in the case of human beings, express a lack of sexual interest.

A quite different pattern of behavior but with analogous consequences, has been studied by Stanford anthropologist Arthur P. Wolf. [12] There the behavioral pattern was a Chinese custom called *sim pua*, in which peasants sought to insure that their offspring would have a mate, the solution of which was to bring a prospective mate into the household at a very early age. But the solution was

frustrated by the same evolutionary theory of avoiding marriage with close relatives that had prevented the kibbutz children from mating.

The *sim pua* children did marry when their time for marriage arrived because the parents insisted upon it, but the behavioral result was strong resistance by the wedding pair: early divorce, and reduced child bearing. In this instance, the cultural rules of the parents had apparently countered the force of biological evolution.

The underlying proposition that all living organisms, including homo sapiens, seek to reproduce may be axiomatic, but the meaning of reproduction needs some qualification. It is not necessarily the most numerous offspring that measures reproductive success. It is that the offspring should also be reproductive. Male and female human beings both seek reproductive success, but their ways or strategies of reproduction differ. This brings us to what may be perceived as a controversial insight: the differences between men and women. The differences begin with the recognition that reproduction is the driving force in all living beings, for the ability to reproduce is, after all, the definition of life. Of course, the people of the modern industrial world can and do employ contraception, thereby managing to separate sexuality from reproduction. That separation, of course, is a very good example of how an evolutionary drive, that has become culturally disadvantageous for human beings, can be reshaped and even restrained by human action. Today we are likely to think that sex is a drive in itself since we can easily separate it from reproduction. Yet reproduction is the evolutionary goal; sex is simply its means.

Although reproduction among human females in industrial societies has been recently and rapidly reduced, the evolutionary basis or reproductive strategies of the sexes remains the same. Women are still the object of men's reproductive goals, though we are more likely to acknowledge that object as a drive for sex. The evolutionary goal for men is to seek females since they have the greater investment in reproduction; they bear offspring. For males the individual cost of reproduction is their investment in competing with males for access to females. Males also invest by attracting females, such as the peacock's lengthy tail or the male bower bird's elaborate nest or the human male's income or support.

Females' reproductive investment, however, differs from that of a male in two ways. One is that, unlike a male, who can increase his offspring with multiple females, a female cannot increase her own offspring by accepting multiple mates. The second burden of females, human and other mammals, is that they must nurture their offspring in their own bodies, experience the dangers of birth, and then feed the offspring from her body. Moreover, as the modern feminist leader, Alice Rossi, has remarked, the endocrinal, hormonal, and psychological relation

between a human mother and her child constitutes a bonding that the lengthy maturation of children depends upon.[13] In short, females do not compete with each other for reproductive success as males do, though females exercise sexual selection, that is, have some choice about the quality of a male's reproductive capacities. (Females can and do deny copulation.)

These very cultural, physiological, and evolutionary connections suggest that the family unit is more a duality with the father as an appendage, in which he is more a source of material support than an integral or inclusive connection to the mother and child. (Among chimpanzees, gorillas, and orangutans – our genetically closest animal relatives – males play no immediate supportive role in assisting with the care of offspring.) Most modern industrial societies see the family as a mother-child-father unit, but not all anthropologists see the so-called nuclear family as a universal.[14] Yet it is surely universal enough for the purposes on which historians and sociologists usually base their studies.[15] Some sociologists, however, do worry that the cohesion of the family seems to be uncertainly explained. It is sometimes said that the nuclear family's purposes or needs are now being performed by other agencies like the community or the state. Under such circumstances, it is argued, the family may gradually weaken or even disappear. Yet, given the evolutionary basis for the mother-child relation, when reinforced by important support from the father, it seems unlikely that the nuclear family could easily disintegrate or seriously weaken.

Once one recognizes the evolutionary basis for the mother-child relation, historians and other students of family life ought to be careful about perceiving the so-called nuclear family as a kind of fragile and merely historical contractual unit rather than as an organic unit built from a long evolutionary history. A feminist and leading sociologist remarked in 1984 that theories of the family are no longer adequate in light of modern evolutionary theory. The differences of the sexes, she continued, "serves the fundamental purpose of reproducting the species. Hence sociological units of analysis such as roles, groups, networks, and classes divert attention from the fact that the subjects of our work are male and female animals with genes, glands, bone and flesh occupying an ecological niche of a particular kind in a tiny fragment of time". These differences, she insisted, "emerged from the long pre-history of mammalian and primate evolution".[16]

Once the differences in the reproductive drives of the sexes are understood, evolutionary theory then offers a possible explanation for two well-known historical phenomona. One is pornography, that is, the production of visual and/or literary depictions of sexual activities that are generally perceived socially as immoral or otherwise impermissable. As is well known, pornography has existed through most of human history and across diverse cultures today and in the past. Generally, it is a human male preference with little attraction for

females. Those differences in attractiveness are closely related to the differential reproductive strategies of the two sexes. Pornography itself is not directed to reproduction, of course; rather it derives from the male strategy, as we have seen, to seek as many opportunities as possible for copulation. The mechanism to achieve that purpose emerges from a hormonally-based drive for sexual arousal and pleasure. Pornography releases that mechanism when no female is available.

Females' lack of interest in pornography similarly derives from their reproductive strategy which is to recognize the costs, both biologically and socially, which female reproduction entails. Immediate or quick arousal is not constructive in a sex where choice is important in assaying the quality of potential offspring. The costs for human females, both physiologically and socially, are high, and there is no reproductive gain for females to compete for males.

Once these differences in strategy in regard to pornography are recognized, one can then see why prostitution is largely a male practice across human history and cultures. Again, it is not reproduction that drives the males, but the sexual urge that evolution has selected as the male's means for achieving reproductive success. Similarly, females' lack of interest in paying male strangers for sex fits in nicely to their reproductive strategy. First of all, prostitution violates a basic principle of the strategy: to have choice in selecting successful offspring. Secondly, it is that lack of choice that causes many prostitutes to feel they are victims rather than partners.

Pornography and prostitution today are often considered socially undesirable as well as being immoral, but both practices persist because the evolutionary source of both of those social disorders, if that is how one wants to describe them, were developed in human beings for quite different purposes, namely reproduction. Prostitution and pornography are excellent examples of how human beings have built cultural patterns (or history) from evolutionary roots, in this case, the drive for reproduction. (We have similarly created pleasurable recreational sexuality from the same evolutionary drive.) Nor are prostitution and pornography the only examples of how an evolutionary practice may end up in a way many today would consider unworthy. One would be the growth of the world's population, which now threatens food supplies and energy resources.

It is worth examining another instance in which evolution is not necessarily progressive or good. The instance is the widespread human lust for sweets, which today, of course, may bring dental decay, overweight, diabetes and other related physical deficiencies. The source of the lust for sweets probably stems from our animal ancestry. Most primates, including human beings, enjoy substances, principally fruits, which contain energizing sugar. Human beings, as we know, began as hominids in Africa and thus picked up that sweet tooth, which was a

valuable taste to acquire when seeking a source of energy. Today, of course, when there are many other sources of energy and less need of it, that sweet tooth pays a price today. That love of sweets which was "intended" by evolution to encourage a supply of energy in order to achieve reproductive success, now may result in just the opposite. On a larger historical scale, that taste for sweetness drove nations to intense commercial competition and even wars as the European search for sugar in the eighteenth and nineteenth century reminds us.[17]

At this juncture it is worthwhile to recognize that the human tastes and actions that I have been describing have been built into our bones, brains, nerves, hormones, and muscles that over millenia natural selection has produced through competition. In a sense, of course, these hereditary movements and actions derive from genes or what today we call DNA. But by the same token, it is just as important to recognize that there are no genes for sweetness, any more than there is a gene that causes your eye to blink when a baseball comes at your head. In both cases that reaction had long been built into your nervous system and muscles through your DNA interacting over time with environment. Nevertheless, one can refuse to eat sugar, just as some people, like the Shakers, have successfully denied the evolutionary drive for reproduction. Or in a more common example, which feminist Alice Rossi has emphasized. Once the process of pregnancy begins, a woman's body begins to prepare the embryo and the woman's body to produce a healthy fetus.[18] But even then, human resolve and intervention can stop the process by actions that go beyond mere individual willfulness.

REPRODUCTION AND THE BEHAVIOR OF WOMEN, MEN, AND PARENTS

Reproduction is the basic drive for all living things, but it does not mean that survival of offspring is necessarily the only goal. Reproductive success could well include ending the life of offspring in order to insure the reproductive success of older ones. Many species of animals are known to do just that. Although evolutionary theory suggests that natural selection would protect offspring so as to provide for reproduction in future generations, parents sometimes kill their own offspring. In their book *Homicide,* Martin Daly and Margo Wilson studied infanticide in sixty societies. In fifty-six, the participants justified filicide on the grounds of bad circumstances for properly rearing children. Parents in twenty-one societies contended that the child was deformed, while people in twenty societies explained infanticide on the ground of doubtful paternity.[19] In short, fitness or reproductive success was the underlying or evolutionary explanation. In principle, that practice could be seen as the evolutionary

basis for human abortion: to maintain resources for the support of extant offspring. It is the same principle, now culturally revised and supported, that partly accounts for the world-wide spread of modern contraception, which, in turn, helps to explain the slowing down of population growth across the modern world.[20]

Evolutionary theory and tradition seem to agree in recognizing differences in the reproductive strategies of men and women. Do these differences extend to child rearing? No one doubts that only women can bear offspring, but many people also believe that rearing children could be shared or exchanged between parents. Altered historical circumstances over the last century or so have increasingly raised that question. No longer is it necessary for a mother to be the only food source or caretaker of an infant. Many mothers now rely upon advances in child care technology, which men can also employ as well. As a consequence, in several industrial societies there has been a movement, principally because of the expansion of women's employment outside the home, for fathers to assume some of the responsibilities of traditional child care. So far, however, the proportion of fathers in all industrial societies who do take on those responsibilities is small.

Alice Rossi, the well-known sociologist and feminist has offered an evolutionary explanation for the continuing predominance of women in the care of children. She recognizes that there are differences between the sexes in regard to reproductive strategies, or as she phrases it, "that there are differences in the ease with which the sexes can learn certain things". Among them was a greater ability of females to care for children. "The female is more closely involved in the reproduction process then the male", she acknowledged. If society wishes to urge shared parental roles, Rossi concluded, it has to accept the mother-infant connection as having "greater emotional depth than the father-infant relationship. "Or else," she continued, society should "provide an institutionalized means for men's compensatory exposure and training in infant and child care in order to close the gap produced by the physiological experience of pregnancy, birth, and nursing."[21]

Rossi's contention is supported by the empirical work of another distinguished sociologist, Melford Spiro. In 1955, Spiro had investigated Israel kibbutzim as a test case of the social construction of sexual differences. A central egalitarian point of the kibbutzim was that women and men should have the same opportunities for the community's work. Because only women could bear, and traditionally reared children, the kibbutz (community) decided that the children of couples would be raised in Children's Houses. Parents would visit their children daily, but the Children's House was the children's home. In that way, women and men alike would be free from child care responsibilities, and women could be free to pursue whatever job they wanted and could perform.

Twenty years later, Spiro reinvestigated the kibbutz only to conclude that its ideal had not worked out as intended. For one thing almost all of the child-care givers at the Children's House were women. For another, women in the kibbutz as a whole filled jobs usually considered traditional for their sex. Spiro was so shocked by the return to traditional ways that he found it "a rather unequivocal exception to our social science axiom . . . that human behavior and motive are primarily, if not exclusively, culturally programmed." One part of Spiro's research in the kibbutzim that particularly shocked him was the resistance of the girls after they had passed puberty to showering and dressing with the boys. As he concluded, the original kibbutz belief that children in the proper learning environment would be sex-blind was proven to be false even in the sexually enlightened conditions in which these children were raised. Consequently, Spiro suggested that "it seems likely that these sex differences (like their sense of sexual shame) were brought about not by culture, but by the triumph of human nature over culture, that is, by motivational dispositions based on sex differences in precultural rather than culturally constituted needs." He then added that if those needs were indeed "precultural", then, "all things being equal, it is probable that they are shared by females in any society."[22]

Spiro's use of the word "precultural" in explaining the kibbutz experience was obviously intended to be ambiguous (if not innovative) in order to avoid accepting a fully evolutionary explanation. A fuller evolutionary explanation for the differences in the child care behavior of men and women would look to the different investment of each sex. For women reproductive success entails the cost of bearing and rearing offspring; while for men the cost is gaining access to females by competing with males and or protecting females. Thus, in the course of evolution males' participation in reproduction has not involved child care, which, instead has been the female's part.

This evolutionary explanation of male reproduction also offers a way of accounting for certain other practices that men engage in historically. One is the harem and the other is polygyny. The traditional explanation for the institution of the harem or polygyny is that male sexual desire is the force behind both; once again, however, that traditional explanation needs to be recognized as only a "proximate" cause, as evolutionists would say, rather than the "ultimate" or evolutionary cause. As already noted, the sex drive is merely a means for reproduction, whereas the evolutionary purpose or ultimate cause is reproductive success, which for males means gaining as many offspring as possible. Polygyny is a much more common phenomenon than the maintaining of a harem, which, after all, incurs high financial costs. Polygyny, like harems, is but another special case of the ultimate cause for the male search

for reproductive success. In both practices, males have access to more females, that is, achieve male reproductive success.

Although the vast majority of modern industrial societies practice monogyny, polygyny is by far the most common form of marriage. About 87% of the 848 societies listed in Murdoch's Ethnographic Atlas (1967) are described as polygynous. (Apparently, however, many societies do not practice it – it is often too costly. Yet in many Islamic societies the number of such marriages is high.) Polygyny, of course, further reinforces the male search for reproductive success.[23]

It is relevant, historically, to recognize that in the United States, Mormon polygyny was sufficiently attractive to some men and women that it had to be legally extirpated by "monogamists" who compelled the federal government to outlaw polygyny. Many Mormons found the practice acceptable, even though at the outset polygyny had not been a practice included in the religion. When the founder, Joseph Smith, added polygyny to the religion, original members as well as those who later joined the church readily accepted the idea of plural wives. An evolutionary explanation would be a drive for male reproductive success; for women's acceptance the explanation would entail the obvious advantage of better resources since male Mormons who had plural wives were usually wealthier than most. Polygyny persists today undercover among some Mormons. (Polyandry, that is, a marriage with more than one husband, as evolutionary theory predicts, barely exists in any society. It neither advances female nor male reproductive success, given the limited number of offspring such a relationship can produce.)

Evolutionary theory offers another way of seeking to account for a striking phenomenon in human affairs: the level of violence among men as compared with that of women cross culturally. Martin Daly and Margo Wilson in *Homicide* thoroughly document the differences across a great spread of geographical locations by showing an amazingly high proportion of homicides between males as against those between females. Homicides between males in Canada (1973–1983) constituted 94% of homicides of all same sex murders; in Belo Horizonte, Brazil the proportion was 97%, in 1982 in England and Wales the proportion was 91.6%; Munda, India (1974) 100%, Congo (1948–1957) 97.5%, Denmark 85.3%, Baden-Würtemburg (1970–1971) 95.9%, Philadelphia (1948–1952) 95.4%.[24] The difference between each of these statistics and 100% is the proportions of women who kill women. As the authors conclude, "*The differences between the sexes is immense and it is universal* [emphasis in original] There is no known human society in which the level of lethal violence among women even begins to approach that among men".[25]

According to evolutionary theory there are two important sources of males' greater violence.[26] One is sexual rivalry among males in seeking access to

females. And that is where the push for violence against males occurs: to fight off rivals for that availability. Competition and violence among males are well known in the animal world as male deer grow horns and other "weapons" for the struggle to acquire access to females. Females possess no need for combat to gain reproductive success though they may seek desirable males as progenitors of their own offspring. By the same token, maintaining that long period of care enhances the violence of the male in order to protect his offspring against threatening males. (Females also enter into the combat scene in that, thanks to sexual selection, females may reject a male for inadequate reproductive capacity and thus open up a further source of conflict among males.) A second source of combat or violence among males is the need to guard against rivals since males cannot be sure that their offspring are truly theirs. (Male lions and some male monkeys have been observed killing the young of females in order to provide opportunity for reproduction.) As recognized earlier, males compete against other males to achieve reproductive success, while females have no need to compete.

That parental investment (that is, caring for the offspring) is the crucial determinant of differential behavior between the sexes is strikingly shown when that investment is reversed. In some species of frogs, birds, and insects, for example, the investment is higher among males than among females. In such species the females are larger and more combative. Seahorses are perhaps the most telling instance of reverse behavior since the male nurtures the young in his abdominal pouch and he is not combative, though females are.

In noticing the differences in the degree of violence among women and men it is worth recallling in twentieth century American history how frequently women, and women leaders like Jane Addams, have been prominent in the peace movements in this country and elsewhere. One is reminded, particularly, of the movement led by women during the 1920s in the United States, soon after their achievement of the vote, to bring about the ratification of the Kellogg-Briand Pact for world peace. Women have often been conspicuous in opposition to nuclear war, the Vietnam war, and other efforts to end violence in the United States and elsewhere. Once again it is important to recognize that we are talking about the sexes in general terms. Obviously, both sexes display individuals who defy the generalization. Some women are more violent than many men, just as some men are more peaceful than many women. It is the central tendency of the reproductive behaviors of the sexes that evolutionary theory seeks to explain.

Another well recognized form of violence may also be explained by evolutionary theory, namely, child abuse by step-parents. In today's society, with frequent divorce and much remarrying, we often know of step-parents and problems between them and their stepchildren. At the same time, social memory

reminds us that such problems are not just recent. Cinderella, Snow White and Hansel and Gretel and similar tales from the Brothers Grimm remind us that stepchildren have left a long memory of abuses.

Evolutionary theory helps to account for those long ago memories as well as more recent connections with step-parents. A step-parent, of course, has no genetic relationship with his or her stepchildren and thus does not have the same biological grounds for helping to rear a child as a biological parent does. (As evolutionary theory reminds us, all parents, human and animal are built to protect and support their offspring above all others.) Statistics of child abuse by step-parents support the genetic argument quite well. In a study made in Hamilton, Ontario, the proportion of instances of child abuse in families with two natural parents to age 16 was less than one case out of a thousand children. Among families of one natural and one step-parent the proportion was much greater. For children of up to four years, the proportion was 13 per thousand; at the age of five to ten years the figure was five per thousand, and at eleven to sixteen years the proportion was 4.8 per thousand. Although one might think that class differences might explain the different proportions between the children of natural parents and those of step-parents, the evidence does not sustain the difference. As Daly and Wilson pointedly note, "the proportion of children living with a step-parent is virtually identical across socio-economic classes in both Canada and the United States".[27]

The purpose here is not to instruct historians on matters relating to the history of step-parents, but to suggest that evolutionary theory can offer clues in accounting for human behavior or motivation. Just as a biographer might want to explore psychoanalytical theory or other forms of psychological theory in studying his or her subject. Psychology is best drawn upon for biographical studies since a psychological theory, like Freud's, for example, generally derives from the analysis of individuals. On the other hand, evolutionary theory is concerned with the species as a whole, and cannot offer any help in accounting for individual behavior.

EVOLUTION FOSTERS COOPERATION

Probably the most familiar motivations historians have derived from their knowledge of evolution have been selfishness and competitiveness. And those negative interpretations on the working of evolution have not been confined to historians. Thomas Huxley, Darwin's "bulldog", as he was called, drew exactly that conclusion at the end of his life in delivering his Romanes Lecture "Ethics and Evolution." The burden of his argument was that evolution, as he under-

stood it, was too selfish to be the basis for an ethical system. Instead, said Huxley, it was morally necessary to have evolution stamped down when thinking about human ethics. "The practice of that which is ethically best – what we call goodness or virtue in all respects", Huxley insisted, "is opposed to that which leads to success in the cosmic struggle for existence. In place of ruthless self-assertion it demands self-restraint; in place of thrusting aside, or treading down, all competitors, it requires that the individuals shall not merely respect, but shall help his fellows; its influence is directed not so much to the survival of the fittest, as to the fitting of as many as possible to survive. It repudiates the gladiatorial theory of existence", he exclaimed.[28] "Let us understand, once and for all that the ethical progress of society depends, not on imitating the cosmic process, [that is, evolution] still less in running away from it, but in combating it".[29] Even modern biologists like the preeminent evolutionary theorist, George C. Williams, still agree with Huxley. "Nothing resembling the Golden Rule or other widely preached ethical principles", Williams remarked, "seems to be operating in living nature. It could scarcely be otherwise, when evolution is guided by a force that maximizes genetic selfishness".[30]

Yet Darwin himself contended that cooperation and even morals resulted in part in the drive for fitness, that is, reproductive success. The Russian naturalist and philosophical anarchist Peter Kropotkin went even farther than Darwin in finding cooperation as a significant part of evolution as evidenced in his well-known book *Mutual Aid* (1902).

More recently, biologists and social scientists have begun to recognize cooperation and even self-sacrifice in animal in addition to human beings. As one biologist remarked twelve years ago, there are more than 150 species of birds in which some individuals spend part or all of their lives helping others to rear offspring.[31] Other scholars point out that a chimpanzee will endanger himself to protect another, or a ground squirrel will warn others of the presence of a predator even at the risk of endangering itself. By and large the explanation for these examples of helpful or "altruistic" behavior is that the helpers or risk-takers are related to those they are saving. Their "sacrifice", in short, is based not on altruism but in helping relatives who share the risk-takers genes. By the same token, what appears to be selfless behavior on the part of human beings is justified by the genetic closeness of the persons involved. A brother is more likely to save a sister than a stranger or perhaps even a cousin. Indeed, genetic closeness can account for the degrees of family adhesion in most societies. Social scientists have often recognized the cooperative basis of human affairs running from Aristotle's definition of man as a social animal to a modern political scientist's book on *The Moral Sense*[32] as well as in Jane Mansbridge's well-known article "The Rise and Fall of Self-Interest in the Explanation of Political Life".[33]

Mansbridge does not draw upon evolutionary theory in raising doubts about self-interest, a primary motivation proclaimed by political scientists as well as by historians. (James Wilson is a rare exception when he draws upon evolutionary theory in *The Moral Sense*.). In one sense, though, no student of politics better captures the present recognition of the genetic basis of human sociality than David Hume and he possessed no knowledge of modern genetics! "A man naturally loves his children better than his nephews", Hume wrote, "his nephews better than his cousins, his cousins better than strangers, where everything else is equal".[34] Evolutionary theory now tells us that siblings, on average, share half of their genes from their parents, and a quarter of them from their nieces and nephews, and only one-eighth of those from cousins. This is, of course, another example of how the feelings of relationships can be explained by evolutionary theory. It also explains why nepotism has long persisted and will continue to exist though resisted because people who are genetically related are more likely to secure offices, favors, or jobs. It also helps to explain why in the past empires have used eunuchs, that is, without families, in securing the undivided loyalty of their closest supporters, just as celibacy may be a source of loyalty for a religious order.

Evolutionists have also uncovered another form of cooperation among animals. It is called "reciprocal altruism" or rephrased less technically as "you scratch my back and I'll scratch yours." Unlike cooperation among family members, "reciprocal altruism" does not depend upon a genetic connection. It is behavior based upon feelings unrelated to genetics. Students of chimpanzee and bonobo [once called pygmy chimpanzees] behavior have recognized it. Perhaps the strongest evidence of cooperative behavior that lacks genetic basis among animals is the sharing of blood (that is, food) by vampire bats which are not related.

The impetus behind these examples is to qualify the seemingly unrelenting selfishness often found by many observers of evolution or natural selection. And qualification is needed since many other observers have long understood and appreciated the presence among human beings of cooperation and even selflessness. As some ethologists remind us, selfishness cannot be the hall mark of humanity when we reflect on the concern for others and the benevolence encountered within our species. As the well-known ethologist Frans de Waal has shown in *Good Natured: the Origin of Right and Wrong in Human and Other Animals* there are many examples of cooperative and friendly relations among chimpanzees and bonobos. Despite the well-known objection of anthropomorphism by some ethologists, de Waal accepts in animals many of the cooperative behavior patterns that are common among human beings. How could open and unrestricted competition "ever explain the concern for others and the benevolence

encountered in our species?" de Waal asks. Natural selection, de Waal continued, does not necessarily deny kindness, any more than that "birds and airplanes appear to defy the law of gravity yet are fully subjected to it . . ." Or put another way, morality has evolved as a way of achieving reproductive success.[35]

Another zoologist, Matt Ridley, sees cooperation and sociality as human characteristics, though he draws his support from his recognition of human universals rather than from animal behavior. "Human beings have social instincts," he concludes. "They come into the world equipped with predispositions to learn how to cooperate, to discriminate the trustworthy from the treacherous, to commit themselves to be trustworthy, to earn good reputations, to exchange goods and information and to divide labor."[36]

Among historians, as I have remarked before, universals are not the approach that historians emphasize when they do their work. They are primarily searching for differences among peoples and tend therefore to ignore universals. Yet when evolutionary theory points to similarities or universals among peoples, historians might gain some fresh insight into the fundamental nature of human beings, despite the presence of differences among people. After all, evolutionary theory treats all members of our species as basically alike. That is why evolution does not recognize racism as a result of evolution. It is worth remembering, too, that cultural or historical arguments have long justified almost any kind of discrimination against peoples or nations. At the same time, evolutionary theory admittedly does not describe human beings as harmonious or selfless, though as a theory and practice it may not be as insistent upon selfishness or competiton as historians sometimes believe.

Some modern students of evolutionary theory emphasize the limits that evolution seems to place upon women, such as their lack of risk-taking, which is said to be derived from their lack of violence as compared with that of men. Some theorists of evolution have even concluded that the prevalence of the "glass ceiling," which has been described as limiting women's opporunities in achieving upper mobility in business managment, is a consequence of evolutionary development.[37]

As a result of such interpretations of evolutionary theory, many historians often misunderstand or even reject evolution in thinking about history or social policy. As one historian recently remarked to me in discussing evolutionary theory, it is nothing more than a form of "Social Darwinism." The fear is that evolutionary theory is essentially deterministic since it accounts for certain kinds of behavior. And it is true that evolutionary theory seeks to explain how our biological history has shaped our behavior in limited, but also important ways. Evolutionary theory is a scientific concept that explains biological actions, but it does not tell us how we should behave. Evolutionary theory tells us that

human beings, like other animals, must reproduce, nevertheless almost all human societies culturally restrict reproduction to the extent of denying or limiting even that strong biological drive, as for example, in the case of marriage. Today the evolutionary drive for reproduction has been limited in another way by separating sexuality from reproduction through the introduction of contraception. Even something like the high level of violence by men can be shaped by values in a man-made cultural environment. Look at the low level of violence by men in Iceland as contrasted with the high level in New York City, yet men in both societies are programmed for the same evolutionary basis for male violence. In sum, an evolutionary behavior pattern can be altered or controlled by human values. But before those adjustments can be made, historians and other humanists need to understand the human behaviors that evolution has bequeathed us.

NOTES

1. In: Henry Adams, *The Degradation of the Democratic Dogma.* (New York, 1920), 267–311.

2. For an introduction into cultural behavior in animals see John Tyler Bonner, *The Evolution of Culture in Animals.* (Princeton, 1980).

3. The Nobelists were Karl von Frisch, who worked on the language of bees, Konrad Lorenz and Nicholas Tinbergen, both of whom worked on instinctual behaviors.

4. See *New York Times*, March 30, 1999. p. D3.

5. Clifford Geertz, *Interpretation of Cultures* (New York, 1973), 36–37, 53.

6. See Paul Ekman, *Annual Review of Psychology*, *30*, 527–554.

7. Walter Burkert, *Creation of the Sacred. Tracks of Biology in Early Religions* (Cambridge, Mass., 1996) 33.

8. Alfred Kroeber, *An Anthropologist Looks at History*, T. Kroeber, ed., (Berkeley, 1963) 212-213, 204.

9. Carl G. Hempel, The Functions of General Laws in History. *The Journal of Philosophy*, *39* (1942), 35–48.

10. See a selection of the literature: Jerome H. Barkow, Leda Cosmides, John Tooby, (Eds), *The Adapted Mind; Evolutionary Psychology and the Generation of Culture* (New York, 1992); John B. Beckstrom, *Darwinism Applied*; *Evolutionary Paths of Social Goals* (Westport, Ct., 1993); Jack Hirschleifer, *Economic Behavior and Adversity* (Chicago, 1987); Allan Mazur, "Biosociology" In: J. E. Short (Ed.), *The State of Sociology* (Beverly Hills, Ca, 1981); Robert J. McShea, *Morality and Human Nature: A New Route to Ethical Theory* (Philadelphia, 1990); Roger Masters, *Beyond Relativism: Science and Human Values* (Hanover, NH, 1993); Mary Maxwell, *Morality Among Nations; An Evolutionary View* (Albany, NY, 1990); Randolph Nesse and George C. Williams, *Why We Get Sick. The New Science of Darwinian Medicine* (New York, 1994); Robert J. Richards, *Darwin and the Emergence of Evolutionary Theories of Mind and Behavior* (Chicago, 1987); W. G. Runciman, *The Social Animal* (London, 1998); Elliott White, (Ed.), *Sociobiology and Human Politics* (Lexington, MA, 1981); James Q. Wilson, *The Moral Sense* (New York, 1993).

11. Joseph Shepher, *Incest: A Biosocial View* (New York, 1983).

12. Arthur P. Wolf, *Marriage and Adoption in China*, 1854–1945 (Stanford, Ca., 1980). His early explorations appeared in "Childhood Association, Sexual Attraction, and the Incest Taboo: A Chinese Case". *American Anthropologist*, *68* (Aug., 1966), 833–898.

13. Alice S. Rossi, "Gender and Parenthood." *American Sociological Review*, *49* (Feb. 1984).

14. One of the objectors is Lila Leibowitz, *Females, Males, Families; A Biosocial Approach*, (North Situate, MA, 1978).

15. Some ethologists also recognize some reason for seeing a persistence of human males associating with the mother-child dyad: the dimorphism of the human sexes. Male gorillas and chimpanzees, like male elephant seals differ hugely from females because their investment is in size and strength, not assistance in nurturing offspring. On the other hand, among gibbons and siamangs, which come close to mating for life, the sexes are about equal in size. Male and female human beings differ in size, but only slightly compared to gorillas or elephant seals. This slight dimorphism of human beings suggests that there is some male investment in caring or assisting with offspring or females.

16. The author was Alice Rossi in her presidential address to the American Sociological Association in 1984.

17. See the delightful and informative study, Sidney Mintz, *Sweetness and Power* (New York, 1985).

18. Alice S. Rossi, "A Biosocial Perspective on Parenting." *Daedalus*, *106* (1977), 4–5; *ibid*, "Gender and Parenthood." *American Sociological Review*, *49* (Feb. 1984) 9; ibid., "The Biosocial Side of Parenthood." *Human Nature*, *75*.

19. Martin Daly and Margo Wilson, *Homicide* (New York, 1988), 58–59.

20. See Max Singer, "The Population Surprise, "*Atlantic Monthly*, 284 (August, 1999), 21–25.

21. Quoted in Carl N. Degler, *In Search of Human Nature* (New York, 1991), 295.

22. Melford E. Spiro, *Gender and Culture* (New York, 1980), 97–100. Since Spiro wrote, the role of women in child care in the kibbutzim has been well recognized and partly explained in Eliezer Ben-Rafael, *Crisis and Transformation; The Kibbutz at Century's End* (Albany, NY, 1997), 61–71.

23. Daly and Wilson, *Homicide*, 132.

24. *Ibid.* 147–148, The authors display, all told, thirty-five different locations in which male violence has been statistically measured.

25. Daly and Wilson, *Homicide*, 146.

26. It is frequently remarked that the source of violence among males as compared to that of females is the greater presence of androgen hormones or testosterone. As noted earlier, evolutionary theory draws a distinction between "proximate causes" and "ultimate causes", which are not unlike immediate and long term causes in standard historical study. In evolutionary theory a proximate cause is one that brings an immediate action such as the hormones that cause aggressive action. The ultimate cause, that is, the evolutionary cause, is why was the hormone developed to produce that action? Natural selection thus used the hormone to develop the behavior necessary for the male individual to compete for reproductive success.

27. *Ibid.* 87. See also their article in R. Geller and J. Lancaster, (Eds), *Child Abuse and Neglect: Bio-social Dimensions* (1987).

28. James Paradis and George C. Williams, *Evolution and Ethics* (Princeton: 1989), 139–140.

29. *Ibid.*, 141.

30. *Ibid.*, 195.

31. J. R. Krebs and N. B. Davies, *An Introduction to Behavioral Ecology* (2nd ed., Sunderland, MA, 1987), 268–278.

32. James Q. Wilson, *The Moral Sense* (New York, 1993).

33. Jane J. Mansbridge, (Ed.), *Beyond Self-Interest* (Chicago, 1990), 3–22.

34. David Hume, *A Treatise of Human Nature* (Buffalo, 1992 [1739]), 238.

35. Frans de Waal, *Good Natured; The Origins of Right and Wrong in Humans and Other Animals* (Cambridge, MA, 1996), 11–12.

36. Matt Ridley, *The Origins of Virtue; Human Instincts and the Evolution of Cooperation* (New York, 1997), 249.

37. See, for example, Kingsley Browne, *Divided Labours; An Evolutionary View of Women at Work* (London, 1998) and the soon to be published *The Decline of Males* by Lionel Tiger.

EVOLUTION IN INTERNATIONAL RELATIONS? FROM SOCIAL DARWINISM TO EVOLUTIONARY THEORY IN THE STUDY OF INTERNATIONAL POLITICS

Vincent S. E. Falger

INTRODUCTION

The question to what extent neo-Darwinian evolutionary theory and concepts have been accepted in the field of International Relations (hereafter IR) could be answered shortly: not in the mainstream thinking – however diverse that may be in itself – and hardly in the periphery.[1] So far, IR theorists apparently do not feel addressed by the results and perspectives that are accumulating in other disciplines, in particular in evolutionary psychology and anthropology. On the other hand, evolutionary scientists do not pay special attention to the field of International Relations although they may be positively interested in politics and political behaviour in general. In this chapter I hope to make clear that the missing link is the lack of realization on the IR side that what is happening in the evolutionary approach of human behaviour is not less important than attempts to overcome the shocks of post-Cold War politics and the increasing influence of economics on world politics. For this purpose I will first look in existing IR literature for traces of themes that come nearest

Evolutionary Approaches in the Behavioral Sciences, Volume 8, pages 91–118.

ISBN: 0-7623-0769-2

to (neo-) Darwinian thinking. Then I will inventory modern evolutionary theory in International Relations. Common themes at the level of presumptions in IR literature of all times are: (a) 'nature' and 'human nature', (b) causes of inter-group violence and war, and (c) change or development or evolution of political structures.

'NATURE' AND 'HUMAN NATURE'

Aristotle developed a 'modern' conception of human beings as *political* animals by nature because they are group members gifted with rationality, and inclined to cooperation, with the city-state as a result. In the Middle Ages, however, 'nature' was seen by the philosophers of the time as a divine creation and as a given. Humans needed divine rules to live a virtuous life and to save their soul, because they are sinful and potentially disturb the divine creation. Machiavelli and Guicciardini, the late medieval, early Renaissance Florentine humanists, turned away the attention from salvation of the soul towards security of the state. Society changed from a natural order of things into a man-made one, or, as Knutsen (1997, 51) puts it, "by replacing God with the state [Machiavelli and Guicciardini] also replaced nature with society."

In the Modern Age, from the 16th century on, rationality, sovereignty, the interstate system (1648 Peace of Westphalia), territoriality, and international law began to dominate secular thinking about international politics. Among many other 17th century publications, Thomas Hobbes' *Leviathan* (1651) no doubt exerted a prominent influence by its naturalistic conception of society's original 'state of nature' (the lack of authority results in anarchy) in which war is omnipresent, "a Warre, as is of every man, against every man." Also Hobbes' other naturalistic analogy, *homo homini lupus*, was taken seriously until deep into the 20th century. Hedley Bull's concept of the 'anarchical society' for international relations in total, expresses the foundation of the Realist school of thought, the most important and in many ways still dominant conceptual framework in IR. As long as the world is divided into territorial states that do not recognize and tolerate external interference in internal affairs, this anarchical character will remain the main aspect of human behaviour in the largest of all groups, international society. This does not imply automatically Hobbes' *bellum omnium contra omnes*, as we now know that his conception of human nature is very one-sided, but the *structure* of the international system still depends thoroughly on what Roger Masters (1964) called a 'primitive' political system in which self-help dominates.

So, one line of IR theorizing, leading far back to the Antiques and Rationalists, has deep roots in conceptions of 'human nature'. "The absence of

a supreme authority does not mean that human sociability is removed from international society altogether," as Knutsen writes very much to the point, but "[r]ather, it means that the sociable nature of humanity is de-emphasized in international actions, and all allowances are made for a larger share of egotistical, unilateral behaviour" (Knutsen, 1997, 3).

In the 19th century we can see that the *Zeitgeist* of 'nature red in tooth and claw' also has room for 'progress' or evolution. Although Darwin's legacy has been, in a way, stigmatized by false interpretations of, in particular, natural selection theory which we now know as social Darwinism (but could better be called Spencerianism), it nevertheless is also the start of a new way of thinking about IR. War as a 'natural' phenomenon in which 'superior' countries beat 'inferior' ones of course had racist presuppositions, but was not typically right-wing politics. Karl Marx, who never formulated a distinct theory of IR, saw the English occupation of India as a sign of progress: a feudal country was forced into modernity by British capitalism, and this was a necessary precondition for ultimate preparation to the end-state of world history, the upraising of the poor and the establishment of socialism. We know how much Marx originally was impressed by Darwin's work and that the great revolutionary saw Darwin's theory as the proof of the foundation of his own dialectical method applied to human history.[2] We also know that this admiration and inspiration was certainly not reciprocal.

In the same vein of dialectical progress, but without in the end devastating the status quo, belong Walter Bagehot's *Physics and Politics* (1872) and the full-blown social Darwinists Herbert Spencer, Benjamin Kidd, William Sumner, and the Germans Friedrich Ratzel and Gumplowitz. War between states was seen as the ultimate test in which the strongest would rightly survive, and the coupling of *is* and *ought* resulted in the well-known apology, and sometimes heroification of war. Vulgarized Darwinism as political legitimization proved very useful for many thinkers and politicians in the Western world.[3] One of the most influential of these has been the Jena professor of zoology Ernst Haeckel. Haeckel, the prominent promotor of Darwinism in Germany[4], developed his monist philosophy in which the superior Germans would prove their quality, under the condition that they would fight alien influences, from liberal political doctrines to non-Aryan elements in the German population. The combination of Haeckel's pure Aryan *Volk* idea with Ratzel's organicist *Lebensraum* conception – of course by means of healthy struggle between groups – led to a social theory that would be a fruitful inspiration for later biologistic tendencies in national socialism (Gasman, 1971; Marten, 1983 and Stein, 1987).

Less influential – less devastating, one may also say – was Darwin's inspiration of Russian biologists and social theorists.[5] In particular, Darwin's central

theme of the struggle for existence and the criticism it invited made a lasting impression, and the best known title is Petr Kropotkin's *Mutual Aid: A Factor in Evolution* (1904). In this book, the cornerstone of Kropotkin's anarchist philosophy, the author demonstrates that the instinct of human sociability evolved into morality.[6] Contrary to Darwin and Marx, Kropotkin argued that cooperation rather than conflict lays at the root of the human historical process. His argument ranged from mutual aid among animals, via idem among savages and barbarians, to mutual aid among ourselves, where the best examples are found in rural communities and villages. In a way, Kropotkin's struggle for the village as the basic human form of social organization and against the state, makes clear that: (a) Darwinism has been an ideological tool in the hands of very different and often opposing politicians and social philosophers, and (b) that the idealized harmony of the small-scale community is inversely related to the chaotic and non-hierarchical structure of international relations. So, it is not surprising to find direct and, even more, indirect influences of evolutionary theory in International Relations. However, it would be better to see these influences in a fundamental way: they helped to put a scientific basis under predominantly political philosophies, *ergo* it is of a highly pseudo-scientific character. At best, it illustrates a *looking for* a rationalist social philosophy; at worst, it is a scientifically unacceptable mixing of facts (or better, supposed facts) and norms, resulting in a direct and severe abuse of science by secular political ideologies of all colours.

Has the field of IR changed since World War II with respect to evolutionary theory? Taking for granted the details of the pre-war discussion on behaviourism in social science, we can summarize the period 1945-c.1990 roughly as follows: 'human nature' as the explicit or implicit basis of several important treatises of Realism declined from a central but unquestioned given to an out-of-date and nearly totally forgotten factor under the influence of the Standard Social Science Methodology (see note 13). Representative is the fate of Hans Morgenthau's classic text *Politics among Nations. The Struggle for Power and Peace* (1st ed. 1948, 6th and last ed. 1985, brief ppb ed. 1993). In this book Morgenthau not only pretends to present an empirical and pragmatic theory of international politics, but also states that his theory concerns the nature of all politics.[7] He sees the history of modern political thought as a contest between two schools that differ fundamentally in their conceptions of the nature of man, society and politics. The idealist school believes in human rationality, universal moral principles, essential goodness and infinite malleability of human nature. Education and reform, and incidentally the use of force, may adjust for the failures of social life. The Realist school, on the other hand, believes "that the world, imperfect as it is from the rational point of view, is the result

of forces inherent in human nature. To improve the world one must work with those forces, not against them. This being inherently a world of opposing interests and of conflict among them, moral principles can never be fully realized, but must at best be approximated through the ever temporary balancing of interests and the ever precarious settlement of conflicts" (Morgenthau, 1967, 3). Realists see in a system of checks and balances a universal principle for all societies, which can be illustrated with historic data. Morgenthau presents six principles of political realism, of which the first three are most important for this chapter.

First, "political realism believes that politics, like society in general, is governed by objective laws that have their roots in human nature. In order to improve society, it is first necessary to understand the laws by which society lives" (Morgenthau, 1967, 4). Morgenthau expressly states that "[h]uman nature, in which the laws of politics have their roots, has not changed since the classical philosophies in China, India, and Greece endeavored to discover these laws" (p. 4).

Second, Realism presumes in political reality a kind of rational outline which guides the statesman in choosing among alternative options. It is the concept of interest, defined in terms of power, that provides the link between reason trying to understand international politics and the facts to be understood.

Third, "Realism does not endow its key concepts of interest defined as power with a meaning that is fixed for once and for all. The idea of interest is indeed the essence of politics and is unaffected by the circumstances of time and place" (p. 8). Morgenthau agrees with Max Weber's observation that not ideas, but (material and ideal) interests create 'images of the world' which have often "served as switches determining the tracks on which the dynamism of interest kept action moving." For Morgenthau this meant that the goals of political action may cover "the whole gamut of objectives any nation has ever pursued or might possibly pursue" (p. 9). The same applies to the concept of power:

> Its content and the manner of its use are determined by the political and cultural environment. Power may comprise anything that establishes and maintains the control of man over man. Thus power covers all social relationships which serve that end, from physical violence to the most subtle psychological ties by which one mind controls another. Power covers the domination of man by man, both when it is disciplined by moral ends and controlled by constitutional safeguards, as in Western democracies, and when it is that untamed and barbaric force which finds its laws in nothing but its own strength and its aggrandizement (p. 9).

In the third chapter of his book Morgenthau presents important refinements of his conception of political power, which he calls one of the most difficult and controversial problems of political science. Here we read his famous dictum "International politics, like all politics, is a struggle for power. Whatever the

ultimate aims of international politics, power is always the immediate aim" (p. 25). It is in line with his statement that "the struggle for power is universal in time and space and is an undeniable fact of experience" (p. 31).

Even though Morgenthau is aware of anthropological studies that show that "certain primitive people seem to be free from the desire for power," this does not imply that the state of mind of these people and the conditions under which they live "can be recreated on a worldwide scale so as to eliminate the struggle for power from the international scene" (p. 31). In his view it would be useless and even self-destructive to free one or more peoples of the earth from the desire for power while leaving it extant in others. For, "[i]f the desire for power cannot be abolished everywhere in the world, those who might be cured would simply fall victim to the power of others" (p. 31). Recognizing that this standpoint can be seen or judged as an objection to progress and reform, Morgenthau then presents his most fundamental argument to *start* with human nature as the foundation of power politics:

> Though it is true that certain social arrangements and institutions have always existed in the past, it does not necessarily follow that they must always exist in the future. The situation is, however, different when we deal with not social arrangements and institutions created by man, but with those elemental bio-psychological drives by which in turn society is created. *The drives to live, to propagate, and to dominate are common to all men*. Their relative strength is dependent upon social conditions that may favor one drive and tend to repress another, or that may withhold social approval from certain manifestations of these drives while encourage others. Thus, to take [an] example only from the sphere of power, most societies condemn killing as a means of attaining power within society, but all societies encourage killing of ennemies in that struggle for power that is called war (p. 31; emphasis added).

In the original text the crucial, italicized sentence bears a footnote which deserves extra attention: "Zoologists have tried to show that the drive to dominate is found even in animals, such as chickens and monkeys, who create social hierarchies on the basis of the will and ability to dominate. See e.g. Warner Allee, *Animal Life and Social Growth* (1932), and *The Social Life of Animals* (1938)" (p. 31, note 5).

In defending the autonomy of the political sphere against other modes of thought, Morgenthau as a Realist pleads for a pluralistic conception of human nature and seems to hint at the difference between man and animal. "Real man is a composite of 'economic man', 'political man', 'moral man', 'religious man', and so on. A man who is nothing but 'political man' would be a beast, for he would be completely lacking in moral restraints. A man who was nothing but 'moral man' would be a fool, for he would be completely lacking in prudence" (p. 13).

Digging deeper into Morgenthau's conception of human nature, it is worth remembering that "the questionable parallel with animal societies," as the

contemporary critic of Morgenthau, Barrington Moore already called it, referred to books of his Chicago collegue, the zoologist W.C. Allee (1932, 1938). In particular *The Social Life of Animals* (1938) is interesting, since Allee is stressing that "the tendency toward struggle for existence is balanced and opposed by the strong influence of the co-operative urge" (Allee 1938, 186), which he calls an equally fundamental biological principle (p. 209; see also pp. 6–16, 217). Morgenthau must have neglected this important aspect of Allee's explanation of sociality on purpose. Most probably he disagreed with the conclusion of Allee's chapter on some human implications, interestingly enough mainly about the relations between nations![8] Although Allee sees many men as aggressive animals, recognizes a similarity of the drive toward dominance in human social hierarchies and those of chickens and other animals (see Morgenthau's above mentioned note 5 of Chapter 3 in *Politics among Nations*), and is aware of the fact that man appears to be relatively easy to lead into mass combat, he denies that war making is instinctive (Morgenthau holds the same position). But Allee's younger collegue must have been thinking of the idealism which was so *en vogue* in the United States at that time, when he read that "[t]here seems to be no inherent biological reason why man cannot learn to extend the principle of co-operation as fully through the field of international relations as he has already done in his more personal affairs" (p. 217).[9]

Morgenthau's selective reading of Allee's work makes clear that he was not really interested in the biological or evolutionary foundations of social behaviour as such, but only in solidifying his own skeptical view of mankind. In this respect, Morgenthau's conception of human nature as the origin of power politics can be understood as *pars pro toto* for the Realist school in IR theory.[10] The classic version, however, had to give way to the modernized versions like those of Kenneth Waltz and Robert Gilpin, and one of the main points in which neo-realism distinguishes itself from the classical power political approach is indeed the explicit *dismissing* of human nature as the foundation of a theory of international relations (Waltz 1959, chapter 3, and 1978 as a structural theory). So, in a time that ethology and sociobiology gained more and more attention, exactly *because of* the rising interest in the biological bases of human behaviour, these developments were totally missed by main stream IR theory as well as its critical variants. Biopolitics, which could have functioned as a *trait d'union*, had no influence on IR whatsoever, even if the field's anthologies traditionally paid lipservice to the existence of 'biological' conceptions of war (see e.g. Dougherty and Pfaltzgraff in every edition[11]). Somit and Peterson (1998) are correct to ascribe the few traces of mentioning of 'biological' literature in IR to ethology, not to biopolitics.[12]

CAUSES OF INTERGROUP VIOLENCE AND WAR

In the early 1990s, the downfall of communism in Eastern Europe not only caused an identity crisis in IR as a discipline, but it coincided with the low tide of previously strong Marxist opposition to transplant ideas from genetics and evolutionary biology to the human social sciences (for an overview, see Segerstråle, 2000). In particular, in psychology and anthropology an important impetus from biological-evolutionary theory was demonstrable. Two factors played a role in this process. First, traditionally both disciplines had, because of a greater orientation towards individual behaviour, a more intimate link with the theoretical and experimental side of biology (e.g., physical and cultural anthropology do not have counterparts in sociology and political science). Second, no other field than politics – and by implication political science – had suffered more from the *déconfiture* of social Darwinism, and nowhere more opposition against biologism existed than in the macro-social sciences, where the Standard Social Science Methodology remained virtually unquestioned.[13] A broad third factor may be that IR, tied as it is to macro-social actor presentations, is much more oriented to history, (international) law and economics than to behavioural analysis. The main trend in IR, apart from flirting with all kinds of (de)constructionisms, reveals decreasing interest in political theory and traditional political philosophy, and increasing descriptive-institutional and normative analyses of international politics. The quantitative approach in IR, which was seen as the hallmark of scientific nature in the 1960's and later, lost terrain also. Human rights problems, the historicization of international politics and international political economy took over most of the field, but even under this unpromising and unrewarding constellation, a very small number of IR people have invested time and energy to apply neo-Darwinian evolutionary concepts and biologically derived research methods. Hereafter I will restrict myself to those who are to be counted as IR specialists.

Having discussed the few main trends in the IR literature related to biology up to the early 1990s in this series before (Falger, 1994), I did not detect a fundamental reorientation of leading theorists to a biological-evolutionary paradigm since that time, such as happened in the 1980s in evolutionary psychology. What could have become an important source of inspiration, Shaw and Wong's *Genetic Seeds of Warfare* (1989), did not turn out to function in that way. This book still is, up till now, the only 'biopolitical' treatise of war, traditionally the most central issue in IR. As I wrote earlier, the book's importance is based on the application of Hamilton's inclusive fitness theory for the development of the authors' view of nepotistic altruism. "Inclusive fitness and, more specifically, kin selection also provide an ultimate, evolutionary rationale for anticipating

origins of 'self-sacrifice to the death' " (Shaw & Wong, 1989, 31). If this is applied to less and nonrelated individuals, there is an evolutionary based reason to expect in-group amity and out-group enmity from the early history of social living species, including man, on.

Given considerations of resource competition, the principle of competitive exclusion will in general promote hostility toward members of out-groups. In-group amity and out-group enmity are thus strengthened from two different sides. The resulting group dynamics are essentially those envisaged by Alexander's balance-of-power hypothesis:[14]

> From our perspective, the important point is not whether warfare per se was or was not the singular force in the rise of tribes, chiefdoms, and states. Nor do we feel obliged to argue that groups were in a constant state of warfare with one another. Rather, it is the threat of resource competition, competitive exclusion, and warfare that matters . . . [W]e submit that balance-of-power strategies evolved to help minimize these threats from expanding outgroups (. . .).
>
> In addition, it is important to acknowledge that regardless of the exact process that led to group expansion, larger groups would likely have enjoyed a competitive advantage that led to group expansion, larger groups (everything else being equal). In the evolutionary long run, larger groups would have displaced smaller groups and their members would thus have staked out a larger share of humanity's gene pool. This implies that behavioral predispositions that facilitated group expansion would have been retained and incorporated into more permanent repertoire of individual and group behavior (Shaw & Wong, 1989, 56–58).

Contemporary manifestations of group cohesion and mobilization for conflict, evident in modern nationalism and patriotism, may have acquired cultural traits, but "it is the underlying psychological dispositions or mechanisms, not the cultural traits themselves, which must be understood directly in terms of an evolutionary theory of warfare." (p. 64)

This rigorous ultimate reasoning distinguishes this book from the non-biologically oriented IR literature, typically fixated at proximate causes. But it is not only the book's quite difficult to understand theoretical basis that will deter readers not familiar with behavioural biology; this goes also for the unusual conception of international politics at the time of writing. Much energy is invested in presenting ethnicity as an optimal grouping with powerful mobilization appeal which, in the case of nationalism, "stems not from particular content but from its inclusive fitness logic" (Shaw & Wong, 1989, 143). The authors rightly deny the modern character of nationalism; the association of nationalism with the nation-state is indeed relatively recent, but it is only one phenotypic expression of a deep in-group/out-group structure inherited from human prehistory. So, in accentuating the differences in theoretical perspective via use of an ultimate, evolutionary rationale in the explanation of intergroup

behaviour, Shaw and Wong have indeed made an original contribution to the almost exclusively proximately oriented social science literature on the origin of war.

Unfortunately, they did not take the trouble to link their own work to the more mainstream IR literature, and the result was to be expected: a complete ignoring of this potentially important book in the field. One can wonder whether the approach is too unconventional (also demanding quite another type of intellectual investment than usual in the social science of IR), or whether the authors are relative outsiders in the field, or perhaps they were just one decade too early – the result is a sorry example of how little International Relations as a discipline is interested in the biological-evolutionary approach.[15]

A second book that might exert influence on 'traditional' IR is Johan van der Dennen's two-volume *The Origin of War* (1995), which is, however, no more than a handsomely printed PhD thesis, not published with a well-known publisher and out of print. This restricts the book's impact seriously, although the author is a well-known and prolific writer in academic circles. Until the condensed version of this book is published with a regular, preferably American publisher and gets the broader attention it deserves, we must rely on the relatively few available copies of *The Origin*. The book, even if it certainly is not typical for IR, is an important statement on the evolutionary origin of primitive war, which should be taken seriously by every student of war – ancient and modern wars, 'primitive' and technologically advanced. Modern IR specialists will be surprised to read about war as the evolution of a male-coalitional reproductive strategy. Indeed, the relationship to non-human primate behaviour (in particular of chimpanzees and bonobos) is central to the argumentation that without this evolutionary connection and prehistory, our human tendencies of ethnocentrism, xenophobia and in-group/out-group differentiation are incomprehensible. Far from being reductionistic, van der Dennen argues that we cannot do without evolution and evolutionary theory to explain why: (a) war only occurs with *Hominidae* and *Pan troglodytes*, (b) violent intergroup conflict does not occur with other primates or mammals in general (although they do have substantial interindividual agonistic behavioural repertoires), and (c) explain the sexual dimorphic character of warfare – it is an overwhelmingly male preoccupation. Coalitional psychology is supposed to play an important role, and does so not only with humans but also with chimps, and not only in intergroup conflicts, but also in intergroup cooperation – *politics*, for short. I predict van der Dennen a much more prominent role if he succeeds in publication of his main theses in a 'normal' book. He may have much more success than Shaw and Wong who are disciplinary outsiders and did not enter the minds of IR specialists.[16]

EVOLUTION OF POLITICAL STRUCTURES

Turning to International Relations as a recognized, be it not very old, academic discipline, the last decade was as unreceptive to evolutionary biology as before. Compared to the hundreds of books published in this field every year, one will look in vain for even one title like the Shaw and Wong book. However, within a short time this may change for the better somewhat. Three developments can be reported – but it is too little and too early to present this as a substantive change.

One development has to do with creating a modest niche for IR in the annual conferences of the Association for Politics and the Life Sciences. In September 1998, at the first independent APLS conference in Boston (with E. O. Wilson as the keynote speaker), Peter Corning organized a panel on 'Political Evolution' in which two IR specialists met and presented papers. One was George Modelski (1998b) who discussed his Evolutionary World Politics project, and the other was this chapter's author, commenting on the project (Falger, 1998). As I said many times before, had there not been international conferences like this one, I would have given up my efforts to combine evolutionary biology and IR a long time ago.[17] Modelski and I had great discussions, and agreed to disagree, but the experience was such that I felt stimulated to organize a panel on IR in the next year's edition of APLS's annual conference.

In Atlanta (September 1999), hosted by the Section Political Behavior and Theory, I had the privilege to chair five presentations under the heading 'International Relations and Biopolitics', which most probably was the first gathering of this kind exclusively devoted to IR. The participants all took different routes, but the coherence was clear-cut. Van der Dennen (1999) presented a summary of his main work on the origin of war, Vasquez and Henehan (1999) and Falger (1999) discussed territoriality as a factor in international conflicts, Sterling-Folker (1999) dug into the theoretical aspects of the relationship between IR and evolutionary biology, and very welcome was Long and Brecke's (1999) paper on reconciliation between and within countries, an empirical topic inspired by nearby living and working Frans de Waal.[18] The panel's interdisciplinary discussion was extra stimulated by Amy Parish and Graig Stamford, two primatologists. As usual, time was too short, but as an experiment it was successful. The main outcome is a greatly expanded book project that will become the first biopolitical volume focussed on, and directed at the field of International Relations.[19] If the editors' intention works out, this book will open the door for evolutionary biology in the study of human behaviour at first sight most distant from individual-focussed macro-social sciences. It is important, however, to make clear that the contribution of an evolutionary

IR can only consist of thorough reflection on certain unquestioned presumptions. The application of ethological approaches and methodologies, or the renewed attention for physiological aspects of leadership, on the other hand, could exploit the data-based and even experimental aspects of a more behaviourally oriented empirical IR.[20]

Evolutionary world politics

The 'evolutionary international relations' project would fit perfectly in Somit and Peterson's still too optimistic assessment of a biological-evolutionarily oriented IR (Somit & Peterson, 1998). But under the heading of 'evolution' there is more than just biology, and this leads to the second development in the 1990s to be reported on. As mentioned before, Peter Corning's panel hosted George Modelski, one of the grand old names in International Relations. Totally independent from the biopolitical branch of evolutionary approaches, Modelski has written a complete *oeuvre* on what he calls 'evolutionary world politics' (EWP). It will lead too far to discuss the project in full detail, but this chapter would be incomplete without paying attention to this remarkable achievement. Besides, presenting the main aspects of the project provides a welcome opportunity to discuss some general problems and possibilities of using evolutionary theory in IR.[21]

"An evolutionary approach [of international politics] revives the importance of history," is one of the main and innovative conclusions George Modelski draws from his project. It is truly innovative, because for the first time a systematic analysis of 'international history,' streching back as far as 3,500 BC, has been made on the basis of an evolutionary theory of world politics. The demonstration of historical evidence of the four phases of evolution of modern world politics and the four steps in the evolution of global politics, leads to the well-known long cycle of global politics.[22] Of course, this is not the first cyclical conception of (world) history, but it is the first modern one in which world politics is perceived as subject to (social) evolution.[23] In the following, the central role of evolution and its related aspects in the EWP are analyzed from the conceptually related but for the rest quite different background in biopolitics. There is a clear difference between the social-evolutionary approach of George Modelski's EWP project and the biobehavioural or evolutionary approach of human behaviour in other social sciences, such as psychology and anthropology. The question whether this difference is of any importance depends on the view one holds of the general analysis of human behaviour: if we see ourselves as actors in a grand structure which is the result of one or two hundred generations interacting individuals and groups – the top-down perspective – we might not

be very much interested in the still deeper structure of a bottom-up view, inherent to *biological* evolution. This deepest structure, formed by the evolutionary history of *Homo sapiens sapiens* during some 10,000 generations, manifests itself in the probability with which individual behavioural choices are made. Human nature, as is argued before, is supposed to be relevant also for the study of International Relations.

Ideally, the two perspectives could reinforce each other and could push the EWP project to the center of the modern biological-evolutionary analysis of human behaviour. But first look at the project from the biobehavioural point of view. In their review article on biopolitics, Somit and Peterson (1998) argue that the study of genetic influences on political behaviour is the most significant area of biopolitics. This is so because it reflects on the manner in which our species'evolutionary history has left *Homo sapiens* genetically endowed with certain social and behavioural tendencies. It will be clear that this is the typical bottom-up evolutionary arguing, starting with different individual organisms reproducing offspring. The essence of what happens in evolution could hardly be described more condensedly than as follows: "Sex provides variety by mixing genes from two parents, producing combinations of traits in their offspring that they do not show themselves. Natural selection, in the form of environmental pressures, then takes care of deciding whether these new features are worthwhile. New genetic combinations that give a better chance of survival, or reproduction, tend to come out on top."[24] Here we meet an important theme for comparison between top-down EWP and bottom-up biopolitics.

Modelski, who stresses evolution as the distinguishing feature of the EWP project, adheres to 'social evolution' which is to be distinguished from biological or genetical evolution, even if the structures of both processes are the same. In this particular conception of social evolution there is no need for sociobiology: evolutionary world politics does not posit "a primary causal influence from genetic endowment or biological make-up to human social behavior" (Modelski, 1997a, 2), and "the question is not how much like biological systems are social systems," but instead "the question is: Given that the theoretical basis of biology is micro- and macro-evolution (in the sense that both change and directionality are essential components of evolutionary theory), what additional analogies might there be for the social sciences, if the social system of the human species, *mutatis mutandis*, is viewed as subject to evolutionary processes?" (Modelski, 1996, 327). His conclusion is that there is no need to let have the social sciences an evolutionary theory that is identical in evolutionary biology (see also Modelski & Poznanski, 1996, 316).

From a biopolitical perspective this is a regrettable separating statement which detracts from the parsimonious character that is supposed to be one of the main

attractions of the biological-evolutionary theory. Although Modelski does not deny the possibility of a causal influence from biological make-up to human social behaviour, it is considered to be irrelevant in the analysis of institutions and long-term historical developments. This may be a safe standpoint in modern social science, but it closes the mind for absolutely relevant developments in the biobehavioural sciences. So, where the EWP interests regard the (evolving) institutional structures in which human behaviour of individuals express itself, biopolitics would propose to reverse the picture, and begin with human social behaviour *tout court*. A *biological*-evolutionary approach of human behaviour starts at the bottom. Human behaviour itself is a product of evolution and contains many highly relevant clues which also work out in more institutionalized frameworks, from so-called 'primitive' to modern international-political ones.

Of course, a Darwinian type of evolutionary paradigm for global politics would not deny the importance of the extended phenotypic character of culture and structures, but it would reintroduce emphasis on the individual (human) organism and its behaviour as the unit of selection. Without human individuals there are no social structures left – leaving aside for the moment the possibilities of self-replicating machines.

From a biopolitical point of view the EWP project would be less vulnerable to two other very different points of critique, had it taken the bottom-up line. First, Modelski's evolutionary paradigm has an element of structural determinism in it, in so far that it is completely sex-neutral. The question if there is any difference between male and female humans operating in politics, from domestic to international politics, is evolutionarily very relevant, but is not posed in the EWP project. This accentuates the traditional androcentric quality of most political science, and certainly International Relations theory. What difference it makes to introduce a sex/gender dimension is discussed later on, but for now we can argue that just forgetting about it may indicate an insensitivity towards what evolution is about.[25]

A second point of critique that flows from the absence of a bio-evolutionary behavioural foundation is the paradox of partial ahistoricity of the EWP project. It is an interesting pradox, because even if there are evidently no IR theorists with a wider range of interests in historical development than George Modelski – his work covers the period from the origin of the first cities and states to far into the new millenium (Modelski, 1997b) – the human species is prehistorically apparently not worth mentioning before the Neolithic agricultural revolution. A biopolitical scientist would request serious attention to possible reconstructions of human behaviour models before the period of cumulative civilization, because the behavioural tendencies involved in the group life that

resulted in cities and proto-states have not disappeared suddenly, or became irrelevant, with the rise of 'civilization'. Indeed, an evolutionary approach of international politics does revive the importance of history, but that doesn't begin with the undeniably revolutionary innovation of cities and states. Isolating a particular set of political phenomena – in this case international or inter-group politics – from its prehistory for conceptual reasons is certainly acceptable in a developing research tradition as the one in EWP, but once it is possible, or indeed, desirable, to anchor this research in other fields of research, like evolutionary psychology and anthropology, it would be a waste not to do so. This would even draw the attention to not only the analogy with chimpanzee and bonobo behaviour, but also to the homology of it. In evolutionary biology anthropocentrism is not done. So, the introduction of evolutionary theory in Modelski's IR analysis is laudable, but it is not complete, because the very *basis* is missing. In contrast to what the *auctor intellectualis* of the EWP project is arguing – that the social sciences can do without an other evolutionary theory than his – a biopolitical approach would advise to explicitly pay very much attention to the original edition of it, in fact start with it.

An important matter of discussion in the search for links between the EWP project and biopolitics should be the concept of evolution as used in the project. Recognizing that nobody has the exclusive rights to monopolize the word 'evolution', it nevertheless goes without saying that the *biological* connotation of evolution is most prominent. Modelski himself rightly points at the increasing use of the word 'evolution' as part of titles of books and articles which most of the time does not imply that the evolutionary dimension in such work has more significance than mere semantics. In his own work 'evolution' is taken very seriously, also from a biological point of view, but here in fact only as an *analogy*, a source of inspiration for hypotheses and theorizing. As such this is inspirational, but it also leads one to wonder how representative Modelski's conception of evolution is. Here, the common neglect in social science of the significance of *homology* becomes apparent.

Analogy and homology are important and related concepts in biology. Homology is resemblance in structure and origin, and a homologous feature has the same evolutionary origin and basic form to one in a different organism, but may not have the same final form and function (compare e.g. the skeleton of a bat wing and a human arm). Analogy, according to the *Cambridge Encyclopaedia of Human Evolution* (Jones et al., 1992), is resemblance in function, but not in structure and origin. An analogous structure has a similar function to one in a different organism, but has a completely different evolutionary developmental origin (e.g. the wings of a bat and an insect). A biological approach of a phenomenon, as phrased by the ethologist Niko Tinbergen, implies

a consciousness of the fact that the phenomena – organisms, but also behaviour – independent of how they are caused, serve certain functions and have arisen evolvingly on a historical foundation, in which the phenomena for the fulfillment of the function have adapted to (Tinbergen, 1963). The three resulting questions, causal, functional, and historical respectively, can be condensed into the goal of the modern adaptationist program: recognize certain features of organisms as components of some special problem-solving machinery. "The subject matter of the adaptationist program is structural, physiological, behavioral, or psychological phenotypic features that have been shaped by selection to serve some function." (Symons, 1992, 155). In evolutionary psychology this leads to the presumption that the brain/mind has many functions, that it has been designed by selection to solve very different kinds of problems. Instead of supposing some sort of generalized 'learning' or 'capacity-for-culture' mechanism, human perceptions of, for example, sexual attractiveness are underpinned by many specialized mechanisms which operate according to their own distinctive rules and principles. In the end studies in evolutionary psychology and anthropology provide insight into the human environment of evolutionary adaptedness – at least, that is their generally acclaimed scientific aim (Symons, 1992). It will be clear that from this perspective analogous arguing is considerable less convincing that homologous arguing.

As said before, this is not the place to do full justice to the highly original and voluminous work of George Modelski, who certainly has inspired many IR authors to reflect on evolution, even in its biological connotations. But although he is well-known with Darwinian views, so far he is more a Spencerian social evolutionist. It remains to be seen whether he can function as the necessary bridge between biology and the IR 'orthodoxy'.

Feminization of Man?

If Modelski will not be the bridge over troubled water, then the third development of the 1990s might hint to someone else. His name in the field may be surrounded by hesitations or even rejections, but he has the stature to draw a new generation of IR students over the line. But before mentioning his name, I would first like to draw attention to an important change in, or addition to the classical perspective of IR. Anthropologist John Pfeiffer published in 1969 *The Emergence of Man* which became well-known but was renamed as *The Emergence of Humankind* in its fourth edition (1985). The greater perceptiveness for the role of women in the study of human behaviour reflects not only some political correctness under the influence of feminism, but also the consequences drawn from an evolutionary approach. Here, not socially deconstructed gender

analyses drew increasingly the attention to differences in male and female reproductive strategies, but increasing empirical observation in primatology led to hypotheses on analogy and homology of these differences in humans (see de Waal, 1982 and Betzig, 1986). Feminist and political scientist Ann Tickner was fully right in pointing to the androcentric character of Hans Morgenthau's six principles of realism (Tickner, 1991). The implicit androcentric character of the EWP project, as mentioned before, is said to be neutralized by referring to the structural quality of the empirical basis: *who* exactly plays politics is not supposed to influence the cyclical nature of world politics (Modelski, pers. comm.). Viewed from the top down this may be the case, but it leaves the obvious question unanswered why males so overwhelmingly dominate politics. Top-down this may not be very interesting, bottom-up it is the logical next question after answering why 'Man' is a social species (a fundamental question hardly posed in political science outside political philosophy and biopolitics). Even if it does not change the essentials of EWP – QED – it does put an evolutionary foundation under the project which it doesn't have of itself.

If male domination in the spheres of politics is evident, from earliest historic records on and cross-culturally, then one answer why this is so, is to look at the evolutionary background of these differences in strategy. Generalizing on the basis of differences in reproductive interests, females invest in raising their children (quality) and males in siring them (quantity). Of course this is a very crude picture, but the essence is not only demonstrable in humans but also in many other primate species. In fact 'politics' is a relatively successful behavioural repertoire for attaining or defending a comparatively better position in the competition for scarce resources, sexual partners included.

The male-male competition is the classic setting for politics, as Frans de Waal has described so vividly in his books, and the male-female relation has only recently (and only in some parts of the developed world) come under the influence of reproduction regulation technology in the hands of women, which changed the evolutionary rationale for the first time in human evolution. But instead of pointing exclusively to the cultural quality of the male-female relation, it is also worth while to wonder about Charles Darwin's 'female choice' as an other important selective force – and totally disregarded outside evolutionary biology. Even if the time span since the introduction of chemical contraceptives is evolutionarily absolutely insignificant – at most two generations of women in mostly Western countries have been able to limit the number of children they want – it did open up the much more substantive possibility of female competition in the traditional realms of politics and economics. It would certainly be premature to draw solid conclusions from less than half a century

of emancipatory politics in the West, but so far it does not look like the traditionally male dominated occupational fields – certainly not the top positions – are taken over for more than a few percent by women. There is a quantitative change, but not structural, in the Nordic countries, and so far much less so in other countries of Western Europe and the United States.

However, in one respect a change under influence of women has been recognized and attractively formulated as an evolutionarily inspired analysis of ultracontemporary and future politics, so far mainly of and in the West. Francis Fukuyama - he is the one who might lead IR into the Evolutionary Age if he takes it seriously long enough - is in particular known for his fundamental discussion of the future of history after the victory of economic and political liberalism over communism in *The End of History and the Last Man*.[26] In 1998 he published under the challenging title 'Women and the Evolution of World Politics' a digest of primatology, evolutionary psychology and increasing individualization and democratization of social life. I consider the article to be a very welcome stimulus in International Relations theory in general, the first of its kind outside the expanding framework of feminist analysis.

Starting with the increasingly shared idea among primatologists that male dominance and bonding in chimpanzees and humans are homologous features, and that males and females have different strategies in their social life, Fukuyama draws the conclusion that in no area is sex-related difference clearer than with respect to aggression and violence. However, biological determinism is avoided if it is realized that accepting a genetic basis for sex differences is first of all making a statistical assertion regarding the distribution of certain characteristics. Different 'male' and 'female' bell-shaped distributions will overlap partially, illustrated by the well-known exceptions (so far) of Golda Meir and Margaret Thatcher. Based on the statistical strong correlation between male enjoyment of camaraderie and experience of aggression, Fukuyama argues that "[a] truly matriarchal world would be less prone to conflict and more conciliatory than the one we inhibit now"[27] (Fukuyama, 1998, 33).

The 'feminization of world politics', which will be promoted by more participation of women – as leaders, officials, soldiers and voters – not only articulate women's interests but can also shift the underlying male agenda.[28] Empirical research has shown that in the United States women are consistently less in favor of military action than men, and it can be predicted "that increasing female participation will probably make the United States and other democracies less inclined to use power around the world as freely as they have in the past" (p. 35). This leads to the 'democratic peace' argument of which the essence is the correlation between the degree of consolidation of liberal democratic institutions and interdemocratic peace. Fukuyama calls this "one of the few

nontrivial generalizations one can make about world politics," and although the reasons for this correlation is still a matter of discussion, he points to a factor that has not drawn much attention so far: developed democracies also tend to be more feminized than authoritarian states (in terms of expansion of female franchise and participation in political decision making). "It should therefore surprise no one that the historically unprecedented shift in the sexual basis of politics should lead to a change in international relations" (p. 36).

Another fundamental biological factor will probably influence international politics increasingly: the fall in family size in the economically developed countries make the fewer children a couple has extra costly emotionally, therefore extra weary of body bags resulting from military actions. The recently fought war over Kosovo, which had zero casualties in action on NATO's side, is a perfect illustration of the dilemma of democratic states and governments amidst of the non-democratic, more war-prone ones. In particular in the rich and democratic European countries, with a much lower immigration ratio than in the United States, the demographic composition within a few decennia will be such that a growing elderly population will not be prepared to sacrifice their youngsters in political-military adventures. This will be expressed in the countries' foreign policies. Meanwhile, however, the differences between males and females remain, and this leads Fukuyama to a reconfirmation of his earlier views: "Liberal democracy and market economies work well because, unlike socialism, radical feminism, and other utopian schemes, they do not try to change human nature. Rather, they accept biologically grounded nature as given and seek to constrain it through institutions, laws, and norms" (p. 40). This acceptance is, of course, by implication, and although no reference to Hans Morgenthau is given here, it could have been done!

EPILOGUE

Those who feel inspired by evolutionary theory and use its concepts may think that the discipline of International Relations has a chance to become receptive to neo-Darwinian approaches. The question, however, is how fat a chance. I would estimate a very meager one, if it has to come from IR itself. Tasting the latest 'debate' in the field – the pros and cons regarding Alexander Wendt's *Social Theory of International Relations* (Wendt, 1999), and the forum discussion on this book in *Review of International Studies* – then I think that this is a rephrasement of the older debate between 'idealists' and 'realists', brought up-to-date in a constructivist language. Even sociobiology, human nature and group selection are mentioned by Wendt (p. 321–353), but a book that takes the question whether states view each other as enemies, rivals or friends as a

fundamental determinant, is not built on a theory of individual and social behaviour which is the starting point in biological evolution.[29]

No, the question is not when IR discovers the potentiality of evolutionary theory; the question is when the interdisciplinary evolutionary approach will discover international relations and international politics as a very interesting domain of human behaviour to analyze. Biologists and other scientists are beginning to become interested. Jared Diamond's *Guns, Germs, and Steel: The Fates of Human Societies* (1997) can be seen as a model for a radically new science of evolutionary international relations. I can imagine a quite different IR course with *Demonic Males: Apes and the Origins of Human Violence* (Wrangham & Peterson, 1996), *Coalitions and Alliances in Humans and other Animals* (Harcourt & de Waal, 1992), and *Peacemaking among Primates* (de Waal, 1988), to mention only a few titles. But we should start, of course, with the latest edition of *Chimpanzee Politics* (1998).[30] I am afraid, however, that my faculty wouldn't allow me to, because biological evolution and all that doesn't fit in their orthodox 'constructionist' historic and legal-normative views of yesterday's international relations. And the students? I could try to seduce the new generation with Fukuyama's (1998, 33) discovery:

> Once one views international relations through the lens of sex and biology, it never again looks the same.

NOTES

1. The assessment of Somit and Peterson (1998, 567–568), that International Relations is the only field in political science where a modest recognition of helpfulness of an evolutionary outlook is to be found, is not well-founded enough to convince, nor is the bibliographical approach used in Somit, Seo and Peterson 1994 (see Falger, 1994, 1997, and note 11 below).

2. On the Darwin-Marx connection, see McLellan, 1977, 423–424, and Desmond and Moore 1992, 601–602. Friedrich Engels, who had the scientific background that Marx missed, was much more reserved about Darwinism; see e.g. his *Dialectic of Nature* (1982, 306–308).

3. Overviews in Wiltshire, 1978; Bowler, 1989; Degler, 1991; Crook, 1994 and Hawkins, 1997.

4. For which he was prominently recognized by the post-Second World War East German communist regime. Haeckel not only was a Darwinist, but also a leading anti-clerical revolutionary, and as such one of the ideological supporters of the German Democratic Republic *avant la lettre* (see Krauße, 1984 and Jahn, Löther & Senglaub, 1985).

5. See in particular Vucinich, 1988 and Todes, 1989.

6. Nowadays a modern view, also taken by contemporary biologists like E. O. Wilson and Frans de Waal.

7. The following analysis of Morgenthau's human nature conception as the origin of power politics is taken from Falger, 1997.

8. "When we attempt to apply to human behavior the same methods of analysis that we have used throughout toward other animal groups, we reach most interesting results when we select some phase of reactions of men in which integration has not developed much beyond that found in some semi- or quasi-social animal aggregations which we have been considering in the lower animals" (Allee, 1938, 185–186).

9. However, Allee was not a naive utopian idealist; his hopeful outlook was empirically founded in the historic events of his day, in particular the Spanish civil war (Allee, 1938, 216–217). This, by the way, points to a problem of recurrent character: since Hobbes' analogy between 'nature' and human social behaviour, social scientists quite easily use biologists' information for their own purposes, while scientists, in their turn, selectively point at political phenomena that would confirm their own pretheoretical world views. This referring without firm knowledge from each other's field signals the lack of an integrated theory in social *and* biological science. Evolutionary theory might provide that, it is hoped by the 'holists'.

10. Two other representatives of the Realist school of IR using 'human nature' conceptions are Frederick Schuman and Reinhold Niebuhr. Schuman (1933) even antedates Morgenthau's work and was considered to be one of the most influential authors in the field, but since his 'human nature' refers to the Freudian pleasure, reality and morality principles, he cannot be considered to have been influenced by a biological-evolutionary way of thinking. The same goes for Niebuhr (1940), who was influential as a Christian theologist writing about the ethical consequences of power politics. Niebuhr's use of human nature is constantly mixed up with his religious belief, for example when he criticizes Christian pacifist-rationalists like Bertrand Russell who believe that man is essentially good at some level of his being. Niebuhr may be of interest for the study of normative approaches to IR, but not for biopolitics. The IR author perhaps most interested in biological aspects of human behaviour was Quincy Wright (1965, 1st ed. 1942), not coincidentally the brother of the well-known biologist Sewall Wright. However, though paying much attention to (primitive) war from a biological perspective (he included a chapter on 'animal warfare'), it certainly is not an evolutionary theory of IR, or even of war (see for a discussion of Wright's *magnum opus* van der Dennen, 1995). In his later work, Wright (1955) presents a sophisticated discussion of biology-related topics (in particular Chapters 27 and 31), but these do not function coherently in an evolutionary-theoretical context.

11. Dougherty and Pfaltzgraff's book is by far the most authoritative comprehensive survey of IR theories, and here are the traces of 'biology' and biopolitics respectively: (a) authors: R. Ardrey, D. Barash, Th. Dobzhansky, W. Etkin, K. Lorenz, E. O. Wilson; P. Corning, S. Hines, R. Masters, A. Somit, Th. Wiegele; (b) subjects: Aggression (animal behavior studies, biological explanations, instinct theories of), Animal behavior studies, Darwinian theory, Decision making (psychological components in crisis), Human behavior, Innate behavior, Social Darwinism; no entry on 'human nature'. In the 1997 edition, no biological/biopolitical literature later than 1985 is cited, most dating of the 1970s (the previous edition was published in 1993, the one before that in 1990; hence the conclusion is justified that this main work of reference is reflecting no awareness of any recent developments relevant for this chapter).

12. One small exception to this rule is the discussion about decision makers in international politics from the perspective of physiology by Russett and Starr (1989, 311).

Although influences from biology are known in a much cited book like Axelrod's *The Evolution of Cooperation* (1984), in which the 'cooperation' between the political scientist Axelrod and the prominent 'sociobiologist' William D. Hamilton is an essential part (Chapter 5), this book cannot serve as an indication for a greater interest in biological-evolutionary theory.

13. The SSSM refers to the prevailing opinion in social science that 'culture' builds from scratch ("all contentful features of the human mind are 'socially constructed' or environmentally derived", Cosmides & Tooby, 1992, 221), denying the existence of a universal human nature. Evolutionary psychology in particular does *not* use the SSSM of culture to answer the question what psychological mechanism allows humans to behave so differently from most animals. In fact, it is an evolutionary vision on the classic nature-nurture problem that still is the foundation of conventional or traditional social science. See also Buss 1999 (also very appropriate as a discussion of human nature as a scientific concept in modern evolutionary theory).

14. This refers to Alexander (1979) who assigns as the main function to human groups the protection from predatory effects of other human groups. Less specific for International Relations than Shaw and Wong, but closely related to the themes of nepotism and ethnocentrism are the books by Tatu Vanhanen (1991, 1992, 1999) and the edited volumes by Reynolds et al. (1987) and Thienpont and Cliquet (1999).

15. Appreciating the book's theoretical foundation does not make it immune for fundamental criticism because of its empirical validation, as I reviewed earlier (Falger, 1991).

16. The signs for van der Dennen are better (see for example Modelski, 1998), but it is remarkable that so far his book and other publications are better received by biologists (see Wrangham, 1999).

17. Without wanting to display any self-pity, it is very important to have somewhere a few people who want to listen to what at home is ignored altogether. For example, without institutions like the European Sociobiological Society and APLS, I would have worked in a vacuum, which one cannot survive intellectually very long. However, Somit and Peterson (1998) are right when they write that biopolitics does not create opportunities in academic careers. At best, I am tolerated, after several failed attempts to drive me back into IR orthodoxy.

18. Typical for the current situation in political science, however, is that there are no institutional or even incidental links between the Atlanta political science community and the world's most prominent political scientist of primates working in nearby Yerkes Primate Center.

19. *Evolutionary International Politics: A Biopolitical Perspective* will be edited by Vincent S.E. Falger and John A. Vasquez, and is scheduled for publication with Greenwood Press in 2001. The provisional table of contents (still subject to alterations) is as follows: Part I Orientation in Biology and International Relations. 1. 'The use of ethology and evolutionary psychology in the study of IR' (Frans B. M. de Waal & Vincent S.E. Falger), 2. 'Evolution of coalitionary killing: the imbalance-of-power hypothesis' (Richard W. Wrangham), 3. 'The politics of female dominance in bonobos (Amy Parish), 4. 'Neurotoxins and human behaviour: its relevance for international politics' (Roger D. Masters), 5. 'The origin of war: evolution of intergroup agonistic behavior in animals and Man' (Johan M.G. van der Dennen), 6. 'Theories of IR and evolutionary biology: parallels, problems and prospects' (Jennifer Sterling-Folker). Part II: Classical Issues in IR in New Perspectives. 7. 'Extending sociobiology's explanatory power: the synthetic materialist theory of politics, war and state formation' (Stephen K.

Sanderson), 8. 'Why territorial disputes are war prone' (John A. Vasquez & Marie T. Henehan), 9. 'The limits of globalization: a biopolitical analysis of borders and territoriality' (Vincent S. E. Falger), 10. 'War and reconciliation' (William J. Long & Peter Brecke), 11. 'Biopolitics and Realism' (James N. Schubert). Part III. New Issues in IR. 12. 'The contribution of evolutionary psychology to the understanding of ethnic conflict (David Goetze & Patrick James), 13. 'Kin terms, patriotism, and international politics' (Gary R. Johnson), 14. 'Is humanity destined to self-destruct?' (Lynton K. Caldwell), 15. 'Women and international politics' (Vincent S. E. Falger). Part IV: Epilogue. 16. '"World Politics as a Primitive Political System": 1964 in retrospect' (Roger D. Masters), 17. Conclusion, by the editors, 18. Bibliographic essay (Johan M.G. van der Dennen & Vincent S. E. Falger).

20. Much of Roger Masters' work is clearly relevant for IR, and so is Bert Park (1986).

21. The following is based on and partly taken from a publication under review (Falger, in press.), result of a paper presented at a conference on evolutionary approaches to world politics in Bloomington, Indiana, 4–6 December 1998, hosted by William R. Thompson.

22. In short, EWP/Modelski starts with the hypothesis that world politics is subject to (social) evolution, and that evolutionary world politics involves four basic mechanisms: selection, variation, amplification (reinforcement), and cooperation. The evolution of the world system, and of world politics, is a learning process for the human race. The four phases of the evolution of modern world politics consist of the phases of Preparation (variation), Nucleation (cooperation), Global Organization (selection), and Consolidation (amplification). The basic beat of the system is the generational period of some 25–30 years (the replacement interval) which is one of generational learning. See Modelski, 1996, and Modelski and Poznanski, 1996 for an overview, and Modelski's very useful collection of articles at http://faculty.washington.edu/modelski/

23. A comparable *oeuvre*, of course, is Immanuel Wallerstein's work, but his historic foundation does not antedate the 15th century and does not take biology into account. On the other hand, Modelski is not the only IR specialist interested in very early history; see for example Cioffi-Revilla, 1996 on the earliest origins and evolution of warfare and politics, but remaining safely within the boundaries of the human social sciences. See Hall, 1996 for an overview of 'world-system' history on social evolution. Addendum: after finishing this manuscript, a very interesting collection of articles on the change in the organization of humankind over the past 5,000 years has been published by Robert A. Denemark et al. (eds), 2000.

24. *The Economist*, June 26th 1999, 110. See Colby, 1996 and Futuyma, 1997 for further elaboration.

25. The political prejudice against this, however, is kept alive on purpose when someone like Miles Kahler (1998, 1) writes: "the association of Darwinian evolutionary theory with sociobiology and renewed interest in the biological sources of human behavior appeared to erect another barrier to application of evolutionary ideas elsewhere in the social sciences." I am afraid that this sentiment was clearly dominant at the Bloomington conference, where Kahler presented this paper, and as such is a good indication of the intellectual atmosphere at the frontiers of IR *vis-à-vis* the usefulness of biological-evolutionary theory (see also note 21; the conference as such, however, proved to be an important stimulus for organizing the 1999 APLS panel on IR where two of the Bloomington participants presented papers).

26. Typically for Fukuyama in the early 1990s, he wrote about Man, not 'human'; the book in all its erudition did not indicate that women were also subject to Fukuyama's

Hegelian/Nietzschean fight for *thymos* or recognition. The book's view of Man confirms indeed the classic androcentric view of men in action.

27. Fukuyama continues: "Where the new biology parts company with feminism is the causal explanation it gives for this difference in sex roles. *The ongoing revolution in the life sciences has almost totally escaped the notice of much of the social sciences and humantities* (. . .)", (Fukuyama, 1998, 33; emphasis added; see for a feminist reaction Tickner, 1999 who evaluates Fukuyama's argument as 'deeply conservative', the typical reaction of mainstreamers who are in the end more preoccupied with normative questions than with innovative theorizing).

28. The theme of 'feminization of politics' as such is not new in biopolitics (see Schubert, 1991), nor is the discussion of the role of women in international politics (see, among others, Falger, 1992, Genovese, 1993 and D'Amico & Beckman, 1995), but these titles do not lead to an evolutionary analysis or theory of International Relations.

29. Of course, the border line between *starting* with such a theory and building upon it, and only more or less sympathetically referring to it, is always in the eye of the beholder. Two recent examples of recent IR literature in the second category are the mainly historical (and also based in personal experience) approach with a touch of knowledge of biological-evolutionary preconditions by William Polk (1997), and the book by 'evaluative political realist' Roger Spegele (1996), who even devotes a complete chapter to human nature and its relation to IR. Neither of the books, however, is by far radical evolutionary enough to account for substantial influence. Of the modern IR literature, Fukuyama's 1998 article comes closest to the first category. However, although he displays in his latest book *The Great Disruption* (1999) a significant knowledge of the biological approach of human behaviour, Fukuyama here writes more as a social philosopher than as a theoretician. Besides, as long as it makes sense to make such distinctions the book really cannot be counted to the body of International Relations literature. Nevertheless, so far Fukuyama is my best bet for the bio-evolutionary revolution in IR – if it ever would happen.

30. But it should be followed by informed antidotum, such as Schubert, 1986 and Masters 1989; see O'Neill, 1986 for an older suggestion to change teaching IR fundamentally.

ACKNOWLEDGMENT

This article has been written during a sabbatical leave provided by the Wiarda Research Institute of the Faculty of Law, University of Utrecht.

REFERENCES

Allee, W. C. (1932). *Animal Life and Social Growth*. Baltimore: William and Wilkins.

Allee, W. C. (1938). *The Social Life of Animals*. London: William Heinemann.

Alexander, R. A. (1979). *Darwinism and Human Affairs*. London: Pitman.

Axelrod, R. (1984). *The Evolution of Cooperation*. New York: Basic Books (Chapter 5 written by William D. Hamilton).

Bagehot, W. (1872). *Physics and Politics, or Thoughts on the Application of the Principles of 'Natural Selection' and 'Inheritance' to Political Society*. London: Henry S. King (1876, 4th ed.).

Betzig, L. (1986). *Despotism and Differential Reproduction: A Darwinian View of History*. New York: Aldine de Gruyter.

Bowler, P. J. (1989). *The Invention of Progress. The Victorians and the Past*. Oxford: Basil Blackwell.

Buss, D. (1999). *Evolutionary Psychology. The New Science of the Mind*. Boston: Allyn & Bacon.

Cioffi-Revilla, C. (1996). Origins and Evolution of War and Politics. *International Studies Quarterly*, *40*, 1–22.

Colby, Chr. (1996). *Introduction to Evolutionary Biology*, http://fp.bio.utk.edu/darwin/default.html

Cosmides, L., & Tooby, J. (1992). Cognitive Adaptations for Social Exchange. In: J. H. Barkow, L. Cosmides & J. Tooby (Eds), *The Adapted Mind. Evolutionary Psychology and the Generation of Culture* (pp. 163–228). New York: Oxford University Press.

Crook, P. (1994). *Darwinism, War and History. The Debate over the Biology of War from The Origin of Species to the First World War*. Cambridge: Cambridge University Press.

D'Amico, F., & Beckman, P. R. (Eds) (1995). *Women in World Politics. An Introduction*. Westport, CT: Bergin & Gasvery.

Degler, C. (1991). *In Search of Human Nature*. New York: Oxford University Press.

Denemark, R. A., Friedman, J., Gills, B. K., & Modelski, G. (Eds) (2000). World System History. The Social Science of Long-Term Change. London, New York: Routledge.

Desmond, A., & Moore, J. (1992). *Darwin*. London: Penguin Books.

De Waal, F. B. M. (1982). *Chimpanzee Politics: Power and Sex among Apes*. London: Jonathan Cape (1998 2nd ed., Baltimore/London: The Johns Hopkins University Press).

De Waal, F. B. M. (1988). *Peacemaking among Primates*. Cambridge, MA: Harvard University Press.

Diamond, J. (1997). *Guns, Germs, and Steel: The Fates of Human Societies*. New York:/London: W.W. Norton.

Dougherty, J. E., & Pfaltzgraff, R. L. (1997). *Contending Theories of International Relations*. New York: Harper & Row (4th ed.; 1st ed. 1971).

Engels, F. (1982). *Dialectics of Nature*. Moscow: Progress Publishers.

Falger, V. S. E. (1991). The Missing Link in International Relations Theory? *Journal of Social and Biological Structures*, *14*, 73-77.

Falger, V. S. E. (1992). Sex Differences in International Politics? An Exploratory Study of Coalitional Behaviour in Biopolitical Perspective. In: J. M. G. van der Dennen (Ed.), *The Nature of the Sexes: The Sociobiology of Sex Differences and the 'Battle of the Sexes'* (pp. 171–194). Groningen: Origin Press.

Falger, V. S. E. (1994). Biopolitics and the Study of International Relations: Implications, Results, and Perspectives. *Research in Biopolitics*, *2*, 115–134.

Falger, V. S. E. (1997). Human Nature in Modern International Relations. Part I. Theoretical Backgrounds. *Research in Biopolitics*, *5*, 155–175.

Falger, V. S. E. (1998). Evolutionary World Politics: An Analysis of George Modelski's Work on International Relations Theory and Historical Practice from a Biobehavioral Perspective. Paper presented at the 18th Annual Conference of the Association for Politics and the Life Sciences, Boston, 3–6 September, 1998.

Falger, V. S. E. (1999). The Limits of Globalization: A Biopolitical Analysis of Borders and Territoriality. Paper presented at the 19th Annual Conference of the Association for Politics and the Life Sciences, Atlanta, 2–5 September 1999.

Falger, V. S. E. (in press) Evolutionary World Politics Enriched: The Biological Foundations of International Relations. In: W. R. Thompson (Ed.), *Evolutionary Interpretations of World Politics*. London, New York: Routledge.

Fukuyama, F. (1992). *The End of History and the Last Man.* New York/Toronto: The Free Press.

Fukuyama, F. (1998). Women and the Evolution of World Politics. *Foreign Affairs, 77*(5), 24–40.

Fukuyama, F. (1999). *The Great Disruption. Human Nature and the Reconstruction of Social Order.* New York: The Free Press.

Futuyma, D. J. (1997). *Evolutionary Biology.* Sunderland, MA: Sinauer.

Gasman, D. (1971). *The Scientific Origins of National Socialism.* New York: Elsevier.

Genovese, M. A. (Ed.) (1993). *Women as National Leaders.* Newbury Park: SAGE.

Hall, T. D. (1996). World-Systems and Evolution. An appraisal. *Journal of World-Systems Research, 2*(2), 1–43 (also available at http://csf.colorado.edu/jwsr/archive/vol2/v2_n4.htm).

Harcourt, A. H., & de Waal, F. B. M. (Eds) (1992). *Coalitions and Alliances in Human and other Animals.* Oxford: Oxford University Press.

Hawkins, M. (1997). *Social Darwinism in European and American Thought, 1860–1945. Nature as a Model and Nature as a Threat.* Cambridge: Cambridge University Press.

Jahn, I., Löther, R., & Senglaub, K. (Eds) (1985). *Geschichte der Biologie. Theorien, Methoden, Institutionen und Kurzbiographien.* Jena: Gustav Fischer Verlag.

Jones, S., Martin, R., & Pilbeam, D. (Eds) (1992). *The Cambridge Encyclopedia of Human Evolution.* Cambridge: Cambridge University Press.

Kahler, M. (1998). Evolution and the Units of World Politics: Why We Live in a World of Nation-States. Paper presented at the Conference on Evolutionary World Politics at Bloomington, Indiana, 4–6 December, 1998.

Knutsen, T. L. (1997). *A History of International Relations Theory.* Manchester/New York: Manchester University Press.

Krauße, E. (1984). *Ernst Haeckel.* Leipzig: Teubner.

Kropotkin, P. (1904) [1972]. *Mutual Aid. A Factor of Evolution.* New York: New York University Press (ed. and Introduction by P. Avrich).

Long, W. L., & Brecke, P. (1999). Reconciliation Between and Within Countries: A 'Rational Choice' or a 'Reasoned Instinct'? Paper presented at the 19th Annual Conference of the Association for Politics and the Life Sciences, Atlanta, 2–5 September 1999.

Marten, H. G. (1983). *Sozialbiologismus. Biologische Grundpositionen der politischen Ideengeschichte.* Frankfurt/New York: Campus Verlag.

Masters, R. D. (1964). World Politics as a Primitive Political System. *World Politics, 16*, 595–619.

Masters, R. D. (1989). *The Nature of Politics.* New Haven: Yale University Press.

McLellan, D. (1977). *Karl Marx. His Life and Thought.* Frogmore: Paladin.

Modelski, G. (1996). Evolutionary Paradigm for Global Politics. *International Studies Quarterly, 40*, 321–342.

Modelski, G. (1997a). What is Evolutionary Politics? *The Evolutionary World Home Page* at http://faculty.washington.edu/modelski/

Modelski, G. (1997b). Cities in the Ancient World: An Inventory (−3500 to −1200). *The Evolutionary World Politics Home Page* (see 1997a).

Modelski, G. (1998a). Book Review of *The Origin of War* by J. M. G. van der Dennen. *Politics and the Life Sciences, 17*, 88–90.

Modelski, G. (1998b). Evolutionary World Politics. Paper presented at the 18th Annual Conference of the Association for Politics and the Life Sciences, Boston, 3–6 September, 1998.

Modelski, G., & Poznanski, K. (1996). Evolutionary Paradigms in the Social Sciences. *International Studies Quarterly, 40*, 315–319.

Morgenthau, H. J. (1967). *Politics among Nations. The Struggle for Power and Peace* (4th ed.). New York: Alfred Knopf.

Niebuhr, R. (1940). *Christianity and Power Politics.* Archon Books.

O'Neill, J. (1986). Evolutionary International Relations: A Biopolitical Framework for Teaching World Politics. *Politics and the Life Sciences*, *5*, 44–53.

Park, B. E. (1986). *The Impact of Illness on World Leaders*. Philadelphia: University of Pennsylvania Press.

Pfeiffer, J. E. (1985). *The Emergence of Humankind*. New York: Harper & Row.

Polk, W. R. (1997). *Neighbors and Strangers. The Fundamentals of Foreign Affairs*. Chicago/London: University of Chicago Press.

Review of International Studies. Forum on *Social Theory of International Politics*. *Review of International Studies*, *26*, 123–180.

Reynolds, V., Falger, V. S. E., & Vine, I. (Eds) (1987). *The Sociobiology of Ethnocentrism. Evolutionary Dimensions of Xenophobia, Discrimination, Racism and Nationalism*. London: Croom Helm.

Russett, B., & Starr, H. (1989). *World Politics. The Menu for Choice*. New York: Freeman.

Schubert, G. (1986). Primate Politics. *Social Science Information*, *25*, 647–680.

Schubert, G. (1991). Sexual Politics and Political Feminism. *Research in Biopolitics*, *1*. Greenwich, CT: JAI Press.

Schuman, F. L. (1933). *International Politics. The Destiny of the Western State System*. New York: McGraw-Hill.

Segerstråle, U. (2000). *Defenders of the Truth. The Battle for Science in the Sociobiology Debate and Beyond*. Oxford: Oxford University Press.

Shaw, R. P., & Wong, Y. (1989). *Genetic Seeds of Warfare: Evolution, Nationalism and Patriotism*. Boston: Unwin Hyman.

Somit, A., & Peterson, S. A. (1998). Review Article: Biopolitics after Three Decades – A Balance Sheet. *British Journal of Political Science*, *28*, 559–571.

Somit, A., Seo, K. K., & Peterson, S. A. (1994). The Impact of Biology on the Study of International Politics. *Research in Biopolitics*, *2*, 135–140.

Spegele, R. D. (1996). *Political Realism in International Theory*. Cambridge: Cambridge University Press.

Stein, G. J. (1987). The Biological Bases of Ethnocentrism, Racism and Nationalism in National Socialism. In: V. Reynolds, V. S. E. Falger & I. Vine (Eds), *The Sociobiology of Ethnocentrism* (pp. 251–267). London: Croom Helm.

Sterling-Folker, J. (1999). Theories of International Relations and Evolutionary Biology: Parallels, Problems, and Prospects. Paper presented at the 19th Annual Conference of the Association for Politics and the Life Sciences, Atlanta, 2–5 September, 1999.

Symons, D. (1992). On the Use and Abuse of Darwinism in the Study of Human Behavior. In: J. H. Barkow, L. Cosmides & J. Tooby (Eds), *The Adapted Mind. Evolutionary Psychology and the Generation of Culture* (pp. 137–159). New York/Oxford: Oxford University Press.

Thienpont, K., & Cliquet, R. (Eds) (1999). *In-Group/Out-Group Behaviour in Modern Societies. An Evolutionary Perspective*. Brussels: Vlaamse Gemeenschap/CBGS.

Tickner, J. A. (1991). Hans Morgenthau's Principles of Political Realism: A Feminist Reformulation. In: R. Grant & K. Newland (Eds), *Gender and International Relations* (pp. 27–40). Milton Keynes: Open University Press.

Tickner, J. A. (1999). Why Women Can't Run the World: International Politics According to Francis Fukuyama. *International Studies Review*, *1*(3), 3–11.

Tinbergen, N. (1963). On Aims and Methods of Ethology. *Zeitschrift für Tierpsychologie*, *20*, 410–433.

Todes, D. P. (1989). *Darwin without Malthus. The Struggle for Existence in Russian Evolutionary Thought*. New York/Oxford: Oxford University Press.

van der Dennen, J. M. G. (1995). *The Origin of War. The Evolution of a Male-Coalitional Reproductive Strategy*. Groningen: Origin Press.

van der Dennen, J. M. G. (1999). The Origin of War: Evolution of Intergroup Agonistic Behavior in Animals and Man. Paper presented at the 19th Annual Conference of the Association for Politics and the Life Sciences, Atlanta, 2–5 September, 1999.

Vanhanen, T. (1991). *Politics of Ethnic Nepotism. India as an Example*. New Dehli: Stirling Publishers.

Vanhanen, T. (1992). *On the Evolutionary Roots of Politics*. New Dehli: Sterling Publishers.

Vanhanen, T. (1999). *Ethnic Conflicts Explained by Ethnic Nepotism*. Stamford, CT: JAI Press (*Reasearch in Biopolitics*, vol. 7).

Vasquez, J. A., & Henehan, M. T. (1999). Why Territorial Disputes are War Prone. Paper presented at the 19th Annual Conference of the Association for Politics and the Life Sciences, Atlanta, 2–5 September, 1999.

Vucinich, A. (1988). *Darwin in Russian Thought*. Berkeley: University of California Press.

Waltz, K. N. (1959). *Man, the State and War. A Theoretical Analysis*. New York: Columbia University Press.

Waltz, K. N. (1979). *Theory of International Politics*. Reading, MA: Addison-Wesley.

Wendt, A. (1999). *Social Theory of International Politics*. Cambridge: Cambridge University Press.

Wiltshire, D. (1978). *The Social and Political Thought of Herbert Spencer*. Oxford: Oxford University Press.

Wrangham, R. W. (1999). Evolution of Coalitionary Killing. *Yearbook of Physical Anthropology, 1999*, 1–30.

Wrangham, R. W., & Peterson, D. (1996). *Demonic Males. Apes and the Origin of Human Violence*. London: Bloomsbury.

Wright, Q. (1955). *The Study of International Relations*. New York: Appleton-Century-Crofts.

Wright, Q. (1965). *A Study of War* (1st ed. 1942). Chicago: The University of Chicago Press.

THE EVOLUTION OF LAW AND BIOLOGY

Monika Gruter Morhenn and Margaret Gruter

INTRODUCTION

The law's primary role in modern society is to regulate human behavior. Indeed, the law regulates myriad human activities; birth, marriage, employment, and death all have legal, as well as social and moral implications. Law, then, is more effective when it understands why humans do what they do. Examining human behavior through a biological lens aids lawmakers in developing legal strategies and theories which are informed about human behavior. It is the aim of the law and biology movement to facilitate within the legal academy an understanding of modern insights from the biological sciences.[1] No attempt is made to reason from *is* to *ought*, for example, no attempt is made to reason that because human behavior is a certain way we ought to allow such behavior. Instead, the primary endeavor of the present day law and biology movement is to shed light on what the law is "up against."

Searching for insights into regulation of human behavior is not a novel undertaking. Indeed, celebrated thinkers in philosophy, sociology, psychology, anthropology and history have pursued this task for hundreds of years. The law and biology movement is a continuation of this quest. Such a pursuit is no easy task; for the questions loom large, the answers appear piecemeal and the gap between the social sciences and life sciences remains substantial.

Why have the findings in the biological sciences during the last 100 years, unlike those of the social sciences, had comparatively little impact on legal

Evolutionary Approaches in the Behavioral Sciences, Volume 8, pages 119–134

ISBN: 0-7623-0769-2

thought? A primary contributor to this short-circuit is the existence of several misconceptions that frequently misguide observers.[2] First, the impact of the vastly different vocabulary in the social versus the life sciences should not be underestimated. Lack of training in both "languages" leads many scholars to misunderstand each other even on the relatively rare occasion when scholars from the social and life sciences cross the divide and talk to each other. This divide is illustrated, and indeed arguably even exacerbated, by the virtually universal tendency of universities to house the biological and social science departments and libraries in ways which do not foster cross-disciplinary communication.

Second, the misconception that law and biology somehow equals genetic determinism prevents the law from benefiting from findings from the biological sciences. Yet, as our understanding of the biological underpinnings of human behavior becomes more sophisticated, there is a growing acceptance within the academy and the general public that human behavior is heavily influenced by our biology as well as our physical and social environments. Modern biology provides information which is probabilistic, not deterministic. Further, evolutionary biology tells us that human behavior is complex and influenced by a variety of biological and environmental factors, thus helping us understand how the policy and legal choices that we make are likely to affect human behavior. Most modern thinkers agree that it is not nature or nurture which explains human behavior, but nature AND nurture. The law and biology movement has expended considerable focus on bringing law and the behavioral sciences together, rather than pitting them against one another.

Third, another notable misconception deserving discussion is the Naturalistic Fallacy.[3] As stated above, law and biology does not attempt to reason from description to normative conclusion. Instead, law and biology seeks to inform, with normative questions left unanswered. Leaving normative questions open is not unique to law and biology. For instance, law and economics may indicate that efficiency maximization is achieved under conditions X, Y and Z with no indication or suggestion that conditions X, Y and Z are desirable. Such law and economics analysis is, without question, pervasive and persuasive in American and international jurisprudence.

Having discussed the purposes of law and biology as well as some of the misconceptions surrounding it, the remainder of this article focuses on how the law and biology movement has facilitated discussion and will continue to facilitate discussion between law and the biological sciences in order to contribute to a more informed jurisprudence. To this end, Part I of this article considers the early connections between law and biology, with Part II tracking the rise of the present day biological approach to law. Finally, Part III highlights current areas of research and inquiry in this growing interdisciplinary arena.

I. EARLY CONNECTIONS BETWEEN LAW AND BIOLOGY

Early jurisprudential references to biology focused on the evolution of law as analogous to the evolution of the species. For example, just after the publication of Charles Darwin's *Origin of the Species*,[4] Sir Henry James Sumner Maine published the *Ancient Law*.[5] Although the work has been criticized for its overly-simplistic evolutionary generalizations, it did elaborate upon the concept of "evolution of the law," as previously highlighted by Friedrich Karl von Savigny's social-historical approach to legal evolution. Similarly, John Henry Wigmore[6] published lengthy discussions relating to the evolution of law. Wigmore's focus was on the law as continually changing and adapting to local environments, with biological evolution as a metaphor for change.

Oliver Wendell Holmes, one of America's greatest jurists, wrote of the evolution of legal doctrines, or specific legal rules, in his celebrated book *The Common Law*[7] and later again in his article, *Law In Science and Science in Law*.[8] Indeed, Holmes was greatly influenced by Darwin. Like Wigmore, Holmes' treatment of law and evolution can be characterized as metaphorical, with Holmes bringing the focus to the level of legal doctrines. Much later, discussions of the evolution of law appeared in the work of Robert Charles Clark, who brought the idea of the evolution of law to the statutory level.[9]

Collectively, these early discussions of the evolution of law focus primarily on the analogy between biological evolution and legal evolution. While the present day law and biology movement extends beyond the use of biological evolution as a metaphor for legal evolution, the insights from these early works continue to influence the making and administering of law.

II. THE PRESENT DAY LAW AND BIOLOGY APPROACH

In more recent years, legal thought has benefited from the insights of biology in numerous ways stretching beyond the level of analogy. Recent scholarship centers around the notion that law can be more effective in meeting its ultimate objectives if it is based upon scientifically-informed models of human behavior. Simply put, the effectiveness of a given law is proportionate to the degree to which the functions of such law complement the functions of the behavior that the law attempts to regulate.[10] In short, a more informed jurisprudence is a jurisprudence that has information about "what it is up against."[11]

In *Law in Sociobiological Perspective*, published in 1977, Margaret Gruter presents insights into the origins and evolution of the law, based on studies of

various animal species and early man.[12] Tracking such findings, Gruter discusses how legal behavior[13] evolved from the behavior of the individual living in groups. Group living required that its members respond to various rules for group living. A reasonable assumption is that primitive legal or rule-following frameworks developed in conjunction with individuals living in groups:

> Over a period of hundreds of thousands of years the implicit rules of group life evolved into formal legal rules. During this time the rules governing group life decreed behavior which, in part due to hereditary factors partly learned and transmitted from one generation to the next, served to adapt man to his environment. Once behavior was expressed in legal norms, the norms could be effective only if the individual members of the group willingly obeyed the rules or could be forced to obey them. In sum, the effectiveness of legal norms depended on the legal behavior of the individual.[14]

Further expansions on the law and biology approach took place in a 1981 symposium of the Gruter Institute for Law and Behavioral Research[15] at Monterey Dunes, California. This Monterey Dunes symposium was the first formal convening of scholars in the area of law and biology. Many significant contributions resulted from the symposium, including those by Richard Alexander, Paul Bohannan, John Beckstrom, E. Adamson Hoebel, Bartley Hoebel, Paul MacLean, Hubert Markl, Roger Masters, Manfred Rehbinder, and Richard Schwartz. Indeed, these scholars proceeded to bring fresh ideas to their respective areas of expertise: Alexander shedding light on the evolutionary origins of morality, emphasizing that biology predicts that individuals are likely to have conflicts of interest;[16] Beckstrom discussing the biology of altruism and the effects on law;[17] Bohannan and E.A. Hoebel building the insights of biology into the field of anthropology and law;[18] and MacLean exploring how insights from the brain sciences contribute to a more informed understanding of the evolution of law.[19]

Law and Biology's Presence In Legal Scholarship

Today, much of the scholarship in law and biology is linked to the work of a handful of pioneering thinkers. Following the publication of *Law in Sociobiological Perspective*,[20] and the convening of scholars in 1981 at Monterey Dunes which led to the subsequent publication of *Law, Biology and Culture: The Evolution of Law*,[21] a series of keystone articles flourished, providing breadth and depth to this new field: law and biology. In *Bringing People Back: Toward a Comprehensive Theory of Taking in Natural Resource Law*,[22] William Rodgers identified a gap in law and economics – a lack of means by which to understand or predict individual preferences and where they

come from – that law and biology promised to fill. While Rodgers was not alone in identifying this gap, his subsequent writings on the topic of deception and self-deception in natural resource law place him in the unique position of having identified a void *and* subsequently working to fill it. Soon thereafter, Donald Elliott's publication of *The Evolutionary Tradition in Jurisprudence*[23] illustrated that evolution had long influenced jurisprudence, as indicated by the writings of great jurists such as Maine, Clark, Holmes, and Wigmore. With this evolutionary tradition in view, the notion that jurisprudence would benefit from ongoing exposure to the insights of biology no longer appeared so foreign. Taken together, these publications brought to light the view that jurisprudence already had and would continue to benefit from a more robust and informed understanding of that which it seeks to channel: human nature. With this framework in existence, a platform for further scholarly inquiry was established.

Building on the momentum, three additional books edited by Margaret Gruter, *Ostracism: A Social and Biological Phenomenon*,[24] *The Sense of Justice: An Inquiry into the Biological Foundations of Law*,[25] and *Law and the Mind: Biological Origins of Human Behavior*,[26] began to bridge the gap between the abstract use of biology as a basis for informing the law and specific applications where law is benefited by insights from biology. In *Ostracism*, for example, the experience of an Amish community in rural Ohio is described and used as an example of how groups tend to ostracize those who do not conform to the norms of that group.

Next, *The Sense of Justice* provides a platform for inquiry concerning brain functions, including the role of emotions, involved in making decisions about right and wrong, or fairness. Here, law is analyzed both as a creation of the human mind and as a product of the biological mechanisms that support the human quest for order and justice. Perspectives from an array of disciplines are provided in response to this inquiry. For example, psychiatrist Michael McGuire discusses how neurological mechanisms have evolved to allow primates to monitor and respond to the social behavior of others, a mechanism integral to group living primates; economist Robert Frank focuses on forms of behavior, such as altruism, that cannot be explained using conventional economic models of the rational actor, and discusses how the role of emotions fills in this void; and legal scholar Wolfgang Fikentscher surveys legal anthropology to show that all cultures have a distinct sense of justice, with Western norms of justice being but a single, cultural instance.

Finally, *Law and the Mind* discusses how various mechanisms in the brain affect individuals' legal behavior within particular areas of law, such as environmental law. For example, how might the laws related to recycling benefit from a more informed understanding of human behavioral tendencies?

Law and Biology In Legal Education

Coupled with the entrance of law and biology into legal scholarship was its entrance into legal education. Donald Elliott and Roger Masters pioneered the first law and biology class in legal curricula in 1988 at Yale Law School.[27] Today, biological insights are being applied in numerous law school classes to enrich discussion in myriad legal fields.[28] With the law and biology movement underway, many of the first students of law and biology, such as Owen Jones, a student in Elliott and Master's pioneering class at Yale, are now nurturing this approach with great attention and energy. In 1998, Jones founded the Society for Evolutionary Analysis in Law (SEAL), an organization of law professors devoted to incorporating biology into legal scholarship and education.[29] Through media such as these, law and biology has become an exciting arena within the legal academy, with immeasurable space for further research and scholarship.

III. CURRENT AREAS OF RESEARCH AND SCHOLARSHIP

In recent years, numerous publications in the field of law and biology have spanned a vast array of topics. Indeed, legal scholars have begun to use the insights from evolutionary biology to shed light on a variety of specific legal issues. The rapid growth in this field is compelling. The following discussion outlines just some of the specific legal issues which have been illuminated by the use of a biological lens.[30]

Economics, Property Rights, Security and Risk

Economics, property rights and, more generally, security and risk have been the focus of a vast array of scholarship in law and biology. Indeed, numerous law and biology conferences have been devoted exclusively to informing legal scholars about how biology plays a role in the laws relating to these topics.[31] However, despite the great quantity of scholarship bringing biological insights to laws relating to property rights, regulatory law relating to security and risk, and economics, some observers may question why law and biology has not produced the same quantity of legal scholarship and has not been the topic of as many law school classes as compared to law and economics, which would seem to be at least a first cousin given their shared roots in the "law and . . ." genre of law school curricula.

Similar to law and biology, one might describe the law and economics movement as a search for a basis on which to ground law on concepts more satisfying

than positivist notions or ad hoc intuitions. Under the law and economics approach, legal scholars attempt to employ a more rigorous approach to the law by analyzing law under cost benefit analysis, taking as a given the goal of reducing unnecessary costs.[32] Under this approach, human beings are assumed to be rational, self-interested maximizers. While this is an attractive and appealing approach because it eliminates the softness of positivist bases of jurisprudence, this deductive approach is unable to tell us what preferences are, why they are as they are, and cannot explain certain behaviors such as altruism.[33] As William Rodgers has aptly stated, we need to "bring people back" into legal analysis, for legal theory that is built on false notions of human motivations has led to a "contest of false modeling."[34]

It is in precisely these shortcomings of law and economics where the strength of the behavioral approach to law is highly visible. This is not to say that law and biology is imperialistic, i.e., law and biology does not contend that it can explain everything, or that there is no place for law and economic analysis. Quite the contrary, law and economics and law and biology should be seen not as competitors, but as complements: the one (law and economics) informing the law about what to expect presupposing certain preferences; the other (law and biology) informing the law about what those preferences are most likely to be.

Given the unsurpassed pace at which modern biology is shedding new light in areas including neurobiology,[35] genetics,[36] and advanced reproductive technologies,[37] it is fair to say that for the law and social sciences such as economics to ignore such insights would be to ignore the reality of precisely that which they must necessarily understand: human behavior.

Significantly, as Zak and Denau show in Chapter IX in this volume, there is a growing movement in economics in which insights from behavioral biology are being used to enrich traditional economic modeling and analysis. Thus, scholars in both the law and biology movement as well as in economics may be viewed as pursuing the same goal: striving for a more informed and thereby robust foundation on which to build their disciplines. Given the complementarities available across the law and biology movement and law and economics movement, cross-disciplinary work within these areas promises to be rich and rewarding.

Such cross-disciplinary work is already underway. Significant publications relating to economics and property rights include those by: Robert Frank, using insights about the biology of emotions and other neurobiological mechanisms to better explain phenomena such as altruism, heretofore unexplained by traditional economic theory;[38] Robert Cooter, developing theories about property rights and the biological foundations of constitutions;[39] Paul Zak, setting forth

how laws protecting private property are a necessary condition for economic development,[40] Jack Hirshleifer, using biological insights to explain cooperation and conflict, including war;[41] Dorothy Glancy, examining how biology illuminates pervasive property rights conflicts such as the difficulties frequently encountered with tenancies in common; Wolfgang Fikentscher, weaving behavioral insights into the complexities of antitrust and copyright law;[42] and Michael McGuire and Mark Grady, applying biological insights from the neurosciences to the nature of constitutions and theories of natural law.[43]

In addition, on the topic of security and risk, and regulatory law generally, E. Donald Elliott's work on how to foster voluntary compliance with environmental regulations has incorporated many insights about tendencies for risk-taking and cooperation into discourse on environmental regulations and statutory complexity generally.[44] Still further, Oliver Goodenough has recently begun applying a biological lens to the study of numerous issues confronting businesses, including security and risk at the level of the firm and of the economy in which a firm operates.

Family Law

The field of family law is rife with opportunity to gain from insights from the biological sciences. Indeed, one of the first applied legal endeavors of participants in the law and biology movement was to file an amicus brief in the Baby M surrogate mother case in 1988.[45] The brief set forth for the Supreme Court of New Jersey what modern biology knows about the bonding process between mother and infant, while carefully stating that the normative decisions were not within the purview of the brief. Further applications in the field of family law are found in Owen Jones's articles on rape and on child molestation, which offer compelling insights into understanding and preventing these phenomena.[46] Of additional import are Cheryl Hanna's publications applying biological insights to domestic violence as well as teen violence.[47] Finally, the area of elder law is enriched by Lawrence Frolik's application of biological insights to issues such as the undue influence doctrine.[48]

Environmental and Natural Resource Law

The areas of environmental and natural resource law are the subject of substantial and significant law and biology scholarship. This is due, in large part, to the fact that regulatory law lends itself well to analysis – such of that in law and biology – which ably predicts the tendencies of individuals in the aggregate. Significant contributions in this area include those by William Rodgers that shed light on how behaviors such as deception and self-deception, and cooperation[49] play powerful roles in compliance (or non-compliance) with

environmental regulations. In addition, Donald Elliott, a prolific scholar in this field, illuminates how biological knowledge about host-parasite relationships and symbiosis contribute to an improved understanding of "the problem of the commons" in environmental law.[50] Adding to this rich base of scholarship is Thomas Moellers's work on human nature and environmental protection.[51]

Criminal Law

The area of criminal law has also benefited from biological insights. Worthy of emphasis here is that biology is ill-suited to predict a particular individual's behavior. Further, as stated previously, law and biology does not argue from is to ought; normative decision are outside of its scope. Instead, the goal here is to shed light on tendencies built into human nature on the level of the aggregate. In this vein, numerous publications are significant, including (but by no means limited to) those by Martin Daly and Margo Wilson,[52] Cheryl Hanna,[53] Albert Alschuler, and Oliver Goodenough.[54] Criminal law will also be enhanced if informed about the ground-breaking work emerging out of the neurosciences.[55]

Employment Law and Workplace Behavior

Employment law, and workplace behavior generally, deal directly with human behavior on a daily basis. Recently, Kingsley Browne's *Divided Labours: An Evolutionary View of Women at Work*[56] discusses how biological tendencies, at the level of the aggregate, may play roles in the workplace and labor market generally. In addition, Oliver Goodenough is leading an initiative to research how the behavioral sciences may help us better to understand workplace behavior, hierarchy, and conflict resolution. Recently, Diana Weinhold, Paul Zak and Monika Gruter Morhenn discuss how the biology of risk-taking versus risk-aversion and the way such behaviors affect work choices are likely to impact the labor market.[57]

Conferences and Seminars

Numerous conferences take place every year allowing scholars to participate in bridging the gap between law and biology. For instance, each summer for the past ten years, legal scholars, economists, and scientists from fields such as evolutionary biology, genetics, ethology, the neurosciences, and anthropology have convened in Squaw Valley, California for teaching conferences and symposia hosted and organized by the Gruter Institute for Law and Behavioral Research.[58] These conferences have brought attention to myriad topics aimed at informing the law about biology. In addition, the Gruter Institute has hosted

numerous unique conferences for law professors and federal judges based on a variety of topics relating to law and biology. Conference topics have included security and risk; neurobiology, human behavior and the law; and evolutionary biology and the sense of justice. Another forum in which scholars from law and the life sciences are convening is the Society for Evolutionary Analysis in Law (SEAL). Among other endeavors, SEAL hosts a yearly conference whose climate of cross-disciplinary collaboration assists legal scholars seeking to publish "outside their field."

CONCLUSION

By applying insights from the behavioral sciences to the law, we enable law to be better leveraged to regulate human behavior. As seen above, the use of biological insights in law has grown from the level of analogy to the level of direct application. Today, legal scholarship as well as legal education are bridging the gap between law and biology in order to improve the law's understanding of that which it seeks to regulate. As illustrated above, this arena has enjoyed exciting and ground-breaking interdisciplinary scholarship. Yet in reviewing this exciting young field, it is also apparent that much if its landscape remains unexplored.

NOTES

1. For purposes of this chapter, the terms "biological sciences," "biology" and "evolutionary biology" are used equivalently, and are meant to include insights from evolutionary biology, evolutionary psychology, the neurosciences, ethology, and genetics.

2. For a comprehensive discussion of many of the misconceptions surrounding law and biology, see Owen Jones, *Sex, Culture and the Biology of Rape: Toward Explanation and Prevention*, 87 California L. Rev. 827 (1999); E. Donald Elliott, *Law and Biology: The New Synthesis?* 41 St. Louis L. J. 595 (1997); and Owen Jones, *Law and Evolutionary Biology: Obstacles and Opportunities*, 10 J. Contemp. Health L. & Pol'y 265 (1994).

3. For further discussion of the Naturalistic Fallacy, see e.g., Owen Jones, *Law and Evolutionary Biology: Obstacles and Opportunities*, 10 J. Contemp. Health L. & Pol'y 265 (1994); E. Donald Elliott, *Law and Biology: The New Synthesis?* 41 St. Louis L. J. 595 (1997); and Owen Jones, *Sex, Culture and the Biology of Rape: Toward Explanation and Prevention*, 87 California L. Rev. 827 (1999).

4. Charles Darwin, *Origin of the Species* (1859).

5. Sir Henry James Sumner Maine, *Ancient Law* (1861).

6. See e.g. *Evolution of Law: Select Readings on the Origin and Development of Legal Institutions* (A. Kocourek & J. Wigmore (Eds) 1915–1918) (3 vols.).

7. Holmes, *The Common Law* (1881). It is in *The Common Law* that Holmes introduces his famous passage: "The life of the law has not been logic: it has been experience.

. . . The felt necessities of the time, the prevalent moral and political theories, intuitions of public policy, avowed or unconscious, even the prejudices which judges share with their fellow-men, have had a good deal more to do than the syllogism in determining the rules by which men should be governed." Holmes, *The Common Law*, at p. 5.

8. Holmes, *Law in Science, Science in Law*, 12 Harv. L. Rev. 433 (1899). For a rich discussion of Holmes' and Wigmore's treatment of the evolution of law, see E. Donald Elliott, *The Evolutionary Tradition in Jurisprudence*, 85 Colum. L. Rev. 38 (1985).

9. Clark, *The Morphogenesis of Subchapter C: An Essay in Statutory Evolution and Reform*, 87 Yale L. J. 90 (1977), see also Clark, *The Interdisciplinary Study of Legal Evolution*, 90 Yale L. J. 1238 (1981).

10. Gruter, M., *Law in Sociobiological Perspective*, at 185.

11. Again, it is to be emphasized that normative decisions about whether to structure any particular law "with or against the grain" of human nature is a separate analysis, beyond the scope of the law and biology approach.

12. Gruter, M., *Law in Sociobiological Perspective*, supra note 10.

13. As used herein, "legal behavior" refers to the behavioral response to rules, norms or laws, whether the response is to obey, evade or disobey. For further discussion of legal behavior, see Gruter, M., *Law in Sociobiological Perspective*, supra note 10; and Gruter, M., *Law and the Mind: Biological Origins of Human Behavior* (1991).

14. Gruter, M., *Law in Sociobiological Perspective*, supra note 10, at 185. For related discussions see Masters, R. D. & Gruter, M. (Eds), *The Sense of Justice* (Sage, 1992), Chapter 5.

15. The proceedings of this symposium appear in *Law, Biology and Culture: The Evolution of Law*, Gruter, M. & Bohannan, P. (Eds) (1983 Ross-Erikson) (available through the Gruter Institute for Law and Behavioral Research).

16. Alexander, Richard, *The Biology of Moral Systems* (1987).

17. Beckstrom, John H., *Sociobiology and the Law: The Biology of Altruism in the Courtroom of the Future* (1985).

18. Bohannan, Paul, "Some Bases of Aggression and Their Relationship to Law" and Hoebel, E. Adamson, "Anthropology, Law and Genetic Inheritance." In: M. Gruter and P. Bohannan (Eds), *Law, Biology and Culture: The Evolution of Law*, supra note 15.

19. MacLean, Paul, "A Triangular Brief on the Evolution of Brain and Law." In: M. Gruter & P. Bohannan (Eds), *Law, Biology and Culture: The Evolution of Law*, supra note 15.

20. Gruter, M., supra note 10, at 185.

21. M. Gruter & P. Bohannan, (Eds), supra note 15.

22. 10 Ecol. L. Q. 205 (1982).

23. 85 Colum. L. Rev. 38 (1985).

24. M. Gruter & R. Masters, (Eds). (1986).

25. R. Masters & M. Gruter, (Eds). (1992).

26. M. Gruter (1991).

27. This ground-breaking class was entitled "The Ethology of Law: The Biological Bases of Legal Behavior."

28. Examples include biological insights into: environmental and natural resource law (William H. Rodgers, University of Washington; E. Donald Elliott, Yale University); regulatory and family law [with emphasis on rape and child abuse] (Owen D. Jones, Arizona State University); elder law and ERISA regulations (Lawrence Frolik at University of Pittsburgh); labor and employment law (Kingsley Browne at Wayne State

University); criminal law and teen violence (Cheryl Hanna at Vermont Law School); criminal law relating to rape (Owen Jones, Arizona State University); property law (Dorothy Glancy, Santa Clara University; Jeffrey Stake, University of Indiana).

29. Further information about SEAL is available at: http://www.sealsite.org

30. Due to the impossibility of referencing each of the works that serve to benefit readers interested in law and biology, attached hereto is a list of Collected References, which identifies many, but not all, of the publications useful in this endeavor.

31. Such conferences include: "Evolutionary Biology, Economics, Business and Law," (Squaw Valley, California, 2000); "Consumer Choice/Market Forces in the Selection of Dispute Resolution Procedures" (Georgetown University Law Center, 1998); "Law, Risk and Risk Management" (University of California at Berkeley, 1996).

32. This is not to say that economists are united among themselves in deriving a uniform law and economics approach.. Indeed, there is a voluminous literature in law and economics relating to this very issue. Significant papers in this area include, inter alia, Priest, *The Common Law Process and the Selection of Efficient Rules*, 6 J. Legal Stud. 65 (1977); Rubin, *Why Is the Common Law Efficient?* 6 J. Legal Stud. 51 (1977); Landes & Posner, *The Economic Structure of Tort Law* (1987 Harvard University Press); Priest, Regulating the Content and Volume of Litigation: An Economic Analysis (1986 *Supreme Court Economic Review*, *1*, 163; Hirshleifer, Evolutionary Models in Economics and law, 4 *Research in Economics and Law*, *167* (1982); Posner, *Sex and Reason* (1992).

33. For a rich discussion in this context, see Robert H. Frank, *Passions Within Reason: The Strategic Role of the Emotions* (1988).

34. William H. Rodgers, *Bringing People Back: Toward a Comprehensive Theory of Taking in Natural Resources Law*, 10 *Ecology L.Q.* 205, 225 (1982).

35. See e.g., Joseph LeDoux, *The Emotional Brain: The Mysterious Underpinnings of Emotional Life* (1996); Steven Pinker, *How the Mind Works* (1998); M. T. McGuire, & A. Troisi, *Darwinian Psychiatry* (1998); Michael Gazzaniga, *The Cognitive Neurosciences* (1995); Antonio Demasio, *Descartes' Error: Emotion, Reason, and the Human Brain* (1994); Roger D. Masters & Michael T. McGuire, *The Neurotransmitter Revolution* (1993); Nicholas Humphrey, *A History of the Mind: Evolution and the Birth of Consciousness* (1992).

36. In addition to the recently published plethora of articles relating to the completion of the mapping of the human genome, see e.g. Matt Ridley, *Genome* (2000); Luigi Luca Cavalli-Sforza, *The Great Human Diasporas* (1995); Luigi Luca Cavalli-Sforza, *Genes, Peoples and Languages*, 265 Sci Am. 104 (1991); Richard Dawkins, *The Selfish Gene* (1989).

37. See e.g. Lee Silver, *Remaking Eden: How Genetic Engineering and Cloning Will Transform the American Family* (1998).

38. See Robert R. Frank, *Passions Within Reason: The Strategic Role of Emotions* (1988).

39. See Robert D. Cooter, "Law from Order: Economic Development and the Jurisprudence of Social Norms." In: *A Not-So-Dismal Science: A Broader, Brighter Approach to Economies and Societies* (1999).

40. See Paul J. Zak, Institutions, Property Rights and Growth [Working Paper, Claremont Graduate University].

41. Jack Hirshleifer, The Bioeconomic Causes of War, 19. *Managerial & Decision Economics*, 457–466 (1998). Jack Hirshleifer, *Evolutionary Models in Economics and Law*, 4; *Research in Economics and Law*, 167 (1982); Jack Hirshleifer, *Natural Economy Versus Political Economy*, 1 J. Soc. & Biological Structures 319, 337 (1978).

42. Wolfgang Fikentscher, *Modes of Thought* (1995).

43. Michael T. McGuire & Mark Grady, *The Nature of Constitutions*, 1 J. of Bioeconomics 227 (2000); and Michael T. McGuire & Mark Grady, A Theory of the Origin of Natural Law, 8 J. *Contemp. Legal Issues*, *87* (1997).

44. See E. Donald Elliott, *Environmental TQM: Anatomy of A Pollution Control Program that Works!*, 92 Mich. L. Rev. 1840 (1994); and E. Donald Elliott, *The Biological Basis of Environmental Law: Cooperative Management of Parasite-Host Relationships* [forthcoming, on file with author].

45. Amicus Brief of the Gruter Institute, *In the Matter of Baby M*, 537 A.2d 1227 (N.J. 1988).

46. See Owen D. Jones, *Sex, Culture, and the Biology of Rape: Toward Explanation and Prevention*, 87 California L. Rev. 827 (1999); and Owen D. Jones, *Evolutionary Analysis in Law: An Introduction and Application to Child Abuse*, 75 N.C. L. Rev. 1117 (1997).

47. See Cheryl Hanna, *Can a Biological Inquiry Help Reduce Male Violence Against Women?*, 22 Vermont L. Rev. 333 (1998); and Cheryl Hanna, *Ganging Up on Girls: Young Women and Their Emerging Violence*, 41 Ariz. L. Rev. 93 (1999).

48. See Lawrence Frolik, *The Biological Roots of the Undue Influence Doctrine: What's Love Got to Do With It?*, 57 U. Pitt. L. Rev. 841 (1996).

49. Much of the material illuminating deception and self-deception comes from Frans de Waal, *Chimpanzee Politics* (1989); Frans de Waal, *Good Natured: The Origins of Right and Wrong in Humans and Other Animals* (1996).

50. See E. Donald Elliott, *Environmental TQM: Anatomy of A Pollution Control Program that Works!*, 92 Mich. L. Rev. 1840 (1994); E. Donald Elliott, *The Biological Basis of Environmental Law: Cooperative Management of Parasite-Host Relationships* [working paper, on file with author]; and E. Donald Elliott, Toward a Theory of Statutory Evolution: The Federalization of Environmental Law, 1 J. L. *Econ. & Org.*, *313* (1985).

51. See Thomas Moellers, *Festschrift fuer Wolfgang Fikentscher* (1998).

52. See Martin Daly & Margo Wilson, *Sex, Evolution and Behavior* (2nd ed., 1983); and Martin Daly & Margo Wilson, *Homicide* (1988).

53. See Cheryl Hanna, supra note 47.

54. See Oliver R. Goodenough, *Biology, Behavior, and Criminal Law: Seeking a Responsible Approach to an Inevitable Interchange*, 22 Vermont L. Rev. 263 (1997) (illustrating that it typically does not follow that a criminal should or will be absolved of a crime on the basis of biologically based exculpatory arguments)

55. See e.g., Michael McGuire's work on the effects of serotonin levels on behavior. Masters, R. D. & McGuire, M. T., *The Neurotransmitter Revolution: Serotonin, Social Behavior, and the Law* (1993).

56. (1999); see also Kingsley Browne, *Sex and Temperament in Modern Society: A Darwinian View of the Glass Ceiling and the Gender Gap*, 37 Az. L. J. 972 (1995).

57. See Diana Weinhold, Paul Zak & Monika Gruter Morhenn, *Homework and the Labor Supply of Women* [Working Paper, London School of Economics].

58. The Gruter Institute for Law and Behavioral Research was founded in 1981 by Margaret Gruter. The Institute, a private non-profit foundation, fosters interdisciplinary research to bridge the gap between law, economics, other social sciences and biology. The Institute has published numerous books, articles and teaching materials, funded many research projects and hosted scores of meetings. Further details, including publication and conference lists are available at http://www.gruterinstitute.org

REFERENCES

Alcock, J., & Sherman, P (1994). The Utility of the Proximate-Ultimate Dichotomy in Ethology, 96. *Ethology*, *58*.

Alexander, R. D. (1987). *The Biology of Moral Systems*.

Alexander, R. D. (1979). *Darwinism and Human Affairs*.

Beckstrom, J. H. (1985). *Sociobiology and the Law: The Biology of Altruism in the Courtroom of the Future*.

Browne, K. R. (June 1984). *Biology, Equality, and the Law: The Legal Significance of Biological Sex Differences*, 38. Southwestern L. J.

Browne, K. R. (1995). *Sex and Temperament in Modern Society: A Darwinian View of the Glass Ceiling and the Gender Gap*, 37. Az. L. J. 972.

Buss, D. M. (1999). *Evolutionary Psychology: The New Science of the Mind*.

Buss, D. M., & Malamuth, N. (Eds) (1997). *Sex Power & Conflict*.

Cavalli-Sforza, L. L. (1991). Genes, Peoples and Languages, 265. *Sci. Am.*, *104*.

Cavalli-Sforza, L. L. (1995). *The Great Human Diasporas*.

Cooter, R. D. (1994). Structural Adjudication and the New Law Merchant: A Model of Decentralized Law, 14. *Internat'l Rev. of L. & Econ.*, *215*.

Daly, M., & Wilson, M. (1983). *Sex, Evolution, and Behavior* (2nd ed.).

Daly, M., & Wilson, M. (1988). *Homicide*.

Damasio, A. R. (1994). *Descartes' Error: Emotion, Reason, and the Human Brain*.

Dawkins, R. (1989). *The Selfish Gene* (new ed.).

de Waal, F. (1997). *Bonobo*.

de Waal, F. (1989). *Chimpanzee Politics: Power and Sex Among Apes*.

de Waal, F. (1996). *Good Natured: The Origins of Right and Wrong in Humans and Other Animals*.

de Waal, F. (1989). *Peacemaking Among Primates*.

Eibl-Eibesfeldt, I. (1989). *Human Ethology*.

Elliott, E. D. (1985). The Evolutionary Tradition in Jurisprudence, 85. *Columbia L. Rev.*, *38*.

Elliott, E. D. (1984). Holmes and Evolution: Legal Process as Artificial Intelligence, 13. *J. Leg. Stud.*, *113*.

Elliott, E. D. et al. (1985). Toward a Theory of Statutory Evolution: The Federalization of Environmental Law, 1. *J. L. Econ. & Org.*, *313*.

Elliott, E. D. (1997). Law and Biology: The New Synthesis?, 41. *St. Louis L. J.*, *595*.

Fikentscher, W. (forthcoming). *Cultural Complexity*.

Fikentscher, W. (1995). *Modes of Thought*.

Fikentscher, W., & McGuire, M. (1994). A Four-Function Theory of Biology for Law, 25. *Rechtstheorie*, *291*.

Frank, R. (1988). *Passions Within Reason: The Strategic Role of the Emotions*.

Frolik, L. (Ed.) (1999). Law and Evolutionary Biology: Selected Essays in Honor of Margaret Gruter.

Frolik, L. (1996). The Biological Roots of the Undue Influence Doctrine: What's Love Got to Do With It?, 57. *U. Pitt. L. Rev.*, *841*.

Gazzaniga, M. S. (1992). *Nature's Mind: The Biological Roots of Thinking, Emotions, Sexuality, Language and Intelligence*.

Gazzaniga, M. S. (1995). *The Cognitive Neurosciences*.

Goldsmith, T. H. (1991). *The Biological Roots of Human Nature: Forging Links Between Evolution and Behavior*.

Goldsmith, T. H., & Jones, O. D. (1999). Evolutionary Biology and Behavior: A Brief Overview and Some Important Concepts, 39. *Jurimetrics, 131.*

Goodenough, O. (1998). Biology, Behavior, and Criminal Law: Seeking a Responsible Approach to an Inevitable Interchange, 22. *Vermont L. Rev., 263.*

Goodenough, O. (1999). Information Replication in Culture: Three Modes for the Transmission of Culture Elements Through Observed Action, Proceedings of the AISB '99 Symposium on Imitation in Animals and Artifacts (Sussex, The Society for the Study of Artificial Intelligence and the Simulation of Behavior).

Goodenough, O. (1995). Mind Viruses: Culture, Evolution and the Puzzle of Altruism, 34. *Soc. Sci. Info., 287.*

Grady, M., & McGuire, M. T. (1997). Theory of the Origin of Natural Law, 8. *J. Contemp. L. Issues, 87.*

Gruter, M. (Jan./Feb. 1991). Behavior, Evolution and the Sense of Justice, 34. *Am. Behav. Sci.*

Gruter, M. (1991). *Law and the Mind: Biological Origins of Human Behavior.*

Gruter, M. (1977). Law in Sociobiological Perspective, 5. *Fl. St. L. Rev., 181.*

Gruter, M. (1979). The Origins of Legal Behavior. *J. Social Biol. Struct., 2.*

Gruter, M., & Bohannan, P. (1982). *Law, Biology & Culture: The Evolution of Law.*

Gruter, M., & Masters, R. (1998). Compassion, Common Sense, and Deception: Social Skills and the Evolution of Law. In: *Festschrift for Wolfgang Fikentscher.*

Gruter, M., & Masters, R. (1986). *Ostracism: A Social and Biological Phenomenon.*

Hamilton, W. D., & Howard, J. (Eds) (1997). *Infection, Polymorphism and Evolution.*

Hamilton, W. D. (1996). Narrow Roads of Gene Land: The Collected Papers of W.D. Hamilton: *Evolution of Social Behaviour.*

Hamilton, W. D. (2000). Narrow Roads of Gene Land : The Collected Papers of W.D. Hamilton (*Evolution of Sex, 2*).

Hanna, C. (1998). Can A Biological Inquiry Help Reduce Male Violence Against Women? 22. *Vermont L. Rev., 333.*

Hanna, C. (1999). Ganging Up on Girls: Young Women and Their Emerging Violence, 41. *Ariz. L. Rev., 93.*

Hirshleifer, J. (1978). Natural Economy Versus Political Economy, 1. *J. Soc. & Biological Structures 319, 337.*

Humphrey, N. (1992). *A History of the Mind: Evolution and the Birth of Consciousness.*

Jones, O. D. (2000). An Evolutionary Analysis of Rape: Reflections on Transitions, 11. *Hastings Women's Law Journal.*

Jones, O. D. (1997). Evolutionary Analysis in Law: An Introduction and Application to Child Abuse, 75. *N.C. L. Rev., 1117.*

Jones, O. D. (1996). Genes, Behavior, and Law, 5. *Politics & The Life Sciences, 101.*

Jones, O. D. (1997). Law and Biology: Toward an Integrated Model of Human Behavior, 8. *J. Contemp. L. Issues, 167.*

Jones, O. D. (1999). Law, Emotions, and Behavioral Biology, 39. *Jurimetrics, 283.*

Jones, O. D. (2000). The Nature of Norms: Biology, Morality, and Fukuyama's Great Disruption. *Michigan Law Review.*

Jones, O. D. (1999). Sex, Culture, and the Biology of Rape: Toward Explanation and Prevention, 87. *California L. Rev., 827.*

Jones, O. D. & Strouse, D. (1999). Introduction to the Symposium on Biology and Sexual Aggression, 39. *Jurimetrics, 113.*

Le Doux, J. (1996). *The Emotional Brain: The Mysterious Underpinnings of Emotional Life.*

Masters, R. D. (1989). *The Nature of Politics.*

Masters, R., & Gruter, M. (1992). *The Sense of Justice: Biological Foundations of Law.*

Masters, R., & McGuire, M. (1993). *The Neurotransmitter Revolution: Serotonin, Social Behavior, and the Law.*

McGinnis, J. (1997). The Human Constitution and Constitutive Law, 8. *J. Contemp. L. Issues, 211.*

McGinnis, J. O. (1996). The Once and Future Property-Based Vision of the First Amendment, 63. *Chicago L. Rev., 49.*

McGuire, M., & Troisi, A. (1998). *Darwinian Psychiatry.*

McGuire, M., & Rehbinder, M. (1993). *Biology, Culture, and Environmental Law.*

McGuire, M. (1993). *Human Nature and the New Europe.*

Mealey, L. (1996). Evolutionary Psychology and the Search for Evolved Mental Mechanisms Underlying Complex Human Behavior. In: J. P. Hurd (Ed.), *Investigating the Biological Foundations of Human Morality, 35.*

Moellers, T. (1998). *Festschrift fuer Wolfgang Fikentscher.*

Perry, J. G. (1998). Reciprocity and Criminal Justice: Using Epigenetics to Design a Program, 22. *Vermont L. Rev., 407.*

Pinker, S. (1997). *How the Mind Works.*

Ridley, M. (1993). *Evolution.*

Ridley, M. (1996). *The Origins of Virtue: Human Instincts and the Evolution of Cooperation.*

Ridley, M. (1993). *The Red Queen: Sex and the Evolution of Human Nature.*

Rodgers, W. H., Jr. (1982). Bringing People Back: Toward A Comprehensive Theory of Taking in Natural Resources Law, 10. *Ecol. L. Q., 205.*

Rodgers, W. H., Jr. (1995). Deception, Self-Deception, and Mythology: The Law of Salmon in the Pacific Northwest, 26. *Pac. L. J., 821.*

Rodgers, W. H., Jr. (1986). The Evolution of Cooperation in Natural Resources Law: The Drifter/Habitué Distinction, 38. *U. Fla. L. Rev., 195.*

Rodgers, W. H., Jr. (1989). The Lesson of the Owl and the Crows: The Role of Deception in the Evolution of the Environmental Statutes, 4. *J. Land Use & Envtl., 377.*

Rodgers, W. H., Jr. (1988). The Lesson of the Red Squirrel: Consensus and Betrayal in the Environmental Statutes, 5. *J. Contemp. Health L. & Pol'y.*

Rodgers, W. H., Jr. (1993). Where Environmental Law And Biology Meet: Of Pandas' Thumbs, Statutory Sleepers, And Effective Law, 65. *U. Colo. L. Rev., 25.*

Ruhl, J. B. (1996). The Fitness of Law: Using Complexity Theory to Describe the Evolution of Law and Society and its Practical Meaning for Democracy, 49. *Vanderbilt Law Review, 1407.*

Sapolsky, R. (1997). *The Trouble with Testosterone and Other Essays on the Biology of the Human Predicament.*

Stake, J. E. (1990). Darwin, Donations, and the Illusion of Dead Hand Control, 64. *Tul. L. Rev., 705.*

Tiger, L. (1999). *The Decline of Males.*

Tiger, L., & Fox, R. (1979). *The Imperial Animal.*

Tiger, L. The Manufacture of Evil: Ethics, Evolution, and the Industrial System (1987)

Tooby, J., & Leda, C. (1992). The Psychological Foundations of Culture. In: J. H. Barkow et al. (Eds.), *The Adapted Mind: Evolutionary Psychology and the Generation of Culture 19, 23.*

Trivers, R. (1985). *Social Evolution.*

Wright, R. (1994). *The Moral Animal: The New Science of Evolutionary Psychology.*

DARWINISM AND PHILOSOPHY

Jean Lachapelle and Michael Ruse

INTRODUCTION

Evolutionary theorizing is a child of the eighteenth century, and from its very beginnings right down to the present it has always been more than pure science. Most prominently and obviously, evolution has served as a kind of secular religion substitute (Ruse, 1996, 1999b, 2000b). The earliest evolutionists like Erasmus Darwin (1789, 1791, 1794-1796, 1803) in England, and Jean Baptiste de Lamarck (1809) in France, were explicit in seeing a form of progressionism embedded in evolutionary change taking us from the simple to the complex – from the monad to the man. Through this progress, they believed that one gets a kind of upward sweep to human history which matches and should replace the Providentialism of the Christian religion. The coming of Charles Darwin's theory through natural selection undoubtedly put evolutionary theorizing on a more secure empirical basis (Ruse, 1999a). But for Darwin and his fellows – most notably Thomas Henry Huxley and Herbert Spencer in England, and Ernst Haeckel in Germany – evolution continued as more than something purely scientific (Richards, 1987, 1992). For all of these men it provided a substitute for the reactionary Christianity of their day, a substitute which they believed was more in tune with the urbanized, industrial society that Europe had then become, and that the rest of the world (America particularly) was shortly to follow.

Through the twentieth century things remained essentially the same, right up to the present day at the beginning of a new millennium. One finds that prominent evolutionists, notably Edward O. Wilson (1975, 1978, 1992, 1998) in

Evolutionary Approaches in the Behavioral Sciences, Volume 8, pages 135–160.

ISBN: 0-7623-0769-2

America and John Maynard Smith (Maynard, Smith & Szathmary, 1995) in Britain, continue to regard evolution as more than mere science, being in some sense rather a full and satisfying alternative world picture. There is little surprise therefore that evolutionists have thought that their ideas would apply across the board, reaching to and having implications for many other areas of human inquiry. Not the least of these is philosophy, an outreaching which has not always been welcomed by the philosophical community itself (Ruse, 1998). Nevertheless, notwithstanding many crudities in evolutionary attempts to throw light on major philosophical questions, we believe that there is much in evolutionism which bears profitably on the major concerns of philosophers. In some sense, it really should matter that we humans are the end products of a long slow process of development rather than the favoured creations of a good God on the Sixth Day.

In this chapter, therefore, we will consider some of the major moves which have been made in the name of evolution into the domain of philosophy. Traditionally, one divides philosophical problems into those dealing with *epistemology*, or the theory of knowledge, and those dealing with *ethics*, or the theory of morality. In this discussion we shall follow this division, turning first to questions about what one can know, and then secondly to questions about what one should do. We shall argue that within these divisions there is a parallel subdivision: both epistemology and ethics have seen (what we regard as) somewhat old-fashioned, and ultimately not-very-profitable, attempts to address the major issues; then both epistemology and ethics have seen (what we regard as) more profitable attempts to solve the problems of knowledge and morality. In a way, we think that these subdivisions reflect attitudes about evolution. The incorrect or less profitable ways rely on old-fashioned, nonDarwinian approaches to evolutionary thought. The more successful ways use Darwinism itself to the full. But whether we are right in this surmise must be for readers themselves to judge.

Evolutionary Epistemology

The problem of epistemology is the problem of knowledge. Generally today the most secure form of knowledge is believed to be scientific knowledge, and it is certainly on this kind of knowledge that traditional "evolutionary epistemologists" have focused their attention (Bradie, 1986). Their approach ostensibly starts with Charles Darwin's theory of evolution through natural selection (Darwin, 1859). Briefly this theory starts with the fact that there is always a potential population explosion, and that this explosion cannot be matched by sustaining food supplies and available space. There will therefore

inevitably be a *struggle for existence*, with more organisms being born than can possible survive and reproduce. Populations of organisms are never entirely uniform. New variations or characteristics of an entirely random nature are constantly being injected into populations. Hence, therefore, in the struggle for existence the competing organisms will be different, and it is the Darwinians' claim that these differences have a crucial deciding effect on the chances of success in the struggle. In particular, on average, those that do survive and reproduce will be different from those that do not and the differences are involved in the survival and reproduction. Hence, overall, there will be a kind of differential reproduction brought on by the struggle. This is known as *natural selection*, or the survival of the fittest. It is the Darwinians' claim that, given enough time, this differential reproduction leads to full-blown organic change, and that this organic change is of a particular kind: specifically, organisms develop characteristics which aid in the struggle. Such characteristics are known as *adaptations*. The hand and the eye are popular examples.

What the epistemologist does now is build on this theory, analogically applying it to culture, specifically to the culture of science. As Van Fraassen (1980, 40) once put it "The success of current scientific theories is no miracle. It is not even surprising to the scientific (Darwinist) mind. For any scientific theory is born into a life of fierce scientific competition, a jungle red in tooth and claw. Only the successful theories survive." One has different theories constantly being thrown up in the course of scientific inquiry. One person, for instance Descartes, suggests that everything is made of small particles, that there is no void, and that motion is a matter of these particles rubbing against each other. Thus one gets the planets revolving around the sun, in ellipses, as described by Kepler. Another theory, that of Newton, agrees that everything is made of small particles, but argues that there are gaps or a void between the particles and that there is action at a distance: particles attract each other with a force known as gravity. On the basis of this theory, Newton likewise is able to derive the Copernican picture of the universe with planets going around the sun in ellipses. One had two theories, just as analogously one has two organisms, and they have to struggle between themselves for success. Eventually, one theory proves itself superior in some way to the other theory: in our example, the Newtonian theory commends itself to scientists rather than the Cartesian theory, and so we get success of Newtonianism. There has been a struggle and a consequent selection. What the evolutionary epistemologist argues is that this is an ongoing process and that eventually it leads to full-blown change.

One finds this kind of evolutionary epistemology, although not by that name, right back to the time of Darwin. For instance, Darwin's bulldog – Thomas Henry Huxley – proposed precisely this view of science.

> Now the essence of the scientific spirit is criticism. It tells us that whenever a doctrine claims our assent we should reply, Take it if you can compel it. The struggle for existence holds as much in the intellectual as in the physical world. A theory is a species of thinking, and its right to exist is coextensive with its power of resisting extinction by its rivals (Huxley, 1893, 229).

More recently, the classic exposition of this analogical form of evolutionary epistemology is that provided by the philosopher of science Stephen Toulmin. He writes as follows:

> In talking about the development of natural science as "evolutionary", I [am not] employing a mere *façon de parler*, or analogy, or metaphor. The idea that the historical changes by which scientific thought develops frequently follow an "evolutionary" pattern needs to be taken quite seriously; and the implications of such a pattern of change can be, not merely suggestive, but explanatory (Toulmin, 1967, 470).

In Toulmin's opinion, what happens is that new ideas are thrown up, these get into the pool – into scientific discourse – through journals and books and so forth. These are then sifted through by scientists and their students. Some prove useful and are taken up, others are rejected, and so we have a kind of selection leading to an evolutionary change. Summing up his position Toulmin writes as follows:

> Science develops . . . as the outcome of a double process: at each stage, a pool of competing intellectual variants is in circulation, and in each generation a selection process is going on, by which certain of these variants are accepted and incorporated into the science concerned, to be passed on to the next generation of workers as integral elements of the tradition.

Looked at in these terms, a particular scientific discipline – say, atomic physics – needs to be thought of, not as the contents of a textbook bearing any specific date, but rather as a developing subject having a continuing identity through time, and characterized as much by its process of growth as by the content of any one historical cross-section . . . Moving from one historical cross-section to the next, the actual ideas transmitted display neither a complete breach at any point – the idea of absolute "scientific revolution" involves an over-simplification – nor perfect replication, either. The change from one cross-section to the next is an *evolutionary* one in this sense too: that later intellectual cross-sections of a tradition reproduce the content of their immediate predecessors, as modified by those particular intellectual novelties which were selected out in the meanwhile – in the light of the professional standards of the science of the time (Toulmin, 1967, 465-466).

EVOLUTIONARY EPISTEMOLOGY CONSIDERED

An analogy is an analogy. It is neither right nor wrong. Rather it is useful and insightful, or not. If someone finds this analogy, between evolutionary change

and the change of scientific knowledge, to be something which is insightful then no amount of criticism is going to detract from this.[1] On the other hand, one might point out that, even if useful, there are dangers lurking. In certain senses, the analogy is no true analogy. This essentially is our position. We cannot deny that there has recently been some very good history of science written by people who are enthused by this analogy. Robert J. Richards, in his magisterial work *Darwin and the Emergence of Evolutionary Theories of Mind and Behavior* (1987), has used this analogy to look very perceptively at the development of Darwinian theories of the mind over the past two and a half centuries. Likewise, the philosopher David Hull (1988) in his major work *Science as a Process* used the analogy to consider the taxonomic wars of the 1960s and 70s, when the new theory of cladism appeared on the scene and eventually pushed aside rivals like traditional evolutionary taxonomy. However, for all the good work that has been done using this analogy, we are troubled by some of the clear disanalogies between the Darwinian picture of evolutionary change, and the change of scientific theories. So much so that we are inclined to conclude that, for all the apparent Darwinism of the science being referred to, in fact the science is of a kind quite different from that envisioned by Charles Darwin, at least in his *Origin of Species*. In respects, it owes more to the thinking of other evolutionists, Darwin's contemporary Herbert Spencer, and others.

The main and much-discussed point of disanalogy is in the nature of variation (Ruse, 1998). For the Darwinian, variation must be random. This is not to say that the new variations or characteristics coming into a population are uncaused. In fact, today, we know a great deal about the causes of the variation. But rather that the variations are random in the sense of not appearing according to need. Suppose for instance the population is faced by a new predator. There is no guarantee in traditional Darwinism that a variation will appear which protects the population against the predator. For instance, a camouflage variation or a new method of fighting back. In fact, if anything one expects that such variations will not appear, and indeed it has been a major concern of Darwinians since the *Origin* to show how the randomness of variation does not necessarily stop or negate the effects of natural selection. But be this as it may, the essential point of the Darwinian scientific position is that variation does not occur according to need. Hence, in a real sense, one does not expect Darwinian evolution to be directed in some overall sort of way. There may be a limited points of improvement, but overall a direction cannot be there. Darwinian evolution is essentially non-progressive.

However, as everybody knows when it comes to science, although sometimes there may be serendipitous discoveries where variations as it were come by

chance (the classic case supposedly is Alexander Fleming's discovery of penicillin which came about purely by accidental contamination of cultures), in most cases the variations of science are anything but random. They are rather the result of genius, and of a great deal of effort being applied in order to speak to perceived needs or questions. The classic case of such a directed variation, if one might so call it, is of Darwin's own discovery of natural selection. Thanks to the notebooks that we have still, we know that Darwin became an evolutionist around March of 1837, but did not discover natural selection until the end of September of 1838. We know however that Darwin worked long and hard during those eighteen months to come up with some kind of mechanism of evolutionary change. He read a great many books; he studied the work and success of animal breeders and plant breeders; he talked at length to many different people; and much more. Darwin was a genius, but he did not achieve his discovery just by chance. It was the end result of a great deal of careful thought. Darwin had a problem, namely how was he to explain evolution – more than this, how was he to explain evolution in the direction of adaptation – and eventually Darwin came up with a solution. His variation, which he introduced into the scientific community in 1859 in the *Origin of Species*, was anything but random (Ruse, 1999a).

The consequence of all of this of course is that one expects to find that scientific knowledge is not simply meandering, going nowhere. It is rather something which is highly directed, in particular it is highly directed towards an understanding or knowledge of the real world. Towards truth, if you will. It may be that Karl Popper (1959) is right that we can never get to the truth, at least we can never know that we have got to the truth, but the general opinion would be that scientific knowledge is in fact moving towards the truth. Related to this, one finds that there are going to be other significant disanalogies between a purely Darwinian type of scientific theory of evolution, and a cultural form of evolution. In particular, if one is working towards the truth then, at some level, one expects to find that, even if not all roads lead to Rome, there will be considerable convergence of different ideas on the same fundamental premises. In fact, it is well known that a significant mark of success in science is when one overarching hypothesis can explain and unit many disparate areas (Whewell, 1840).

Again, Darwin's theory is the classic case of such a unification, known technically as a consilience of inductions, but it is by no means the only one. Isaac Newton, for instance, achieved his great successes by showing that terrestrial motion and celestial motion can be explained by the same premises. More recently in the last century, the great revolution in geology – when it was shown how everything can be explained by moving plates, on which the

continents rest – was hailed as a great advance forward in primary reason because geologists were able to show how knowledge from so many different areas – biogeography, earthquakes, volcanoes, and much more – are all united under the same hypotheses (Ruse, 1989).

In other words, what one expects to find and what one does find in science is that unification and coming together is a major phenomenon. However, in Darwinian biological theory, although one does get something analogous, particularly in plants, in fact this coming together (hybridization) is, if anything, a kind of degenerate form of evolution. It is certainly not something which is widespread in the animal world, and more than this is eventually something which cannot lead to altogether new forms. What one is doing is reworking what one already has, rather than moving forward to something new. So, here again one has a major disanalogy between the biological and the cultural.

This is obviously not all that there is to be said on this matter. One might perhaps argue that the apparent breakdown of the analogy between science and culture is indicative of something else, less a disanalogy and more a misunderstanding by us of the culture of science. One might argue that, rather than looking for some kind of ultimate truth through a progressive development of science, one should think of science itself as being much more relativistic or pragmatic. Much as one might say organisms are fit, but only fit for the particular time and place. As is well known, in recent years, this kind of relativistic move has been made by many philosophers of science. Notoriously, this or something like this was the position of Thomas Kuhn in his *Structure of Scientific Revolutions* (1962), and it is interesting and surely pertinent to note that Kuhn in fact was strongly committed to an evolutionary model of change. However, such relativism is by no means universally accepted by philosophers. We will simply leave the discussion at this point, noting that if indeed one is to adopt this traditional evolutionary epistemology, then at the very least one is going to have to accept certain implications for the nature of science – in particular, about its relativistic and nonobjective status. These may be conclusions which are acceptable, welcomed even, but they are conclusions which will have to be grasped in some form or another. Moreover, it is by no means the case that all philosophers find this conclusion congenial.

DARWINIAN EPISTEMOLOGY

There is another approach to epistemology using evolutionary theory. This is one which has been favoured by a number of scientists and philosophers in recent years. Basically, this approach starts with the fact that the scientist, the knower, is himself or herself a product of evolution through natural selection.

That is to say, the way or ways in which we think and know are themselves shaped by evolutionary biology, and in particular by the mechanism of natural selection. There is a real world out there in which we humans live, and with which we have to deal. In order to act successfully in this world, that is to succeed in the struggle for existence, we humans have certain adaptations which enable us to get about and function successfully. Some of these adaptations obviously are the basic mammalian adaptations, for instance those to do with digestion, excretion, sex and so forth. However, we humans (if not uniquely at least very particularly) have taken up the path of knowledge, and we have developed large brains in order to think about the world and to deal with it (Ruse, 2000b). That is to say we – unlike the ants, for example – do not just work by instinct. We work by intelligence, by thinking about problems, by solving them, and so on and so forth. Obviously much of this is social and indeed some evolutionists today would argue that all of our knowledge ultimately relates back to sociality. But, be this as it may, the Darwinian evolutionist does want to claim that the way that we think is adapted so that we might survive and ultimately reproduce. (Excellent modern introductions to the science we presuppose are Pinker, 1997 and Mithen, 1996.)

Now, how as thinking beings are we to deal with this world of ours? Fairly obviously, we have got to be able to make inferences about the world, inferences which in some sense we can speak of as leading to the truth. This is not to say that we are never deceived, never deceived systematically, but that ultimately the inferences must add up to something relatively close to the truth. Otherwise, we would fail in our attempts to survive (Ruse, 1998, 2000a). What kinds of things are we going to have in order to think about the world truly? What the Darwinian evolutionist argues is that the brain or mind is constructed in such a way that we have certain dispositions to think about things, according to a truth-conferring methodology (Lorenz, 1941; Popper, 1972, 1974; Lumsden & Wilson, 1981, 1983). These are dispositions, and innate in some sense. The Darwinian evolutionist is not going to deny John Locke's major criticism of innate ideas, but claim rather that we have a propensity to think in certain sorts of ways as we grow and develop in our particular culture. Because we all think about things in these certain sort of ways, even if they may manifest themselves in different ways in different cultures, our minds must ultimately have some kind of shared deep structure, to use the terminology of the linguist Noam Chomsky (1957).[2]

What form do these innate dispositions or propensities take? The Darwinian epistemologist hypothesizes that they will probably be the basic rules of logic, of mathematics, and of reasoning generally (Quine, 1969; Sober, 1981; Ruse, 1998). We stress that no one is saying that every child will be born knowing

that either an object is A or not A, nor will such a child be immediately aware that one plus one equals two.[3] However, the Darwinian evolutionist does think that there will be a propensity to think in these sorts of ways, and such an evolutionist relates this back to our past history. Thus, in the Pleistocene let us say, the protohuman who has been given one object in return for two objects will be less successful in life's battles than the protohuman who is aware that having been giving two objects the just and proper return is two objects likewise. Similarly, to take the question of unification or consilience, the protohuman who spots various clues – let us say blood stains, growling in the undergrowth, footprints and the like – and who puts together these various items to infer that quite possibly there are tigers still about is going to be better suited in life's struggles than the proto-human who, having spotted the bloodstains, and footprints and growling and so forth, simply concludes that tigers are a hypothesis, and not a fact.

The great American philosopher Willard Van Orman Quine puts his finger on the problem thuswise:

> One part of the problem of induction, the part that asks why there should be regularities in nature at all, can, I think, be dismissed. *That* there or have been regularities, for whatever reason, is an established fact of science; and we cannot ask better than that. *Why* there have been regularities is an obscure question, for it is hard to see what would count as an answer. What does make clear sense is this other part of the problem of induction: why does our innate subjective spacing of qualities accord so well with the functionally relevant groupings in nature as to make our inductions tend to come out right? Why should our subjective spacing of qualities have a special purchase on nature and a lien on the future?
>
> There is some encouragement in Darwin. If people's innate spacing qualities is a gene-linked trait, then the spacing that has made for the most successful inductions will have tended to predominate through natural selection. Creatures inveterately wrong in their inductions have a pathetic but praise-worthy tendency to die before reproducing their kind (Quine, 1969, 126).

DARWINIAN EPISTEMOLOGY CONSIDERED

Just as with traditional evolutionary epistemology, many have queries and criticisms to be made of the more Darwinian form of epistemology which we are proposing here. Even though we are arguing for a literal application of Darwinism to problems of knowledge, rather than simply an analogy, critics feel that there are so many problems that our position must at best be vaguely hypothetical, and at worst little more than vacuous or pious hopes. Most particularly, the critics point to the fact that modern science is a great deal more than just a few crude rules of thumb. They argue that, when one considers modern science, no one genuinely could want to relate this directly to Darwinian theory

(Polkinghorne, 1991). Take modern population genetics or quantum theory or some such thing. No one could say that the person who knows about quantum mechanics is at an adaptive advantage over the person who is ignorant on the subject. Indeed, one might often want to suggest that the truly knowledgeable are, if anything, less adapted from a Darwinian point of view than the less knowledgeable. People often express the worry that those who spend all their time studying and so forth spend too far too little time on the basic needs of life as opposed to those who devote all their attentions to eating, drinking, and having sex. More specifically, take for instance the examples of Charles Darwin and Gregor Mendel. It is notorious that Darwin was thoroughly confused over the questions of heredity, whereas the Moravian monk Mendel was (unbeknownst to everyone) working on what proved to be the truly profitable and true principles which govern inheritance. Yet Mendel was a Catholic priest and died childless, whereas Darwin had a typically large Victorian family. Biologically speaking, Darwin was better adapted than Mendel, and yet culturally speaking one would have to go in reverse.

This is a serious objection that should not be minimized. However, one can surely go some way towards answering it (Ruse, 2000a). No one is saying that modern genetics or quantum theory is itself directly adaptive. But, genetics and quantum theory do not just spring out of nowhere. They are rather things which are built or constructed (without saying that they are fictions) out of simpler parts. Why should one accept something like genetics? Simply because Mendelian genetics obeys the basic principles of logic and arithmetic in the light of empirical studies. One starts in genetics with the fairly simple, for instance with Mendel's two laws. Then by fairly elementary mathematics one works to something like the Hardy-Weinberg law. Finally, one adds further complications until one gets the sophisticated theories of today. The Darwinian epistemologist therefore would certainly agree that at some level modern science moves up into culture and may well have no direct adaptive value. (Although, surely one could say that the great success of humans today is a function of the science, so one does not want to say there is no adaptive value whatsoever.) But, what the Darwinian epistemologist would also want to say is that the original premises and reasoning of even the most sophisticated science are grounded in fairly simple rules of thumb, and these can indeed be given a Darwinian backing. So it is just not the case that a literal approach to Darwinism throws no light on our science at all.

There are other points that one might want to raise, but let us conclude this section by just quickly speaking to, if not entirely answering, one concern, namely that at some level one is caught in a circularity. Is one not assuming that there is a real world out there, and that Darwinism in some sense maps it

properly? Then, having done this, one takes the theory and turns it back on itself to look at the way that we think. But surely the way that we think, according to the Darwinian, is something which has already been assumed by the Darwinian! In other words, we are assuming that Darwinism works and turns out true theories, namely the theory of evolution through natural selection. Then, having done this, we take the theory of natural selection, and turn it back on itself to prove that in fact Darwinism works: that when we have the theory of natural selection, it is true inasmuch as it is grounded in Darwinian adaptation. If this is not a circular argument, then we do not know what would be. (This argument has been raised most persistently recently by the Christian fundamentalist philosopher Alvin Plantinga (1991, 1994, 1995). He stresses again and again that one is using Darwinism to turn back on itself to prove Darwinism. This, he triumphantly insists, is a circularity. He believes that there is no such circularity in accepting that everything comes directly from God with His guarantee firmly written on it.)

We would agree that there is indeed a circularity here. What we would deny is that this is a vicious circularity which renders everything worthless. Rather, we think that here one has one of those fairly common epistemological feedback situations: one has a certain amount of knowledge, one uses this, and then this feeds back in some way to give us improved support. Certainly, we agree that at some level one is having to make assumptions about the way that things work, and then one is using these assumptions to look at the very ways that they do work. But we would argue that this is not a fault of ours. It is something inherent in the system. As Otto Neurath used to say, we are all rather in the position of a man afloat on a boat which is leaking. We cannot get to the shore and pull our boat up to repair it. We must stay on the boat and do our best to repair it as we continue to sail. To do otherwise is to sink and die.

More literally, we are humans and we have to do the best with what we have. We have come up with the theory of evolution through natural selection. What we now do is apply this theory. If it works, then we continue with it. If it does not, then in good Popperian fashion we either reject it or we modify it. There is nothing to stop us, and much to encourage us, in applying this theory to ourselves, and this is what the Darwinian epistemologist does. If the picture seems to make a coherent whole, then so far and so good. If it does not, then we must revise. What the Darwinian evolutionist says is that, as best we can make out we do come up with a picture which hangs together, and this at some level is the best that we can do, or ever hope to do. Admittedly, this means that ultimately the truth is going to be a matter of coherence of beliefs rather than correspondence to some absolute reality. But the Darwinian epistemologist would argue that this if anything is a virtue of his or her position, rather than

a failing. Many philosophers, most notably recently Richard Rorty (1979), have argued that the notion of truth as correspondence simply does not work. (Although, as Hilary Putnam (1981) has perceptively noted, we all have a yearning for truth as correspondence.) The best that we can hope for, argue these philosophers, is some kind of coherence theory of truth. The Darwinian not only agrees with this but does his or her part to show exactly how truth as coherence will be functioning. In a sense, therefore, one has a kind of truncated epistemology, but the Darwinian epistemologist argues that this is all that one can have or should ever hope to have.

EVOLUTIONARY ETHICS

We come now to the questions of ethics or morality. What ought one to do? Why ought one to do that which one ought to do? There is a traditional form of evolutionary ethics which goes back even before Darwin, very much taken up and developed and exploited by Herbert Spencer. Somewhat misleadingly this is known as "social Darwinism", although in respects perhaps it is better known as "social Spencerianism" (Richards, 1987; Ruse, 2000a, b; Maienschein & Ruse, 1998). Essentially what is argued here is something very much akin to what is argued in traditional evolutionary epistemology. One has an evolutionary process going on in the real world. One has a struggle for existence between organisms and the ultimate triumph of certain forms which are selected and go on to a brighter evolutionary future. Likewise by analogy in the world of humans, the world of culture, one has a kind of struggle and kind of natural selection going on. Socially, one has different people all striving to get the same thing. It might be individuals fighting over a crust of bread, or firms over customers, or nations over territory, or some such thing. There is a clash and a struggle in a very real sort of sense. What the traditional evolutionary ethicist, the social Darwinian, argues is that not only does this struggle occur, but that it is a natural thing. Far from denying it or wanting to prevent it, we should cherish it and want it to continue. At least, we should not want to stand in its way. Hence, the social Darwinian argues that there is a struggle and that the victory will go to that which is the better or the best.

In socioeconomic terms, one finds that firms struggle for resources, and then for customers and for profits and the like. The social Darwinian argues that this is a right and natural form of human expression that we should cherish, and that we should recognize that in some sense the end result is good. The winner is better or leaner or stronger – in some sense superior – to the loser, and unless we allow this to happen things will go sadly amiss. Fitness therefore is not simply a descriptive term, but an evaluative term. Life is a struggle, and

the fitter not only are the successful but are the better. Listen to Darwin himself who somewhat cagily tried out the heated waters of social Darwinism.

> With savages, the weak in body or mind are soon eliminated; and those that survive commonly exhibit a vigorous state of health. We civilised men, on the other hand, do our utmost to check the process of elimination; we build asylums for the imbecile, the maimed, and the sick; we institute poor-laws; and our medical men exert their utmost skill to save the life of every one to the last moment. There is reason to believe that vaccination has preserved thousands, who from a weak constitution would formerly have succumbed to small-pox. Thus the weak members of civilised societies propagate their kind. No one who has attended to the breeding of domestic animals will doubt that this must be highly injurious to the race of man. It is surprising how soon a want of care, or care wrongly directed, leads to the degeneration of a domestic race; but excepting in the case of man himself, hardly any one is so ignorant as to allow his worst animals to breed (Darwin, 1871, 1, 168).

Later in the century, particularly under the influence of Spencer, social Darwinians became much stronger in their claims. For instance, the early American sociologist William Graham Sumner was a great enthusiast for struggle and the consequent results.

> Let it be understood that we cannot go outside of this alternative: liberty, inequality, survival of the fittest; not-liberty, equality, survival of the unfittest. The former carries society forward and favors all its best members; the latter carries society downwards and favors all its worst members (Sumner, 1914, 25).

SOCIAL DARWINISM CRITICIZED

There are many who would criticize social Darwinism. Indeed, from its earliest years it has been an object of attack and amongst philosophers it has become the philosophical equivalent of a bad smell. It is not simply something which is wrong but is something which is ideologically tainted in some way. Even today, one finds that social Darwinism is generally taken to be the apotheosis of how not to do philosophy in general, and how not to apply science to philosophy in particular. So, let us say right at the beginning that we think that for all its faults (and we believe they are many), social Darwinism has been given something of an unfair press. If indeed you look at the actual writings of traditional social Darwinians, starting with Spencer, you find that far from simply endorsing some kind of crude *laissez-faire* socioeconomic position, their thinking is in many respects subtle and commendable.[4]

Take the basic claims of the social Darwinian about how we should behave. (In other words, take what is generally called by philosophers the *normative ethics* of social Darwinism.) It is true that one often finds expressions or sentiments of enthusiasm for extreme libertarian philosophers, and indeed also for various militaristic views of struggle between nations. All too typically, we

regret to say, Adolph Hitler picked up on some of these ideas, and incorporated them into his Nazi tract, *Mein Kampf* (Bullock, 1991). However, such sentiments are far from universal. Herbert Spencer himself was certainly no militarist. Indeed, he deplored the struggle between nations thinking that this was a barrier to free trade (as indeed of course it is). He thought that if one is to have some kind of unfettered social system, then far better that swords be beaten into plough shares and that people learn to cooperate rather than to compete.

Again, one finds that there have been social Darwinians who have preached socialism or even anarchism. The co-discoverer of natural selection, Alfred Russel Wallace (1900), was always convinced that the way to promote evolution is by working together. In this he was joined by the Russian Prince Peter Kropotkin (1902), who argued that there exists naturally between animals and extending up to humans a form of natural sympathy: what he labeled "mutual aid." If this is allowed to go unfettered, it will lead to a kind of harmony amongst humans rather than competition. It is true that people like Wallace and Kropotkin are not usually called social Darwinians, but their logic was very much in the tradition of Spencer and his more immediate followers.

Jumping right up to the present, one finds that a kind of social Darwinism is endorsed by a great many evolutionists today, most notably by Edward O. Wilson (1984, 1992), the entomologist and sociobiologist. He argues that humans have evolved in symbiotic relationship with nature and unless we recognize this and promote nature – in particular, unless we promote biodiversity – we humans will wither and die. Quite literally, Wilson thinks that a world of plastic would be deadly for humankind. Because of this, Wilson argues strenuously that we ought to preserve such an oasis of biodiversity as the Brazilian rain forest and he has spent much time in recent years working precisely to this sort of end. Admittedly, this is all far from anything dreamed of by Spencer or by Darwin or by any other Victorian, but it is very much in the tradition of conventional evolutionary ethics. Hence, it is just not the case that social Darwinism simply can be faulted because it leads to undesirable policies or norms.

There is however a criticism which is philosophically more deadly. This is one which was made much of by G. E. Moore in 1903 in his *Principia Ethica*, although in fact the form of the criticism goes back to pre-evolutionary times and to David Hume (1978). This is the criticism that the traditional ethicist moves illicitly from claims about matters of fact to claims about matters of obligation. One goes incorrectly from "is" statements to "ought" statements. Hume wrote:

> In every system of morality, which I have hitherto met with, I have always remark'd that the author proceeds for some time in the ordinary way of reasoning, and establishes the being of a god, or makes observations concerning human affairs; when of a sudden I am

> surpriz'd to find, that instead of the usual copulations of propositions, *is*, and *is not*, I meet with no proposition that is not connected with an *ought*, or an *ought not*. This change is imperceptible; but is, however, of the last consequence. For as this *ought*, or *ought not*, expresses some new relation or affirmation, 'tis necessary that it shou'd be observ'd and explain'd; and at the same time that a reason should be given, for what seems altogether inconceivable, how this new relation can be a deduction from others, which are entirely different from it (Hume, 1978, 469).

In similar spirit, Moore argued that traditional evolutionary ethicists commit what he called the "naturalistic fallacy". One is going from descriptions to prescriptions. One is in some sense trying to derive the good from natural properties and that is just illicit.

> It may be true that all things which are good are *also* something else, just as it is true that all things which are yellow produce a certain kind of vibration in the light. And it is a fact, that Ethics aims at discovering what are those other properties belonging to all things which are good. But far too many philosophers have thought that when they named those other properties they were actually defining good; that these properties, in fact, were simply not "other", but absolutely and entirely the same with goodness (Moore, 1903, 10).

The general criticisms of Hume and Moore have been much discussed by philosophers. There are those who claim that Hume did not really mean what he appeared to say, and there are others who argue that Moore's naturalistic fallacy may be naturalistic but it is not genuinely fallacious. Here it would be inappropriate to spend time on the general discussion, so let us say simply that we do ourselves accept that there is something to the criticism. There is a difference between claims about fact and claims about obligation. (As you will see in a future section, we want to make something of this very distinction, turning it back on itself in a way.) However, we would note also that traditionally evolutionary ethicists usually are supremely unconcerned when charges about the naturalistic fallacy are raised. An unconcern which leads one to suspect that it is not simply a question of philosophical insensitivity at work here, but rather that there are deeper concerns and commitments of the traditional evolutionary ethicist. When faced with charges that one should not go from the way that evolution operates to the way that evolution should operate, traditionally evolutionary ethicists respond by agreeing that normally the naturalistic fallacy is indeed fallacious, but by insisting also that evolutionism is an exception. They point to the fact that in science one frequently goes from language of one kind to language of another kind. One goes from talk about molecules buzzing around in containers at high speeds to talk about temperature and pressure. Likewise, they argue that although there is certainly a change from the language of fact to the language of obligation, sometimes – specifically on the one occasion of evolution – this transition is legitimate. Traditional evolutionary ethicists agree that normally one cannot go from fact to obligation, but argue that in the case

of evolution this is permissible. Edward O. Wilson (1994, 1998) therefore continues to insist that, since we have evolved in symbiotic relationship with other organisms, it is right and proper to promote this relationship and this obligation should guide our actions.

Why would the traditional evolutionary ethicist or social Darwinism want to make an exception in the case of evolution? There is a simple and straightforward answer. It is our old friend progress once again. The traditional evolutionary ethicist does not see evolution as a slow, meandering process going nowhere. Rather, he or she sees it as directed, going from the simple to the complex, from the monad to the man. Hence, the traditional evolutionary ethicist believes that, as evolution progresses, some form of value is ever increasing in the process (Ruse, 1998, 2000a). Because of this, it is our moral obligation not to stand in the way of such value development. To the contrary, we should encourage it and perhaps even promote it. Hence, Wilson (who is very strongly explicit in his commitment to progressionism) argues that evolution has brought about things of value, namely humans, and that if humans go extinct things of value will be lost. Therefore it is our moral obligation to promote the well-being of humans, not just now but for the future, and the way in which we can and must do this is by promoting biodiversity. Without biodiversity, all is lost.

The position now makes sense, but is it true? Here, let us simply say that we are not convinced that it is. We are far from certain that the course of evolution is progressive. It is true that evolution has led from blobs to humans, and we are the first to assert that we value humans over blobs. But we value humans over blobs because we are human, and because we read into evolution our values, namely that humans are superior to blobs. It is not something that we read simply from the fact of evolution. It is part of our interpretation. Inasmuch as we understand evolution as progressive, we are doing this because we have read in the progress. To conclude therefore that evolution is progressive, and that hence we have a basis for value, would indeed be viciously circular. The question is not whether one can understand evolution as progressive, but whether evolution, particularly evolution from a Darwinian perspective, forces upon us a belief in progress. We are not convinced that this is so. A theory which has as its center the notion of the survival of the fittest surely does not necessarily favour humans over others. It is true that at the moment we humans are doing well; but a good case could be made for saying that certain forms of parasite, for instance the AIDS virus, are if anything doing even better. Certainly, in terms of evolutionary success, the AIDS virus is far superior to the great apes, but on any value scale would one want to put the AIDS virus above the chimpanzee? We rather think not.

We should say that in this criticism we are far from alone, or indeed the first. To his great credit, Thomas Henry Huxley was one who spotted this hidden

substructure to traditional evolutionary ethics. Against his old friend Herbert Spencer, Huxley insisted in his last and greatest essay, *Evolution and Ethics*, that far from thinking that the products of evolution are to be valued, often it is our moral obligation to fight against them.

> For his successful progress, throughout the savage state, man has been largely indebted to those qualitites which he shares with the ape and the tiger; his exceptional physical organization; his cunning his sociability, his curiosity, and his imitativeness; his ruthless and ferocious destructiveness when his anger is roused by opposition.
>
> But, in proportion as men have passed from anarchy to social organization, and in proportion as civilization has grown in worth, these deeply ingrained serviceable qualitites have become defects (Huxley & Huxley, 1947, 63-4).

Here we applaud Huxley, and conclude that although normatively there may be much to be said for traditional evolutionary ethics, at the level of foundations or what is known traditionally as *metaethics*, the edifice is built on sand.

DARWINIAN ETHICS

As in epistemology, we ourselves offer an alternative – a literal Darwinian position – as opposed to the kind of metaphorical or analogical position that we have just been considering (Ruse, 1986, 1994, 1998). Humans are the products of natural selection, and above all humans are social beings. In order to be successful social beings, we must therefore have adaptations which enable us to engage socially. We have to have various adaptations to deal with parasites, as was shown by the unhappy experiences of isolated populations when first exposed to western missionaries. More pertinently for our discussion, we have to have certain kinds of ways of getting on together. These ways of getting on together will involve what biologists call "altruism", although note that this is a metaphorical use of the term rather than necessarily a literal sense. Biological altruism is some kind of engagement between organisms which ultimately pays off for the individual.[5] The best case of biological "altruism" that we know of occurs between the social insects. There we find sterile workers spending all their lives devoting themselves to the well-being of the nest and to the future success of their fertile sisters.

As William Hamilton pointed out in the early 1960s, the sterile workers behave in this "altruistic" way because ultimately it is of reproductive benefit to themselves (Hamilton, 1964a, b). Because of the peculiar breeding system of the hymenoptera, to which most of the social insects belong, females are more closely related to sisters than they are to offspring. (In the hymenoptera, females have both a mother and father, whereas males have only mothers. This means that sisters have a 75% relationship, whereas mothers and offspring have

only 50%.) Hence, to use Richard Dawkins's (1976) metaphor, the "selfish genes" of the hymenoptera dictate that they are benefitting themselves better by becoming non-reproductive and by raising sisters, than being reproductive and raising daughters. This is biological "altruism," although in no sense would one want to say any literal altruism is involved. No one is saying that the hymenoptera are behaving as Mother Theresa, and nobly sacrificing their own good for the benefits of others.

What we argue as Darwinians is that humans are highly "altruistic." What we argue also as Darwinians is that the "altruism" of the ants would not do in the case of humans. They are innately programmed to behave as they do; no learning is required. Such programming works just fine so long as nothing goes wrong; but if something untoward does happen, then problems arise and it may well be that wholesale slaughter will fall on members of the hymenoptera nest. For the hymenoptera this does not matter too much. A queen ant for instance has literally millions of offspring. She can afford to lose a few hundred thousand, if something goes wrong. Humans, on the other hand, have adopted the evolutionary strategy of having just a few offspring but looking after them carefully. We cannot afford to lose offspring every time something goes wrong. Hence, the way of being innately "altruistic" is not for us.

What we argue as Darwinians is that, as in epistemology, natural selection has set up in us certain innate dispositions. In the social realm, these motivate us to act in reciprocity or to be "altruistic" with our fellow humans. When we come to consciousness and as we develop ways of thinking within our culture, it does not mean that we will be blindly programmed always to act in such an "altruistic" sort of way. It does mean that we will be innately disposed to think that we *should* behave in such a way, even though this is not say that we necessarily always *will* obey the conscious calling of our innate dispositions. We argue therefore that natural selection has given to humans certain innate dispositions, not simply to interact but to think that we ought to interact with others. In other words, what we argue is that in order to make us biological "altruists" natural selection has made us literal or genuine altruists.[6]

You might question why it is necessary that we be genuine altruists. For biological "altruism", would it not be enough that we all be simply self-interested calculators? Of course, much of the time we are just this. When we go out and make purchases, the exchange of goods for money is not something which is done out of benevolence. It comes ultimately out of self interest. This self interest does not surprise us as Darwinians. If we were not innately self serving or selfish, we would not function as successful beings. If every time food was available we simply gave it away, or every time a prospective partner or sexual opportunity arose we simply ignored it, we would soon fail in the

struggle for survival and reproduction. Necessarily, we must be self regarding or selfish in some sense. However, as social beings we need something to break through this to enable us to work "altruistically" with others.

What we argue is that natural selection has given us a sense of moral altruism to do this. Thus, suppose I am faced with a situation where I have a certain amount of food and a companion who has no such food. Should I eat it all myself or should I share it? Naturally, my inclination is to hoard it and look after myself. However, at the same time I feel kicking in a sense of moral obligation that I ought to share. We do not say that a person always will share but we do say that this sense of obligation will bring people to share on many occasions. And the point we would make is that this is of selective advantage to the individual. The person who shares is much more likely to get something given back in turn, when he or she is without. The person who refuses to share is probably the person who is going to go hungry on other occasions and therefore overall to suffer. What we argue, therefore, is that altruism is something which is brought about by natural selection. In other words, we argue that humans are made into moral beings simply because of the dictates of evolutionary biology.

DARWINIAN ETHICS CONSIDERED

Interestingly, despite a general reluctance to embrace traditional social Darwinism, an increasing number of professional ethicists responds favourably to the kind of argumentation that we have been giving in the last section (Mackie, 1978; Murphy, 1982). In fact, now often using the methods and techniques of game theory, considerable effort is being put into the devising of theories which will match and explain particular aspects or idiosyncracies of human moral behaviour (Skyrms, 1998; Frank, 1988, 1999). Yet, at the same time, many (including those who react favourably to the biology) will argue that such a position as we have sketched can offer only a truncated picture of ethics. Supposedly, it fails to address the truly basic questions of ethical inquiry.

Critics allow that a Darwinian approach such as we have adopted can explain why humans behave in the way that they do and can perhaps go as far as to explain why humans have the kind of normative sentiments that they do. It is agreed that humans come to accept that one should promote people as ends or the general happiness or some such thing, because natural selection has worked upon these humans. It is agreed therefore that selection does not necessarily promote just pure Darwinian greediness, but can rather speak to questions of cooperation and fairness, and justice and so forth. However, critics argue that, when it comes to questions about justification – not why we behave as we do,

or why we think we should be good, but why we really ought to be good – such a naturalistic approach as we have taken fails entirely (Rawls, 1980; Woolcock, 1993). To get to the question of justification for morality, the metaethical question, one must invoke other arguments. Or else rest with a gap or with a gap jumped illicitly through committing the naturalistic fallacy.

However, while we are sympathetic to these kinds of objections we are not convinced that they are well taken.[7] We think indeed that perhaps a Darwinian approach such as we have sketched can speak to the ultimate questions of morality. Or more particularly, our approach can speak to the ultimate questions of morality by showing that these questions ultimately are unanswerable! (Note the parallel here with our embracing coherentism rather than hardline correspondence, when dealing with matters of epistemology.) There is certainly a tradition in ethics which argues that, at some level, morality is simply a matter of emotions or sentiments and there are no foundations beyond these (Hudson 1983). This basically is the position that we are taking. What we would argue is that once you have given a Darwinian account, you see that there is no underlying basis to morality (Ruse & Wilson, 1985, 1986). Morality is nothing more, although we would add nothing, less than a set of emotions put into place by our biology in order to make us good cooperators, or (to use the language of biologists) good "altruists". In this sense therefore we are emotivists or, as supporters of our position are sometimes labeled, ethical skeptics. (Note that the skepticism is not about morality or normative morality as such, but about the foundations. We are skeptical about metaethical justification.)

However, we are more than just traditional emotivists. Many find emotivism unacceptable because it simply does not seem to be the case that moral claims (like "rape is wrong") are simply expressions of emotion. When one says "rape is wrong" one means that rape *really truly* is wrong, and that this is something binding not only on one individual but on all individuals. Emotions seem not to have that kind of status. If one says "I don't like spinach", another might respond that one ought to eat the spinach nevertheless because it is good for one. But no one would say that the emotion as such is wrong, nor would anyone want to say that the emotion (or its opposite) should be generalized. That is, that everybody ought to like spinach or some such thing. One might say it would be more useful if everybody did like spinach, and one might try to change people's minds. But ultimately, there is no moral justification here. Only a pragmatic justification or some such thing like that.

We see the difference at issue here as being a reflection of the is/ought dichotomy of David Hume. To say "I don't like spinach" is a claim of fact; whereas to say "I ought not rape" is a claim of morality or obligation. What we argue, however, is that in order to make us good cooperators or biological

"altruists", our evolutionary biology has filled us with sentiments not just of emotion, but of obligation. This means that although our moral claims are based on emotion, they have to be emotions of a particular kind: emotions which involve obligation. Since, in order to make us believe in obligation, it is necessary to make us feel that there is in fact some universal standard or criterion above and beyond our emotions, what we also argue that although morality is just emotion – it is subjective rather than objective – in order to make morality work it was important for our biology to make us think that morality is more than just emotion.

In other words, it is an important part of our biology that we "objectify" our emotions when it comes to morality (Mackie, 1977, 1979). When we say "rape is wrong", this is in fact just a sentiment. However, its meaning is more than just a sentiment. Its meaning is of something which has some sort of objective referent. Thus, having accepted Hume's is/ought dichotomy, what we argue is that, in order to make us good cooperators, biology has given us morality. But morality is of a form different from simple sentiments, although ultimately morality is based on nothing more than sentiments. If you like to put it provocatively, morality is an illusion put in place by our genes in order to make us good cooperators. (Note, however, the illusion is not about believing in morality. The illusion is that morality has some kind of objective or universal referent.)[8]

What about the objection that even if moral sentiments are emotions – What else could they be? – this is not to say that that is all there is to morality. What if one objects that there will be some kind of objective foundation, which the evolutionist necessarily misses? After all, we sense the railway engine bearing down upon us through organs of evolutionary adaptation. This is not to deny the reality of the engine. Why should it not be the case that, in some way we sense or intuit morality, but this is not to deny that there is some ultimate reality that the Darwinian ethicist seems to be denying? (See Nozick (1981) for such a point.)

Ultimately, we admit that one cannot deny the possible existence of a morality out and beyond us, any more than the Darwinian epistemologist can deny the existence of a real world out and beyond us – a real world that our coherence theory does not capture fully, if at all. However, what one can say is that, as in the case of the real world, the objective morality no longer seems quite as relevant and pressing as many suppose. As social beings, we have need of something which gets us beyond our biological selfishness, and the way in which we do get beyond this selfishness is through our moral sentiments. These we use and need, whatever ultimate reality might be. It seems to us quite possible that ultimate reality (with respect to morality) could be something completely different from anything that we believe in, and yet we would function as we do quite happily

irrespective. Suppose, for instance, ultimate reality in the moral sense demands that we hate each other or some thing as that. We would still have evolved as we have, as social beings thinking that we ought to cooperate and to like each other and get along, whatever those ultimate dictates might be. In order to argue that the ultimate dictates are going to have some effect, the critic of the Darwinian position must show in some way how these ultimate dictates are going to impinge on us as social beings. This has never been done, and quite frankly we are not at all sure how one might set about doing this. So, we conclude: even if there be an ultimate reality with ultimate moral standards these are irrelevant to us as the products of Darwinian evolution. (Somewhat more strongly, we would probably conclude that an ultimate moral reality that has no effect on us humans is a contradiction in terms. So truly, we are atheists about ultimate moral reality, not agnostics.)

CONCLUSION

These then are some of the approaches which have been taken by philosophers and philosophically minded biologists towards the major problems of philosophy. We cannot pretend that today an evolutionary approach to philosophy, even a Darwinian approach, is one which finds universal or even majority favour amongst professional philosophers. What we would ask is that our readers judge our ideas and arguments on their merits rather than from a democratic head count. We ourselves would re-emphasize that the most important thing we know about human beings is that we are the end products of a long, slow process of evolution through natural selection. We believe that this has to be relevant, even to the most important questions of epistemology and ethics. We fully admit that we may not have the details right, but we are convinced that we are taking the right approach. We hope that having read what we had to say, you will feel some sympathy and inclination to join with us on our quest.

NOTES

1. It might be argued that the relationship between biological evolution and cultural evolution, the latter as represented by the development of science, is no analogy but a literal extension of evolution from one domain to another – just as the evolution of humans although undoubtedly different in respects from the evolution of other organisms is no analogy but a literal extension from one area or domain to another. Indeed, one of us (Lachapelle, see references) prefers this reading whereas the other (Ruse) prefers to stay with the notion of analogy. This is an important distinction which should be discussed. But not here. It is enough to note this difference and to recognize that for the points we would make in this discussion either understanding is legitimate.

2. Chomsky notoriously is no friend of Darwinism. People who have taken his or related ideas and are more sympathetic to our position include Pinker (1994, 1997), Cosmides (1989), and Lumsden and Wilson (1981, 1983).

3. The question of innate mathematical knowledge is a hot area at the moment, and as philosophical naturalists who think that their thought must conform to and be guided by empirical science, we appreciate that new discoveries might clarify the extent to which mathematical understanding is part of our biological heritage. On these points see Butterworth (1999) and Devlin (2000).

4. Of course, if you simply define social Darwinism as something which implies struggle and the superiority of capitalism and of Protestant white males, then clearly it is an ethically dubious position. But if (as we do here) you take it in a generic sense of something using evolution as a basis or guide to morality, then evaluations must be more nuanced, and this holds even if – especially if – you confine discussion to those traditionally taken to be social Darwinians (like Herbert Spencer).

5. "Altruism" is one of those terms which is used in a somewhat loose way in the literature and where one can finds quite different understandings from person to person. Wilson (1975), for instance, understands (biological) "altruism" as involving sacrifice by the altruist towards (or for the benefit of) others. This is fine as far as it goes, but the point is that evolutionary biologists today would always assume that this sacrifice is for the benefit or end of a greater biological good for the altruist. No one does something for nothing. We are simply making this point explicit in our usage. See Ruse (2001) for more details.

6. We do not imply that human sociality is simply a function of biological "altruism," which is achieved only by literal altruism. Clearly being selfish and devious can serve our social ends as well. Our point rather is that "altruism" brought about by sociality is (an important) part of the packet. On the dark side to human nature, "original sin", see Ruse, 2000a.

7. We are sympathetic to the qualification that on its own Darwinian ethics does not speak directly to such issues as the morality of abortion or the evil of capital punishment. No high level ethical theory does this, on its own. It is always necessary to add (often empirical) specific sub-premises to get specific conclusions. I should love my neighbour as myself, but should I stick a knife into this person's chest? Yes, if I am a heart surgeon about to replace a valve. No, if I am a robber about to take a wallet.

8. Honesty forces us to admit that one of us (Lachapelle) is considerably less enthused by the use of the word "illusion" in this context than is the other (Ruse). The more timid (or cautious) member of our pair agrees that there is no objective referent, but points out that there is no objective referent for the blue of the sky and yet we do not speak of "illusion" in such a case. The more adventuresome (or reckless) member of our pair stresses the fact that we are led by our biology to think of morality (unlike other emotions) as objective even though it is not, and hence welcomes the use of "illusion" in this case. Those of our readers who tend towards timidity are invited to use a different term in this context.

REFERENCES

Bradie, M. (1986). Assessing evolutionary epistemology. *Biology and Philosophy*, *1*, 401–460.

Bullock, A. (1991). *Hitler and Stalin: Parallel Lives*. London: HarperCollins.

Butterworth, B. (1999). *What Counts: How Every Brain is Hardwired for Math*. New York: Simon and Schuster.

Chomsky, N. (1957). *Syntactic Structures*. The Hague: Mouton.

Cosmides, L. (1989). The logic of social exchange: has natural selection shaped how humans reason? Studies with the Wason selection task. *Cognition, 31*, 187–276.

Darwin, C. (1859). *On the Origin of Species*. London: John Murray.

Darwin, E. (1789). *The Botanic Garden (Part II, The Loves of the Plants)*. London: J. Johnson.

Darwin, E. (1791). *The Botanic Garden (Part I, The Economy of Vegetation)*. London: J. Johnson.

Darwin, E. (1794–1796). *Zoonomia; or, The Laws of Organic Life*. London: J. Johnson.

Darwin, E. (1803). *The Temple of Nature*. London: J. Johnson.

Dawkins, R. (1976). *The Selfish Gene*. Oxford: Oxford University Press.

Devlin, K. J. (2000). *The Math Gene: Why Everybody Has It, But Most People Don't Use It*. New York: Basic Books.

Frank, R. (1988). *Passions Within Reason*. New York: Norton.

Frank, R. (1999). *Cooperation through emotional commitment*. Unpublished.

Hamilton, W. D. (1964a). The genetical evolution of social behaviour I. *Journal of Theoretical Biology, 7*, 1–16.

Hamilton, W. D. (1964b). The genetical evolution of social behaviour II. *Journal of Theoretical Biology, 7*, 17–32.

Hudson, W. D. (1983). *Modern Moral Philosophy*, 2 ed. London: Macmillan.

Hume, D. (1978). *A Treatise of Human Nature*. Oxford : Oxford University Press.

Huxley, T. H. (1893). The Coming of Age of 'The Origin of Species'. In: *Darwiniana: Collected Essays* (pp. 227–243). London: Macmillan.

Huxley, T. H., & Huxley J. S. (1947). *Evolution and Ethics 1893–1943*. London: Pilot.

Kropotkin, P. ([1902]1955). *Mutual Aid*, (Ed., A. Montague). Boston: Extending Horizons Books.

Kuhn, T. (1962). *The Structure of Scientific Revolutions*. Chicago: University of Chicago Press.

Lachapelle, J. (1999). Culture and evolutionary explanations. Unpublished Ph.D. thesis. University of Guelph.

Lachapelle, J. (2000). Cultural Evolution, reductionism in the social sciences, and explanatory pluralism. *Philosophy of the Social Sciences, 30*(3), 331–361.

Lamarck, J. B. (1809). *Philosophie zoologique*. Paris: Dentu.

Lorenz, K. ([1941]1982). Kant's Lehre vom a priorischen im Lichte geganwartiger Biologie. '*Blatter fur Deutsche Philosophie', 15*, 94–125. Translated and reprinted as: Kant's doctrine of the 'a priori' in the light of comtemporary biology. *Learning, Development, and Culture; Essays in Evolutionary Epistemology* (pp. 121–143) Ed., H. C. Plotkin. Chichester: Wiley.

Lumsden, C. J., & Wilson, E. O. (1981). *Genes, Mind, and Culture*. Cambridge, Mass.: Harvard University Press.

Lumsden, C. J., & Wilson, E. O. (1983). *Promethean Fire: Reflections on the Origin of Mind*. Cambridge, Mass.: Harvard University Press.

Mackie, J. (1977). *Ethics*. Harmondsworth, Mddx.: Penguin.

Mackie, J. (1978). The law of the jungle. *Philosophy, 53*, 553–573.

Mackie, J. (1979). *Hume's Moral Theory*. London: Routledge and Kegan Paul.

Maienschein, J., & Ruse, M. (Eds) (1998). *Biology and Ethics*. Cambridge: Cambridge University Press.

Maynard Smith, J., & Szathmary, E. (1995). *The Major Transitions in Evolution*. New York: Oxford University Press.

Mithen, S. (1996). *The Prehistory of the Mind*. London: Thames and Hudson.

Moore, G. E. (1903). *Principia Ethica*. Cambridge : Cambridge University Press.

Murphy, J. (1982). *Evolution, Morality, and the Meaning of Life*. Totowa, N.J.: Rowman and Littlefield.

Nozick, R. (1981). *Philosophical Explanations*. Cambridge, Mass.: Harvard University Press.

Pinker, S. (1994). *The Language Instinct: How the Mind Creates Language*. New York: William Morrow.

Pinker, S. (1997). *How the Mind Works*. New York: Norton.

Plantinga, A. (1991). An evolutionary argument against naturalism. *Logos*, *12*, 27–49.

Plantinga, A. (1994). *Naturalism defeated*. Unpublished Manuscript.

Plantinga, A. (1995). Methodological naturalism. In: Van der Meer (Ed.), *Facets of Faith and Science*. Lanham, Md.: University Press of America.

Polkinghorne, J. (1991). *Reason and Reality: The Relationship between Science and Theology*. Philadelphia: Trinity Press.

Popper, K. R. (1959). *The Logic of Scientific Discovery*. London: Hutchinson.

Popper, K. R. (1972). *Objective Knowledge*. Oxford : Oxford University Press.

Popper, K. R. (1974). Darwinism as a metaphysical research programme. In: P. A. Schilpp (Ed.), *The Philosophy of Karl Popper*. Vol. 1. (pp. 133–143). LaSalle, Ill.: Open Court.

Putnam, H. (1981). *Reason, Truth, and History*. Cambridge: Cambridge University Press.

Quine, W. V. O. (1969). *Ontological Relativity and Other Essays*. New York: Columbia University Press.

Rawls, J. (1980). Kantian constructivism in moral theory. *Journal of Philosophy*, *77*, 515–572.

Richards, R. J. (1987). *Darwin and the Emergence of Evolutionary Theories of Mind and Behaviour*. Chicago, Ill.: University of Chicago Press.

Richards, R. J. (1992). *The Meaning of Evolution: The Morphological Construction and Ideological Reconstruction of Darwin's Theory*. Chicago: University of Chicago Press.

Rorty, R. (1979). *Philosophy and the Mirror of Nature*. Princeton: Princeton University Press.

Ruse, M. (1986). Evolutionary ethics: a phoenix arisen. *Zygon*, *21*, 95–112.

Ruse, M. (1989). *The Darwinian Paradigm: Essays on its History, Philosophy and Religious Implications*. London: Routledge.

Ruse, M. (1994). *Evolutionary Naturalism: Selected Essays*. London: Routledge.

Ruse, M. (1996). *Monad to Man: The Concept of Progress in Evolutionary Biology*. Cambridge, Mass.: Harvard University Press.

Ruse, M. (1998). *Taking Darwin Seriously: A Naturalistic Approach to Philosophy*. Second ed. Buffalo, N.Y.: Prometheus.

Ruse, M. (1999a). *The Darwinian Revolution: Science Red in Tooth and Claw*, 2nd ed. Chicago: University of Chicago Press.

Ruse, M. (1999b). *Mystery of Mysteries: Is Evolution a Social Construction?* Cambridge, Mass.: Harvard University Press.

Ruse, M. (2000a). *Can a Darwinian be a Christian? The Relationship between Science and Religion*. Cambridge: Cambridge University Press.

Ruse, M. (2000b). *The Evolution Wars*. Santa Barbara, California: ABC-CLIO.

Ruse, M. (2001). Altruism: A Darwinian naturalist's perspective. In: J. Schloss (Ed.), *Altruism*. New York: Oxford University Press.

Ruse, M., & Wilson, E. O. (1985). The evolution of morality. *New Scientist*, *1478*, 108–128.

Ruse, M., & Wilson, E. O. (1986). Moral philosophy as applied science. *Philosophy*, *61*, 173–192.

Skyrms, B. (1998). *Evolution of the Social Contract*. Cambridge: Cambridge University Press.

Sober, E. (1981). The evolution of rationality. *Synthese*, *46*, 95–120.

Toulmin, S. (1967). The evolutionary development of science. *American Scientist*, *57*, 456–471.

Van Fraassen, B. (1980). *The Scientific Image*. Oxford: Oxford University Press.

Wallace, A. R. (1900). *Studies: Scientific and Social*. London: Macmillan.

Whewell, W. (1840). *The Philosophy of the Inductive Sciences*. London: Parker.

Wilson, E. O. (1975). *Sociobiology: The New Synthesis*. Cambridge, Mass.: Harvard University Press.

Wilson, E. O. (1978). *On Human Nature*. Cambridge, Mass. Cambridge University Press.

Wilson, E. O. (1984). *Biophilia*. Cambridge, Mass.: Harvard University Press.

Wilson, E. O. (1992). *The Diversity of Life*. Cambridge, Mass.: Harvard University Press.

Wilson, E. O. (1994). *Naturalist*. Washington, D.C.: Island Books/Shearwater Books.

Wilson, E. O. (1998). *Consilience*. New York: Vintage Books.

Woolcock, P. (1993). Ruse's Darwinian meta-ethics: a critique. *Biology and Philosophy*, *8*, 423–439.

EVOLUTIONARY THEORY AND POLITICAL THOUGHT

Joseph Losco

INTRODUCTION

> Today, in the thinking of citizens and social scientists alike, the deeply held assumption is that culture has severed for good the linkage between human behavior and biology. The conviction is that human beings in their social behavior, alone among animals, have succeeded in escaping biology. The irony is heavy here. For that belief is accompanied by another deeply held conviction that human beings, like all other living things, are products of the evolution that Charles Darwin explained with his theory of natural selection. The irony is almost palpable inasmuch as Darwin entertained no doubt that behavior was as integral a part of human evolution as bodily shape.
>
> Carl Degler (1991, p. viii)

For over a third of a century now, a core group of political scientists has been promoting the importance of the life sciences for their discipline. Of course, the use of biological concepts and approaches goes back much further. Aristotle applied some of the same techniques in his political queries as he did in his study of nature. In modern times, Morley Roberts (1938) is credited with reawakening interest among students of politics in the life sciences and with coining the term "bio-politics." Current sustained interest, however, is traceable to the work of Albert Somit who noted that important findings employed in ethology and pharmacology might very well allow us "to account for political phenomena for which we as yet have no satisfactory explanation" (1968: 550). Shortly thereafter, Thomas Landon Thorson (1970) asserted that Darwinian

Evolutionary Approaches in the Behavioral Sciences, Volume 8, pages 161–180.

ISBN: 0-7623-0769-2

biology, not physics, was the appropriate paradigm for the study of politics because biology, unlike physics, takes the concepts of time and change seriously. "Whatever deserves the name of political science must start with a conception of the human animal in nature and must conclude with an account which will deal with the whole phenomenon of politics, the choosing and creating, as well as the reacting and responding" (p. 178). Since Thorson, the march of followers looking to the life sciences to fulfill the promise of a science of politics has been steady, if not much noticed.

Among disciplinary specialists, political theorists have long shown the most interest in evolutionary approaches to the study of politics (Losco & Baird, 1982; Losco, 1996). This is not surprising given the Aristotelian roots of naturalistic studies of politics and the propensity of theorists to be drawn to epistemological issues. In a survey conducted for this paper,[1] political scientists closest to the life sciences movement overwhelming selected the works of political theorists or theoretical works in political science as having the greatest impact in broadening the appeal of evolutionary concepts for the study of politics. While based on a rather small N, the results are suggestive of the views of political scientists closest to developments in the life sciences. When asked to name one seminal work by a political scientist that has had the greatest impact on mainstream political science, three works were named most often: Roger Masters' *The Nature of Politics*, Robert Axelrod's *The Evolution of Cooperation*, and James Q. Wilson's *The Moral Sense*. It is perhaps no accident that these three volumes represent the sub-field of political theory, broadly construed.

Clearly members of the politics and life sciences community believe an evolutionary perspective is important. 100% of respondents categorized as important (17%) or very important (83%) the use of evolutionary concepts in political science. Yet, only one respondent believed evolutionary concepts have had *any* impact on mainstream political science to date. 70% believe it has had

---

1. In early 2000, a questionnaire was sent to members of the Association for Politics and the Life Sciences identified by department affiliation as political scientists. This Association, begun in the late 1970s, is the pre-eminent professional association for those interested in life science approaches to the study of politics and it publishes an internationally-circulated semi-annual journal. Sixty three questionnaires were mailed. While there may be more than 63 political scientists out of the total of approximately 350 Association members, questionnaires were sent only to those who clearly designated themselves in the official Association Listing as political scientists through department affiliation. Twenty-four completed responses were returned for a response rate of 38%. While the results cannot be considered definitive, they are suggestive of the ways those most active and interested in life science approaches perceive the current state of the political science discipline and the impact the life sciences may have upon it.

very little impact; 13% believe it has had no impact at all. Three respondents (13%) were unable to tell if an impact had occurred. Despite what most of these committed individuals saw as virtually no progress, 48% nevertheless continue to express the belief that it is likely that evolutionary concepts and methods will become a central feature of the political science discipline in the future. 22% believe this prospect is unlikely or not very likely at all. The remainder remain neutral about these prospects. As to where in the discipline change is likely to occur, a consensus seems to have emerged on the most likely impacted areas. 86% believe the study of political psychology and behavior will benefit most from incorporation of evolutionary perspectives while 74% believe philosophy and ethics will so benefit. (Respondents were allowed to select more than one listed sub-field.) Other beneficiaries in descending order were international politics and warfare studies (52%), gender and ethnic studies (48%), public policy (35%), empirical theory (30%), law (26%), and institutional analysis (26%).

While these results are too sparse for wide generalization, a few features of the current landscape seem clear. Even those who would appear to be the most knowledgeable and ardent supporters of the use of evolutionary ideas for the study of politics believe it has had very little impact in the heart of the discipline. Even so, they remain optimistic that their labor will bear fruit. Finally, consistent with the earlier studies mentioned above, there appears to be consensus that one of the most likely areas of growth will continue to be political theory and philosophy.

The remainder of this chapter is an attempt to isolate and explore areas of consensus that are beginning to appear among authors utilizing evolutionary ideas in political theory. Space does not permit an exhaustive review of the literature which has become quite voluminous. Still, much of the work has been contributed by a core of prodigious authors whose work will be represented and whose names appeared prominently in survey results. We will concentrate here on common themes among political theorists employing an evolutionary perspective that are crystalizing at the dawn of the new millennium.

THE NATURALISTIC FALLACY AND MORAL SENSIBILITY

Theorists who advance a portrait of the human animal as one exhibiting a natural moral sensibility are often reminded of the naturalistic fallacy (Hume, 1740; Moore, 1903). But more and more, advocates of a biologically-based view of human nature are insisting that the formal limits of the fallacy do not preclude a naturalistic ethic, one based on some understanding of natural desires and emotions that evolved to enhance reproductive fitness.

Most forcefully presented by Hume, the naturalistic fallacy holds that attempts to derive "ought" statements from "is" statements are untenable. Purely descriptive premises cannot yield prescriptive conclusions. Attempts to bridge this gap with the help of evolutionary biology, modern critics like Leslie Paul Thiele assert, are justifiably doomed to philosophic marginality on grounds of formal logic and for ideological reasons as well. As Thiele reads Hume, "Regardless of the facts at hand . . . the unaided vehicle of reason is incapable of delivering us to the moral realm. To arrive at this destination, Hume insists, reason must be fired up with the fuel of sentiment or passion" (1999: 7). In short, naturalistic observations represent one type of statement (empirical) while ethical statements represent a different and incommensurable type (prescriptive).

Yet, beyond the formalistic logic that informs this fallacy, the field of ethics is replete with examples of philosophers who assemble factual evidentiary material to justify moral positions; not the least of these is Hume himself. James Q. Wilson (1993) notes this strange occurrence in Hume's treatment of justice in which Hume grounds this convention (conceived as a human invention to make man more secure in his property) on our "natural duty" to take care of one's children. "Here Hume derives an 'ought' statement from an 'is' statement scarcely eight pages after asserting that this cannot be done," recounts Wilson (p. 238). He continues:

> Lest his readers think this was a careless slip of the pen, he repeats his conclusions in many other places. Benevolence – that is, being "humane, merciful, friendly, generous" – "universally express[es] the highest merit." These sentiments are "natural," and even though they are fainter than self-love, "it is impossible for such a creature as man to be totally indifferent to the well or ill-being of his fellow creatures, and not readily, of himself, to pronounce, where nothing gives him any particular bias, that what promotes their happiness is good, what tends to their misery is evil." Moral sentiments are "so rooted in our constitution and temper, that without confounding the human mind by disease or madness, 'tis impossible to extirpate and destroy them" (p. 238).

Roger Masters, too, questions the absolute division between fact and value that characterizes the is-ought split. Arguing that biology can help us understand the concept of human flourishing much as the study of medicine leads to an appreciation of health, Masters (1989) asserts:

> I am, of course, aware of the so-called naturalistic fallacy which has been so often condemned by logicians and methodologists. But when the doctor prescribes a treatment, we don't normally object that this practice bridges the logical distinction between the facts of diagnosis and the values of health. Biology, like medicine, has much to teach us about our species and the risks we confront by ignoring the natural basis and consequences of our habits of life (p. xv). [See also, Chisholm, 1999: 23 for a similar perspective from a student of biology.]

Larry Arnhart (1998) has explicitly postulated a naturalistic ethic based on evolved natural desires. He identifies twenty natural desires that are universal

to all human societies. These include the desire for a complete life, parental care, sexual identity, sexual mating, bonding, friendship, social ranking, justice as reciprocity, political rule, war, health, beauty, wealth, speech, practical habitation, practical reasoning, practical arts, aesthetic pleasure, religious understanding, and intellectual understanding. Like Aristotle, upon whose work he builds, Arnhart postulates that these desires are rooted in our biological heritage and that fulfillment of these desires contributes to a good and complete life. As a result, these natural desires *ought* freely to be expressed and *ought* to be allowed full expression in society. Of course, innumerable historic and social factors may contribute to the emergence of social customs or laws that frustrate these desires. While prudence dictates caution in challenging customs that short-circuit natural desires, a naturalistic ethic nevertheless gives us firm grounds for rejecting such customs as inconsistent with natural human aspirations, grounds which relativistic and nihilistic alternatives do not supply,

Elsewhere, Arnhart (1995) uses his model to indicate how particular *oughts* can be derived from the natural desires uncovered by evolutionary biology. Consistent with Aristotle's pre-evolutionary naturalism, Arnhart believes that modern evolutionary naturalism, while respecting some differences between the sexes in natural propensities, does not justify the claim of one sex to moral or intellectual superiority (1995: 149) In a discussion of feminism and primatology, Arnhart argues that a proper understanding of the natural evolved desires of women can yield ethical conclusions about such practices as female circumcision that are not merely based on culture. He argues such practices are simply *wrong* since they frustrate the natural desires and aims of women even if they were created by men exercising their own "natural" tendency for domination:

> In judging a customary practice like female circumcision, we must understand the intricate interaction of nature and history. We must consider the particular historic conditions that would support such a custom and make it difficult to change. Yet, we must also consider the universal nature of men and women – the natural repertoire of needs and powers – that might either be expressed or frustrated by such a custom. Perhaps the natural tendency of males to want to control the sexuality of women to secure their paternal investment explains why some men would adopt such a practice. Yet the fact that it interferes so severely with the natural needs of women indicates that the human costs are so high for both men and women that its putative benefits do not justify it (1995: 155–156).

The late Richard Shelly Hartigan (1995) argues that while it may be true (à la Hume) that moral statements are grounded in convention and custom while facts are the product of reason and sense perception (p. 93), it is also probable that human custom has developed at least partly in response to the ways in which the human animal attempted to cope with his physical and social environment over evolutionary time. The tradition of "natural law," which he

argues predates its Medieval forefathers in one form or another and informs the very root of Western philosophy, was one attempt to supply universal, if not absolute, rules of guidance based on natural human propensities. The Decalogue was yet another: "Generally, the Decalogue, minus the first three commandments dealing with God, constituted the most evident principles of the natural law. These 'social commandments' – how men should live together – were revealed in history by God as a kind of reinforcement which, however, natural reason could attain by itself" (1995: 92). The Ten Commandments, Hartigan observes, reflect the human desire for order and suggest proven means for navigating a course toward our 'natural' ends. He posits the Ten Commandments state "clearly the 'do's' and 'don'ts' which, when followed, have assured human cooperation, stability, and predictability of behavior – the basic ingredients for survival and propagation" (1988: 167).

> Filtered though various ancient and modern translations, and reduced to their bare essentials, the social commandments deal with the basic issues of life, lineage, and property protection, mutually maintained through the assistance of truth-telling. In varied formulations, these are the core principles of every human society even today. May we project these principles backward and presume that they also formed the warp and weft of our very earliest forebears? I believe so. (1995: 100)

Recent discoveries in social biology, especially kin selection and reciprocal altruism, Hartigan believed, help us understand the roots of human morality as revealed in traditional sources like natural law and the Decalogue. These discoveries have "opened the door for a truly scientific understanding of the biological, evolutionary basis of human morality by showing how altruistic behavior among kin actually promotes individual best interest" (1995: 99). Still, he concedes that knowledge of the natural constituent elements of human behavior and ethics may not vitiate the ongoing need for belief in supernatural sanctions.

Political theorists like Masters, Arnhart, Hartigan, and, to a lesser extent, James Q. Wilson have based much of their work on the findings of students of biology who express similar views about the potential of the evolutionary sciences for uncovering the basis of human morality. Most notable, perhaps, is the work of E. O. Wilson:

> In the process, it [the study of human biology] will fashion a biology of ethics, which will make possible the selection of a more deeply understood and enduring code of moral values (1978: 196).

Richard Alexander has blamed the failure of moral philosophy on the failure of practitioners to take biology seriously, noting that "those who have tried to analyze morality have failed to treat the human traits that underlie moral behavior as outcomes of evolution – as outcomes of the process dominated by natural selection that forms the organizing principles of modern biology" (1987: xiv).

Empirical and theoretical work by Trivers (1985), Barash (1977), Dawkins (1976, 1982), Hamilton (1964, 1987), Fox (1989) and others now familiar to those who have followed developments in politics and the life sciences have also contributed to optimism about a science of morality. Thanks to their pioneering work, we are able to understand more clearly the ways in which kin selection and reciprocal altruism helped shape human social behavior. They argue that many of the lessons from evolution are inscribed in the repertoire of emotion and desire that characterize the species today. These attributes have contributed to and help sustain the reproductive fitness of their bearers. If evolutionary biology yields insights into human flourishing, as many of these theorists believe, then knowledge of what behaviors contribute to its attainment cannot be dismissed as morally neutral.

Philosophers, too, have weighed in on the matter of using biological knowledge in the study of ethics. Robert Richards writes: "The first principle of an ethical system can be justified only by appeal to another kind of discourse, an appeal to which factual evidence about sentiments and beliefs is adduced . . . So moral principles ultimately can be justified only by facts. The rebuttal, then, to the charge that at some level evolutionary ethics must be an attempt to derive its norms from facts is simply that every ethical system must" (1993: 127). Similarly, Daniel Dennett asks: "If 'ought' cannot be derived from 'is', just what can 'ought' be derived from? Is ethics an entirely 'autonomous' field of inquiry? Does it float, untethered to facts from any other discipline or tradition?" (1995: 467). Ethics must, he insists, be based on an appreciation of human nature. Evolutionary biology can help us here – though, as Dennett is quick to point out – simple assertions of genetic causation must be avoided.

In regard to Dennett's caveat, it should be noted that biologically-inclined political theorists are not unaware of the pitfalls and perils of their enterprise. They are careful to qualify their work. They understand, for example, that those behaviors that have served as useful moral guides from the standpoint of reproductive fitness in previous environments may not always be adaptive. The emphasis in more sophisticated scholars has shifted from genetic causation to the contextuality of evolutionary solutions to social problems. Certain solutions may work in one environment but not in another (e.g. the evolved human preference for sweets may be counterproductive in the modern world). As Masters notes: "Social biology promises an understanding of both 'special relativity' (why individuals within any society or culture have ethical disagreements, based on their differing perspectives) and 'general relativity' (why different societies and cultures have different laws and customs based on their environment and history). Yet, this new naturalism would account for such cultural variations without falling into the opposite excess of mere relativism, since evolutionary

theory provides a standard against which human behavior can be measured" (1989: 182–183).

Political theorists likewise understand there is no one-to-one relationship between genes and behavior, moral or otherwise. The linkages are many and complicated. And they recognize the difficulty in disentangling the relation between adaptive behaviors and the development of byproducts of adaptive behaviors. Some biological traits may be byproducts of cultural invention rather than the other way around (e.g. large human brains may be a result of social cooperation, not a cause [Masters (1989: 26)]).

Finally, they understand that moral goodness cannot simply be equated with evolution. "Lest this view be dismissed as a return to the discredited view that 'whatever is, is right,' " Masters reminds us, "one should recall Aristotle's emphasis on nature as a criterion of ethical judgment . . . In medicine, no one would maintain that, because a ruptured appendix can occur in the 'natural course of events,' people ought to die of appendicitis" (1989: 183). Those who propose a "naturalistic ethics" are motivated by the notion that there are clues in our evolutionary heritage as to why humans have adopted the particular moral choices that they have and by the conviction that these choices have been consequential for the species. Kin selection offers a clue to the evolution of familial duty. Reciprocal altruism speaks to our sentiments about helping others and the way we feel when others do or do not return our generosity. James Q. Wilson, perhaps, best captures the sense in which political theorists seek further to explore the biological roots – but not simple or causal determinants – of human morality:

> Morality, [Hume] said, rests on sentiment, and of course, generations of students have understood this. But morality does not rest on mere sentiment, because there is nothing "mere" about certain sentiments. There are many ways of knowing; the teachings of the heart deserve to be taken as seriously as the lessons of the mind (p. 238).

THE NATURE OF POLITICAL PERSONALITY

Related to this reassessment of the naturalistic fallacy is a reconsideration of the foundation of human personality. Under reconsideration is both the bifurcation of reason and passion characteristic of modern liberalism and the function of the components of the self.

The psychology that informs liberalism divides the self into two distinct and opposing forces: reason and passion. Armed with the dualism of Descartes and the mechanics of Newton, early liberal theorists created a clockwork image of human personality that dominates the study of politics to this day. Reason, according to some variant of this formulation from Hobbes to Kant, must either

serve the passions (e.g. Hobbes, Hume) or subordinate them to the dictates of reason (e.g. Kant). According to this formulation, the passions are arbitrary and give rise to arbitrary values, the worth of which are matters of personal taste that resists consensus. While reason must struggle to break free of the grip of the passions, it alone is capable of providing an understanding of the world around us and the role we play in it. Kant, of course, takes the most extreme position, severing the moral potency of reason from the bonds of passion. The transcendence of individual desire makes possible the moral life in conformity with the requirements of thought. Even Locke, however, who is more guarded about the efficacy of reason, argues that reason – if we will but consult it – will help the individual resist the urge to license which the passions inspire. Ideas that arise in the mind can at least have force if they spawn an additional "uneasiness" that redirects the will (Locke [1960]: 316). This bifurcation of the human personality into the separate and often warring elements of passion and reason has been the starting point for political analysis in liberal theory. As Unger has noted: "For the last three hundred years, the division of thought and feeling has been the starting point about ideas about human nature. Stated forcefully by the inventors of the liberal doctrine, it has ever since been the beneficiary of a thousand revisions and the target of a thousand attacks. Neither the attacks nor revisions, however, have done away with the original thinking" (1975: 40). Evolutionary theory is beginning to force a reconsideration. (For a more complete discussion of the warring nature of reason and passion in liberal psychology, see Losco, 1995.)

As is evident from the theorists addressed in the previous section, there is a growing awareness that morality is not simply a rational imperative or an irrational urge. If we are to take seriously the notion of a moral sense grounded in evolutionary theory, we must radically reconstruct liberal psychology. This is being done by revising our views of both passion and reason and the role each plays in the human personality.

The study of emotions from a biological perspective is not a new enterprise. Darwin (1872) wrote extensively on the matter, noting the plausible function and operation of emotions for the survival and reproduction of organisms. Darwin used naturalistic observations and cross-species comparisons to study the facial expressions he believed revealed emotional content. These served, he argued, as cues in transmitting and receiving information about the environment which organisms used to accommodate changing circumstances. Emotions evolved in humans during an era of face-to-face interactions as populations gathered in small hunter-gatherer bands. In the context of early social environments, facial displays gave cues as to emotional and motivational states of one's conspecifics and helped coordinate social activities. Ethologists have noted the

import of face-to-face relations in higher order primates (e.g. Lawick-Goodall, 1971) and suggest an evolutionary continuum in using such cues, culminating with the human species (Mazur & Robertson, 1972; Fridlund, 1994).

Evolutionary theorists are increasingly convinced that emotions help play an important role in the elicitation of behaviors that enhance fitness. Trivers (1971), for example, holds that the evolution of emotions like sympathy, guilt, trust, suspicion, and gratitude helped guide behavior in appropriate ways to ensure reciprocity in social relations. A number of models exist that tie emotion to ongoing physiological well-being (e.g. Ekman & Davidson, 1994; Scott, 1980; Plutchik, 1980; Plutchik, Kellerman & Conte, 1979). Using an evolutionary approach, in studying the nature of moral judgment, Allan Gibbard (1990) suggests that social emotions evolved in pairs as a means for coordinating action and response. Emotions "matched" in this way served to ensure more predictable outcomes as individuals interacted with others. "Propensities well coordinated with the propensities of others would have been fitness-enhancing and so we may view a vast array of human propensities as coordinating devices. Our emotional propensities, I suggest, are largely the results of these selection pressures, and so are our normative capacities" (p. 67). The fact is that certain emotions "go together"; they make sense when they are experienced in a coordinated manner. For example, Gibbard (1990) notes the complementarity of guilt and shame as an "adaptive syndrome." Anger which begets guilt may diminish the punishment inspired by an offensive act since guilt signals shame and contrition for the action. In short, this matched pair suggests a conflict-reducing device that can short-circuit potentially fitness-reducing conflict. Similarly, gratitude is more likely to beget benevolence than disdain. The most important consequence of these propositions is, as Gibbard notes, that "human emotions are above all social." They evolved through a process of social adaptation that emphasizes the sociality of the species, contradicting the individual and separate portrait of the passions advanced by liberal theory.

None of this is meant to imply that emotions are independent of reason or culture or that emotions serve as an innate moral compass that always hits its mark. Emotional arousal and display rarely occur in a cognitive vacuum. Cognitions play a major role in the expression and triggering of emotions. As Plutchik (1980) notes, cognition interacts with emotion to inform it until equilibrium with the environment is established. For example, in the response of fear, a negative stimulus must be interpreted as threatening before the sequence of physiological and motor responses we associate with the subjective feeling of fear can be experienced. Cognition continues to monitor the situation and helps sustain the appropriate physiological activity (e.g. alertness, readiness, etc.) until the threat has diminished or disappeared. This is not to imply that

cognition can generate and sustain an emotive response alone. Increased blood flow, a quickened heart rate, and tensing of the muscles can rarely be "willed" by cognitive manipulation alone. However, the example does suggest that cognition and affect are intimately interactive.

We can add to the complexity of our understanding of emotions by citing the developmental and cultural factors that must also be taken into account in rendering a full understanding of emotional response. Obviously, individuals with developmental handicaps may not be able to express the full range of human emotions or may come to express them in ways deemed inappropriate. Social learning also impacts the exhibition of emotion. American females are taught to "hold their anger," while males may be encouraged to display aggression, at least in certain arenas (Low, 2000.) But not all emotional displays are capable of equal control by socializing mechanisms. Management of overt behaviors and the spoken word, for example, are more easily controlled than facial expression (Ekman, 1980: 89). It is more difficult to condition a fearful response to a happy face than a happy face to a fearful one (Lanzetta & Orr, 1980). In controlled experiments, "leakages," have been found to override culturally transmitted instructions for display.

There is ample cross cultural evidence of uniformity of some emotional displays. Eibl-Eibesfeldt (1972) reports identical facial displays among children who were born sighted and blind. Ekman (1980) reports that some emotional displays like smiling in infants occurs almost from birth with little inducement from parents. Pictures taken of people from a number of diverse cultures were accurately interpreted by members of other cultures with a high degree of precision (Ekman, Friesen & Ellsworth, 1972; Ekman, 1973). In one experiment, Japanese and American subjects were exposed to stress-inducing and stress-reducing films while videotapes were made of their facial expressions. Examiners who reviewed the taped expressions found a better than .90 correlation in facial movement shown by Japanese and American subjects. This led Ekman to conclude: "We believe this is definitive evidence that facial expressions of emotion are not limited to simulations but do occur spontaneously as well" (1980: 94).

Roger Masters and Dennis Sullivan (1987) have shown that citizens tend to interpret emotional displays by political leaders in a fairly uniform manner across cultures. In studies conducted in the United States and France, subjects were exposed to videotaped sequences of political leaders. Controls were used for the amount of visual and audio information communicated. In assessing three major dimensions of facial displays (happiness/reassurance, anger/threat, and fear/evasion), the researchers found "viewers decode these three kinds of displays accurately, report distinct emotional responses to each type, and have

physiological responses during the display consistent with the verbal self-reports of emotion" (1987: 885). Masters go on to show that the categories by which we tend to interpret emotional displays are tied to major categories of evolutionary function identified by ethologists in studies of human and non-human animal populations.

Our views on human cognition are changing as well. No longer can we view reason as a cool, detached force directed by abstract principles of logic. Instead, evolutionary theory teaches that reason is infused with pre-patterned algorithms originating in the need to enhance fitness opportunities in early environments. Cosmides and Tooby (1992), for example, have shown the persistence of these algorithms in experiments in which subjects repeatedly abandon the requirements of formal logic to pursue a strategy of catching cheaters in games involving social exchange. The authors have tested a number of algorithms derived from a consideration of social situations likely to have faced our Pleistocene ancestors and have recorded similar empirical successes. Contrary to the traditional vision that reason is impartial, content-blind, and general purpose, the authors conclude that "human reasoning is governed by a diverse collection of evolved mechanisms, many of which are functionally specialized, domain-specific, content imbued, and content-imparting...this research supports the view that the human mind imposes contentful structure on the social world, derived from specialized functional design inherent in its evolved cognitive architecture" (pp. 220–221).

What do we make of findings like this? As students of biopolitics are aware, even if we could demonstrate that emotions evolved to serve some fitness enhancing function, we cannot go on to conclude that they will continue to do so. Neither can we conclude that emotions represent moral positions. My emotions may produce a desire to hurt someone who has hurt me. That does not make my positions morally right. However, at the very least, we can conclude that evolution has equipped most normal humans with a moral sensitivity, that we react in predictable emotional ways to the responses of others, that there are moral consequences to our responses to our own emotional urges and that political theorists need to take emotions seriously in reconceptualizing the political psychology of liberalism. Similarly, we can no longer conceptualize reason as an impartial arbiter of desire and as a contentless computer designed to apply abstract logic. Reason and emotion evolved together to enhance fitness. They cannot be considered separate and/or warring entities as the architects of liberal political theory have advised. While not advocating a return to simple classic political psychology, at least one theorist (Losco, 1995) argues for a more balanced account of reason and emotion, reminiscent of the harmony envisioned by classic authors, rather than liberalism's conflictual account of the political psyche.

THE BIOLOGICAL FOUNDATIONS OF THE STATE

While there exist numerous differences regarding the details of the evolutionary emergence of the state, there is growing consensus that evolutionary biology can help us understand the rise and maintenance of state systems. The explicit use of Neo-Darwinian ideas and concepts gained attention in the early nineteen seventies.

Drawing upon the work of primatologists, Fred Willhoite suggested that dominance patterns evident in both higher primates and humans may serve the vital evolutionary function of reproductive success. Dominance hierarchies, he observed help maintain order in the group and a level of distinction between leaders and followers. These hierarchies "may be an ineradicable feature of the social and political life of mankind" (1971: 638). In a separate article, Willhoite (1976) noted that contrary to the views of classic philosophers there was probably never a time when humans walked the earth as solitary individuals without some patterns of dominance and submission which are the precursors of modern political life.

Somit and Peterson (1997) have taken Willhoite's thesis further, arguing that the primate propensity for dominance hierarchies makes democratic and egalitarian regimes fragile and unstable. Asserting that democratic regimes are a fairly recent phenomenon in human evolution, the authors cross index results from a variety of studies of democratic governments to show that the number of democracies in the world may have peaked in the 1950s with as few as forty-one regimes. Where democracies exists, they argue, a number of ecological factors (internal to the country and external with reference to the world political environment) seem present. Among these are the absence of great disparities in wealth; access to educational institutions; urbanization; adequate social networks; some degree of ethnic, linguistic, and religious unity; prior democratic history, and a predisposing civic culture.

Like Willhoite, the authors argue that dominance hierarchies reflect the Darwinian heritage we share with other primates. As in the case of related primates, dominance hierarchies serve to minimize the conflict that threatens reproductive success and inclusive fitness. Hierarchies make for greater stability. ". . . Stability on balance is conducive to more successful reproduction among members of that society. Hierarchy and social stability, an outgrowth of dominance relations among a social species, function to enhance the likelihood that the individuals who constitute that species will maximize their inclusive fitness" (1997: 54).

At the level of proximate causation. Somit and Peterson attribute our propensity to form hierarchies to an innate tendency toward indoctrinability. Drawing

on work by E. O. Wilson (1978), Barash (1975), Dawkins (1976), and others, the authors contend that evolution has favored the rise of this capacity as a means for regulating group conflict. "But as the size and complexity of the groups increased from hunting-gathering bands of perhaps thirty to forty persons to nation-states numbering millions and hundreds of millions, and as rival interests and ideas began increasingly to compete with each other, indoctrinability has become a fecund source both of intrasocietal ethnic and religious violence and of inter-state hostility, bloodshed, and warfare" (p. 82). While the portrait painted by these authors is bleak, they remind us that the tendency for hierarchy evolved in an earlier world and may no longer be adaptive. Thus, they call for increased civic education to balance our genetic heritage.

Moran (1997) has added an important contextual variable to this portrait, noting that hierarchal arrangements like social class are more likely to create internal conflict (class warfare) only where exit options are limited. This suggests that hierarchies that develop in relatively open political systems will experience greater stability.

While Willhoite, Somit, and Peterson, and Moran describe one side of human relationships within the state, they have not addressed the perhaps more important and more puzzling phenomenon of cooperation which is the hallmark of the modern state. One political theorist (Johnson, 1995) believes that the institution of government that characterizes the modern social and political relationships actually predates human life. ". . . [I]indeed," Johnson writes, "government and politics are probably 3.5 million years old . . . originat[ing] early in the history of life on earth, and . . . found at every level of life's hierarchy, from the simplest cells to the huge nation state in which most humans live today" (p. 244). Throughout evolution, Johnson believes, nature has evolved new forms of cooperation at every level biological organization. Thus, cooperative arrangements among interacting organisms is not unique and should not be surprising. Most life science theorists, however, do not go back as far as Johnson in their analyses of state formation. Indeed, most would probably reject his usage of the terms "politics" and "government" as applied to units of analysis below the individual and the group.

Masters (1989) starts with the insights of biological science to confirm Aristotle's characterization of man as the *zoon politikon*. This does not simply mean, however, that man is inherently social. Humans have a natural affinity for social attributes like bonding but also for agonistic behaviors such as those associated with dominance. Just as the individual from an evolutionary perspective must maintain a balance between various social and self-oriented attributes in order to survive and reproduce, collections of individuals must strive to balance benefits conferred from group efforts with costs, especially as these

groups increasingly involve individuals beyond kin. It is likely that sustenance of larger group affiliation involved the interaction of a variety of strategies including intergroup rivalry and intragroup cooperation in addition to individual behavioral strategies like kin selection and reciprocal altruism (p. 196). The realm in which the strategic blend of these social behaviors takes place is the province of politics which exists at the intersection of ethology and anthropology, where the social and natural sciences meet. (For a critique of Masters views on reductionism and units of analysis as employed in his theory of the state, see Eulau & Zlomke, 1995.)

While Masters holds that evolutionary biology disconfirms a strictly solitary Hobbesian view of human nature, this does not mean that the state is merely an extension of natural sociality. Evolutionary theory only extends sociality via kin selection to a number of related others and via reciprocal altruism only to small number of individuals with whom we have the occasion for face to face contact. Additional environmental and ecological factors must be considered in accounting for the emergence of the state as a device for extended social cooperation.

Using a variety of game theoretic matrices, Masters details several conditions under which the benefits of cooperation outweigh the costs. These are precisely the types of conditions under which the nation state must have emerged, he reasons, and which must continue to function in sustaining large political societies. "It follows," Masters asserts, "that participation in a large scale community or state is possible only under environmental circumstances in which the benefits to subjects outweigh the comparative disadvantages to obeying rulers who gain even more from their powerful positions" (p. 188). These conditions are not always met and occasionally give way to social disintegration or failures of the mechanisms employed by the state (e.g. bureaucracies) for social control (see also Peterson's discussion of "bureaucratic pathologies," 1986).

Robert Axelrod (1986) has also done pioneering work on social cooperation that has a bearing on the evolution of the state. Evolutionary theory provides the logic for the game theoretic approach he employs. "The evolutionary approach dictates that strategies proving relatively effective are more likely to be employed in the future while less effective strategies are dropped" (p. 1109). Using this approach, he is able to predict that in certain conflict situations, the strategy "tit-for-tat" is stable, meaning that one will return in kind the acts directed toward one by an interacting other (1983). Axelrod assumes a "norm" to be established when a stable strategy emerges. In one such game, boldness and vengeance are assigned payoff points. In a reiterated series of computer simulations, it was found that boldness levels fell off rapidly when payoffs for vengefulness were introduced. "Then, gradually the amount of vengefulness

also fell because there was no direct incentive to pay the enforcement cost of punishing the defector . Once vengeance became rare, the average level of boldness rose again and the norm completely collapsed. Moreover, the collapse was a stable outcome" (1986: 1109). While models like Axelrod's are useful in demonstrating the likelihood of cooperation and the conditions which sustain it, they are often criticized as being purely formal and not including mediating personal factors that impact real world applications. This portrait is beginning to change with even more help from evolutionary biology.

Elinor Ostrom modifies Axelrod's strongly rational model in her own theory of community action. She argues that strict rational actor models are only useful in certain prescribed settings (e.g. under strong competition that eliminates players that do not aggressively maximize immediate external values). These conditions, she notes, have not and do not always and everywhere prevail. They are properly employed only as a limiting case of bounded or incomplete rationality. Thus, models like Axelrod's are only one of a member of a family of rationality models rather than the only way to model human behavior (p. 3). Speaking of the sometimes tragic outcomes that are modeled by strictly rational models, she asserts: "We are neither trapped in inexorable tragedies nor free of moral responsibility for creating and sustaining incentives that facilitate our own achievement of mutually productive outcomes" (p. 16).

Ostrom explicitly draws upon evolutionary biology (Hamilton, 1964; Trivers, 1971; and Tooby, 1989, 1992) to modify the concept of rationality, discussing the importance of reciprocity, face to face interaction, trust, reputation, and capacity to learn and invent rules to regulate social conflict. "Models of human behavior based on theories consistent with our evolutionary and adaptive heritage need to join the ranks of theoretical tools used in the social and biological sciences," she concludes (p. 3). For example, she finds that game theoretic models allowing opportunities for face-to-face communication markedly improve the chances for cooperation (p. 7). Ethologists have long argued the importance of personal contact in mitigating conflict. "Our evolutionary heritage has hardwired us to be boundedly self-seeking at the same time that we are capable of learning heuristics and norms, such as reciprocity, that help achieve successful collective action" (p. 2). Thus, evolutionary theory is important in helping us understand how the changes in rational actor preferences may emerge and stabilize. "That humans rapidly learn and effectively use heuristics, norms, and rules is consistent with lessons learned from evolutionary psychology . . ." (p. 10).

Yet, consistent with the other thoughtful theorists we have considered, Ostrom asserts that "humans do not inherit particular reciprocity norms via a biological process. The argument is more subtle" (p. 10). We inherit a sensitivity for

learning norms that increase own long term benefits when confronting social dilemmas with others who have learned and value similar norms. This is not simple reductionism or genetic determinism, but a prescription for the study of a complex and interactive chain of events the elements of which we are only beginning to identify. Still, she is clear about the importance of this avenue of study. "The theory of collective action", Ostrom explains, "is *the* central subject of political science. It is the core of the justification for the state" (p. 1).

CONCLUDING REMARKS

Emerging consensus among biologically-oriented political theorists regarding the limited applicability of the naturalistic fallacy, the nature of the political personality, and the emergence of the state attest to the vitality of political theory under the influence of evolutionary thinking. Still, even supporters of an evolutionary based discipline admit the influence of such thinking upon mainstream political science is minimal. This is puzzling for a number of reasons. One reason is the prominence of many bio-behavioral exponents among the luminaries of the discipline. In recent years, at least three presidents of the American Political Science Association (John Wahlke, James Q. Wilson, and Elinor Ostrom) have produced works utilizing evolutionary concepts and approaches, even though only one (Wahlke) might be considered a self-conscious advocate of biological approaches within the field.

Another puzzlement is reflected in the head-note to this chapter by Carl Degler. Advocates of a purely or primarily culture-based explanation of society and politics, while dominating the social and behavioral sciences for the past half-century for a number of reasons Degler admirably recounts, have failed to come to grips with the inconsistency resulting in both a belief in the process of evolution and a rejection of biologically relevant factors in human behavior. This inconsistency parallels the intellectual tension present in the debate over the naturalistic fallacy discussed above. No doubt some sincerely believe that with the emergence of the mind, human reliance on older, evolved mechanisms of social navigation ended – despite increasing evidence to the contrary (Cosmides & Tooby, 1994). Others remain wed to an "empty organism" view of the human animal as either simpler to model than complex portraits positing internal attributes (see, for example, White's critique of this view, 1992) or more amenable to social engineering for reasons only ideology can explain (e.g. Chomsky's critique, 1975).

Still, as part of a wider discipline that strives desperately for scientific validation, it must be admitted that biopolitics generally has not yielded many demonstrable empirical successes. Albert Somit (1980) correctly sensed the

impatience of political scientists in 1980 when he wrote that the maximum amount of time advocates could continue to prophecie or speculate about the potential utility of biology for the discipline without delivering empirical results was about ten years. Yet, biopolitics advocates were probably not sufficiently realistic about the difficulties they had to overcome in redirecting research in the field. By some estimates (Zisk, 1977), it took approximately thirty years for economic and formal modeling approaches to gain a respectable share of space in the discipline's premiere journal, *APSR*.

As Wilson predicted (1978), those disciplines closest to the natural sciences (anthropology, psychology) have already demonstrated remarkable advances by applying evolutionary thinking as entries in this volume abundantly attest. Those furthest removed from the natural sciences (political science included) are likely to take longer. We should also remember, as Degler (1991) has pointed out, that the legacy of Social Darwinism and the justifiable scorn it engendered runs deep. Three decades after the formation of the Association for Politics and the Life Sciences, there remains only one biopolitics graduate program specifically geared toward political scientists (Northern Illinois University).

Three things, however, seem certain at this juncture: (1) a consensus is growing among political theorists within the biopolitical movement regarding the direction and utility of evolutionary findings; (2) advocates for a biological reorientation probably underestimated the time and effort necessary to reorient mainstream political scientists toward their efforts; (3) if mainstream impatience manages to drive out evolutionary thinking, political science will be the poorer.

REFERENCES

Alexander, R. (1987). *The Biology of Moral Systems*. N.Y.: Aldine De Gruyter.

Arnhart, L. (1998). *Darwinian Natural Right: The Biological Ethics of Human Nature*. Albany, N.Y.: State University of New York Press.

Arnhart, L. (1995). Human Nature-One, Two or None? Feminism and Primatology. In: A. Somit and J. Losco (Eds), *Human Nature and Politics* (pp. 137–166). Greenwich, CT: JAI Press.

Axelrod, R. (1986). An Evolutionary Approach to Norms. *American Political Science Review*, *80*(4), 1095–1112.

Axelrod, R. (1983). *The Evolution of Cooperation*. Cambridge, MA: Harvard University Press.

Barash, D. (1977). *Sociology and Behavior*. N.Y.: Elsevier.

Chisholm, J. (1999). *Death, Hope and Sex*. Cambridge: Cambridge University Press.

Chomsky, N. (1975). *Reflections on Language*. N.Y.: Pantheon Books.

Cosmides, L., & Tooby, L. (1992). Cognitive Adaptations for Social Exchange. In: J. H. Barton, L. Cosmides & J. Tooby (Eds), *The Adapted Mind: Evolutionary Psychology and the Generation of Culture* (pp. 163–228). N.Y.: Oxford University Press.

Cosmides, L., & Tooby, J. (1989). Evolutionary Psychology and the Generation of Culture, Part II: Case Study: A Computational Theory of Social Exchange. *Ethnology and Sociobiology*, *10*, 51–97.

Darwin, C. (1872) [1955]). *The Expression of Emotions in Man and Animals*. N.Y.: Philosophical Library.

Dawkins, R. (1982). *The Extended Phenotype: The Gene As the Unit of Selection*. Oxford: Oxford University Press.

Dawkins, R. (1976). *The Selfish Gene*. Oxford: Oxford University Press.

Degler, C. (1991). *In Search of Human Nature: The Decline and Revival of Darwinism in American Social Thought*. New York: Oxford.

Dennett, D. (1995). *Darwin's Dangerous Idea: Evolution and The Meanings of Life*. N.Y.: Simon and Schuster.

Ekman, P., & Davidson, R. J. (Eds) (1994). *The Nature of Emotions: Fundamental Questions*. New York: Oxford Press.

Ekman, P. (1980). Biological and Cultural Contributions to Body and Facial Movement in the Expression of Emotions. In: A. O. Rorty (Ed.), *Explaining Emotions*. Berkeley, CA: University of California Press.

Ekman, P. (1973). Darwin and Cross-Cultural Studies in Facial Expression. In: P. Ekman (Ed.), *Darwin and Facial Expression: A Century of Research in Review*. N.Y.: Academic Press.

Ekman, P., Fraiesen, W., & Ellsworth, P. (1972). *Emotion in The Human Face*. N.Y.: Pergamon.

Eibl-Eibesfeldt, I. (1972). Similarities and Differences Between Cultures in Expression Movements. In: R. Hinde (Ed.), *Non-Verbal Communications*. N.Y.: Cambridge.

Eulau, H., & Zlomke, S. (1995). Biological Phenomena, Levels of Analysis and Reductionism: Masters, Model of Human Behavior As a Case Study. In: A. Somit & J. Losco (Eds), *Human Nature and Politics* (pp. 173–196). Greenwich, CT: JAI Press.

Fox, R. (1989). *The Search for Society: Quest for a Biosocial Science and Morality*. New Brunswick: Rutgers University Press.

Fridlund, A. J. (1994). *Human Facial Expression: An Evolutionary View*. San Diego, CA: Academic Press.

Gibbard, A. (1990). *Wise Choices, Apt Feelings: A Theory of Normative Judgment*. Cambridge, MA: Harvard University Press.

Hamilton, W. D. (1987). Kinship, Recognition, Disease, and Intelligence: Constraints of Social Evolution. In: Y. Ito, J. L. Brown & J. Kikkawa (Eds), *Animal Societies: Theories and Facts*. Japan Scientific Societies Press.

Hamiltion, W. D. (1964). The Genetical Evolution of Social Behavior. *Journal of Theoretical Biology*, *7*, 1–52.

Hartigan, R. (1995). Natural Law, Nature and Biology: A Modern Synthesis. In: A. Somit & J. Losco (Eds), *Human Nature and Politics* (pp. 85–104). Greenwich, CT: JAI Press.

Hartigan, R. (1988). *The Future Remembered: An Essay in Biopolitics*. Notre Dame: University of Notre Dame Press.

Hume, D. (1740). *A Treatise on Human Nature*, 2nd edition. (Revised by P. Nidditch). Oxford: Claredon Press.

Johnson, G. (1995). The Evolutionary Origins of Government and Politics. In: A. Somit & J. Losco (Eds), *Human Nature and Politics* (pp. 243–305). Greenwich, CT: JAI Press.

Lanzetta, J. T., & Orr, S. P. (1980). Influence on Facial Expressions on the Classical Conditioning of Fear. *Journal of Personality and Social Psychology*, *39*, 1081–1087.

van Lawick-Goodall, J. (1971). *In the Shadow of Man*. Boston: Houghton-Mifflin.

Locke, J. (1689 [1960]). *The Second Treatise of Government*, P. Laslett (Ed.). New York: Mentor.

Losco, J. (1996). Sociobiology and Political Science: A Status Report. In: A. Somit & S. A. Peterson (Eds), *Research in Biopolitics*, Vol. 4 (pp. 151–177). Greenwich, CT: JAI Press.

Losco, J. (1995). The Bifurcated Self: A Life Sciences Critique of Liberal Political Psychology. In: A. Somit & J. Losco (Eds), *Human Nature and Politics* (pp. 105–136). Greenwich, CT: JAI Press.

Losco, J., & Baird, D. D. (1982). The Impact of Sociobiology on Political Science. *American Behavioral Scientist*, *25*(3), January/February, 335–360.

Low, B. (2000). *Why Sex Matters: A Darwinian Look at Human Behavior*. Princeton, N.J.: Princeton University Press.

Masters, R. D. (1989). *The Nature of Politics*. New Haven, CT: Yale University Press.

Masters, R. D., & Sullivan, D. G. (1987). Gut Reactions and the Political Effects of Media. *PS*, *20*, 880–889.

Mazur, A., & Robertson, L. (1972). *Biology and Social Behavior*. N.Y.: Free Press.

Moran, I. (1997). Marxist Gorillas: Nonhuman Primates and the Biopolitics of Class. In: A. Somit & S. A. Peterson (Eds), *Research in Biology and Politics*, Vol. 5 (pp. 57–70). Greenwich, CT: JAI Press.

Moore, G. (1903). *Principia Ethica*. Cambridge: Cambridge University Press.

Ostrom, E. (1997). A Behavioral Approach to the Rational Choice Theory of Collective Action. *American Political Science Review*, *92*, 1–22.

Peterson, S. A. (1986). Why Policies Don't Work: A Biocognitive Perspective. In: E. White & J. Losco (Eds), *Biology and Bureaucracy* (pp. 447–502). Lanham, MD: University Press of America.

Plutchik, R. (1980). A General Psycho Evolutionary Theory of Emotion. In: R. Plutchik & H. Kellerman (Eds), *Emotion: Theory, Research and Experience*, Vol. 2. N.Y.: Academic Press.

Plutchik, R., Kellerman, H., & Conte, H. R. (1979). A Structural Theory of Ego Defenses. In: C. E. Izard (Ed.), *Emotions, Personality, and Psychopathology*. N.Y.: Plenum.

Richards, R. (1993). Birth, Death, and Resurrection of Evolutionary Ethics. In: M. Nitecki & D. Nitecki (Eds), *Evolutionary Ethics*. Albany: State University of New York Press.

Roberts, M. (1938). *Bio-Politics: An Essay on Physiology, Pathology, and Politics of the Social and Somatic Organism*. London: J. M. Dent and Sons.

Scott, J. P. (1980). The Function of Emotions in Behavioral Systems. In: R. Plutchik & H. Kellerman (Eds), *Emotion: Theory, Research and Experience*, Vol. 1. N.Y.: Academic Press.

Somit, A. (1980). *The Literature of Biopolitics*, 2nd ed. DeKalb, IL: Center for Biopolitical Research.

Somit, A. (1968). Toward A More Biologically-Oriented Political Science: Ethology and Psychopharmacology. *Midwest Journal of Political Science*, *12*, November, 550–567.

Somit, A., & Peterson, S. A. (1997). *Darwinism, Dominance, and Democracy: The Basis of Authoritarianism*. Westport, CT: Praeger.

Thiele, L. P. (1999). Evolutionary Narratives and Ecological Ethics. *Political Theory*, *27*(1), February, 6–38.

Thorson, T. (1970). *Biopolitics*. N.Y.: Holt, Rinehart, and Winston.

Trivers, R. L. (1985). *Social Evolution*. Menlo Park, CA: Benjamin/Cummings.

Trivers, R. L. (1971). The Evolution of Reciprocal Altruism. *Quarterly Review of Biology*, *46*, 35–57.

Unger, R. (1975). *Knowledge and Politics*. N.Y.: Free Press.

White, E. (1992). *The End of the Empty Organism*. New York: Praeger.

Willhoite, F. (1976). Primates and Political Authority. *American Political Science Review*, *70*, 1110–1126.

Willhoite, F. (1971). Ethology and the Tradition of Political Thought. *Journal of Politics*, *33*, (August).

Wilson, E. O. (1978). *On Human Nature*. Cambridge, MA: Harvard University Press.

Wilson, J. (1993). *The Moral Sense*. N.Y.: Free Press.

Zisk, B. (1997). The Discipline at Century's End: Ecological Critique. Paper presented at the Annual Meeting of the *American Political Science Association*, Washington, D.C.

BIOPOLITICS IN THE YEAR 2000

Albert Somit and Steven A. Peterson

I. BIOPOLITICS IS . . .

Biopolitics is the shorthand term commonly used to describe the approach of those political scientists who use biological concepts (especially neo-Darwinian evolutionary theory) and biological research techniques to study, explain, predict and sometimes even to prescribe *political behavior*.

That term, however, actually subsumes not one but several "biological" approaches to political phenomena. Almost from its inception in the mid-1960s (Degler, 1991), the biopolitical "movement" marched in several diverse directions. At this juncture, we will simply describe these, postponing specific bibliographic references for a later point in the discussion.

A. Major Lines of Inquiry

Conceptually the most important and most radical – from the standpoint of mainstream political science – approach is that which emphasizes the need to understand the evolutionary origins of the behavioral predispositions which *Homo sapiens* shares with all the other social primates and which significantly affect our political life. Profoundly influenced by ethology and contemporary evolutionary theory, those working in this area insist that we should give proper weight to the role played by *nature* as well as by *nurture* in shaping our social and political behavior.[1] That insistence runs directly counter, of course, to the long-accepted disciplinary paradigm which holds that human political behavior

Evolutionary Approaches in the Behavioral Sciences, Volume 8, pages 181–201.

ISBN: 0-7623-0769-2

is *learned* and that possible genetically transmitted proclivities can and should be ignored.[2] Closely related to this thrust, as we shall see, is another though much smaller group, which seeks to derive ethical and moral guidelines for political decisions from evolutionary theory, in effect, to rest a political philosophy on a "scientific" basis.

A second major strand in biopolitics is concerned with the manner in which physiological factors might impact political behavior, especially at the individual level. Among the most obvious of such possible influences would be illness, age, fatigue, psychopharmaceutical drugs, crowding, pain, sleeplessness, tension and malnutrition. Working along parallel lines is a newly-emergent line of inquiry which is exploring the relationship between neurological function, on the one hand, and attitudes and behavior, on the other.

A third approach has sought to utilize physiological measures as clues to political attitudes and to potential or incipient political behavior. "Body language" is no doubt the best known of these but changes in blood pressure, eye-blink rate. stimulus response rates and hormonal analysis have also been explored as potentially useful indicators. Also involved here are attempts to utilize these measures in conjunction with written or oral survey responses to ascertain *strength* of feeling or belief, as indicative of a possible "intensity" which might not be apparent from a simple oral or written answer.

Having identified the various components which collectively constitute what we call "biopolitics",[3] we must here emphasize a very important aspect of that movement. Although they march in diverse directions, and occasionally even to a different drummer, those in biopolitics nearly unanimously agree on the validity of neo-Darwinian evolutionary theory. And that theory, applied to *Homo sapiens*, leads to only one conclusion: our political behavior, just as with so many other aspects of our life, is often influenced by genetic tendencies which are the legacy of our species' evolutionary history. This, above all, is the central tenet of biopolitics. As almost everyone who has been involved in biopolitics concurs, that enterprise cannot viewed as truly successful until a truly Darwinian perspective has been accepted by our fellow political scientists.

Intellectual Sources

Two of these sources have already been mentioned: neo-Darwinian evolutionary theory and ethology.

By the turn of the 20th century, the task of correcting the weaknesses of earlier Darwinian thought had been largely accomplished. The resulting "neo-Darwinian synthesis" provided, for the first time, a powerful intellectual instrument for understanding how evolution worked to shape both the physical structure (morphology) and the behavior of living organisms.[4] Granted, some

important questions have yet to be fully answered – but that is true of practically every scientific theory.

A second major impetus came from ethology which, in barely a decade, was transformed from a relatively obscure to an almost indecently glamorous specialization. Two ethologists, Konrad Lorenz and Nikolaas Tinbergen became Nobel Laureates, with all of the publicity attendant upon these awards; as our older readers may recall, several ethologically-based books even made the best-seller lists.[5] Although the ethological literature itself focused almost entirely on animal behavior, the moral, not always just hinted, was quite clear.

Third, there was the growing realization that it *was* possible to shape and reshape emotional states, intellectual processes, and even beliefs by manipulating the physiological functioning of the body. Convincing evidence of this came from psychopharmacology, where newly discovered drugs made it possible, in relatively short order, to empty mental asylums of most of their inmates. Additional support came from studies of "brainwashing," or as it was more formally called, "coercive indoctrination." As examination of Chinese and Soviet practices revealed, a systematic combination of physical pain, terror, sleeplessness, fatigue, and malnutrition could not only coerce confessions but could literally effect a change in political belief.[6] And, although it may not fall under the heading of "evidence," science fiction, surely starting with *Brave New World*, increasingly depicted future societies in which these techniques were quite often used to shape and control political behavior.

Finally – and we are talking about the mid-1960s and early 1970s – there was a growing discontent with the "state of the discipline." That discontent was reflected in an extraordinarily nasty, no-holds-barred dispute between advocates and opponents of behavioralism and a resulting sense, among many member of the profession of "a plague on both your houses"; a discontent which contributed to the creation and brief popularity of the New Caucus For Political Science; a discontent which would be voiced, to mention only two of the critics, by Glendon Schubert (1976) and John Wahlke (1976), and which eventually became the central theme of Wahlke's American Political Science Association (APSA) presidential address (1979).

As Wahlke saw it, behavioralism, which had essentially dominated the discipline since the early 1950s, had failed to make appreciable progress towards its professed goal – a "scientific" political science. In fact, he argued, given its assumptions about behavior and its research methodology, behavioralism was incapable of achieving that objective. It was time to try a radically different approach by turning to a discipline which had recently achieved such impressive advances – biology, in general, and ethology and evolutionary theory in particular.

II. PROFESSIONAL RECOGNITION AND ORGANIZATIONAL EMBODIMENTS

Biopolitics received its first formal recognition within political science in 1973 when the International Political Science Association (IPSA) established a Research Committee on Biology and Politics (Research Committee No. 12). Over the ensuing quarter-century plus, RC 12 has been allocated two or more panel sessions at each of the triennial IPSA Congresses and has also organized several international meetings (Paris, 1975; Bellagio, Italy, 1977; Warsaw, 1987; Chicago, 1988; and Alfred, New York, 1996) at which ongoing or projected biopolitical research is discussed and critiqued.

There was considerable debate at the Paris, 1975 meeting over whether those working in the field should establish their own professional association.[7] That issue was eventually resolved in 1983 by the creation of the Association for Politics and the Life Sciences (APLS), with the full accouterments of dues-paying members, a Council, an Executive Director, and a much debated constitution. Two years later, APLS launched its own journal, *Politics and the Life Sciences*.

APLS was officially recognized as an "organized section" by APSA in 1985 and, as a result, was entitled to one or more program "slots" at each of the subsequent annual APSA conventions. It lost that status, however, in 1995 because a growing proportion of its members, whose primary interest tended to be in public policy issues, were drawn from other disciplines and chose not to become APSA members. Since then, APLS has held its own annual convention at the same time and in the same city, but not at the same hotel(s), as APSA. Understandably, this has created a problem, one to which we will later return.

III. THE BIOPOLITICAL LITERATURE: OVERVIEW AND SPECIFIC CONTRIBUTIONS

A. Overview

To be viable, an academic field must develop its own literature; from the very beginning, biopolitics' growth in this respect has been truly remarkable. The first biopolitical review article (Somit, 1972) surveyed some forty plus items. Since then, the literature has burgeoned to considerably more than a thousand papers and articles, numerous books, as well as an annual series, *Research in Biopolitics*, now going into its eighth volume.[8]

The participants in the biopolitical enterprise, moreover, have come from all of the fields or sub-areas of political science; mirroring the diverse subject interests of its authors, the biopolitical literature cuts across all of these fields. Put somewhat differently, those working in biopolitics constitute, in terms of their field specializations, a representative cross-section of the profession. Quite naturally, they have attempted, in their papers and publications, to persuade their mainstream colleagues that biological concepts and techniques can be productively utilized in practically everyone of these fields.[9]

B. Specific Foci

Genetic Influences on Political Behavior

However they differ otherwise, those in biopolitics use the central paradigm of biology – neo-Darwinian evolutionary theory – to study and explain political phenomena. Contemporary evolutionary theory as applied to human social and political behavior runs directly counter to Durkheim's dictum, embraced by the social sciences for the greater part of the twentieth century, that "social facts require social explanation." In other words, human behavior can be explained *only* in terms of social learning, social conditioning, and social situations. To what extent has the Durkheim-based credo been noticeably modified as a consequence of the neo-Darwinian critique?

Lamentably, there is little evidence that evolutionary explanations of political behavior have evoked much interest from, let alone been enthusiastically accepted by, our colleagues in the discipline. We offer three cases in point. First, most ethologists (and sociobiologists) agree that human males and females are subject to somewhat different selective forces and, thus, developed both different physical characteristics and, to an important degree, different behavioral tendencies. Using this as his point of departure, Glendon Schubert (1991) has argued that it is *male* behavior that largely accounts for the violence and aggression that has characterized human politics since the dawn of History. Our species would be much better governed, he contends, if women, rather than men, had a larger role in politics. Schubert's thesis has elicited little attention in political science generally and, because he assumes a genetic difference between the sexes, a mixed reception from feminists.

Roger Masters (1993; see also Masters, 1989) has attempted to derive a naturalistic ethics from evolutionary theory (and see Arnhart, 1998). According to Masters, neo-Darwinism provides a means of resolving the "is-ought" issue that has so long been debated by political philosophers. It is as yet unclear what impact this line of argument has had on the discipline.

Finally, we (Somit & Peterson, 1997) offered a neo-Darwinian explanation for the rarity of democratic polities both historically and even today, during the so-called "Age of Democracy." The explanation, we suggest, is to be found in our species' evolutionary history. Over literally millions of years, natural selection has endowed *homo sapiens*, as it did the other social primates, with an innate "bias" toward hierarchical social and political structures. This innate tendency is a major – but not the only – reason why the overwhelming majority of political societies have been and continue to be authoritarian in nature. Thus far, this Darwinian explanation for the pervasiveness of authoritarianism has failed to garner widespread notice and support.

In short, and probably for the reasons outlined below, efforts to persuade political scientists to adopt a more evolutionary approach have so far fallen onto exceedingly stony soil.

Physiological Factors Influencing Political Behavior

Among the more obvious physiological conditions that would exercise an influence, especially on *elite* behavior, are stress, fatigue, physical or mental illness, hormonal imbalance, pain, drugs (medicinal or otherwise), and alcohol (e.g. Gilbert, 1992). Physiological factors that might significantly affect *mass* political behavior include health status, stress, nutritional level, energy level, and crowding. Literally dozens of studies of possible physiological influences can be found in the biopolitics literature (e.g. Madsen, 1986; Schwartz, Garrison, & Alouf, 1975; Booth & Welch, 1976. See the survey in Peterson, 1996).

Some biopolitical analysts have focused on the functioning of the human brain as important for understanding political decision-making and behavior. George Marcus and colleagues (1998) have attempted to demonstrate a linkage between brain functioning (indirectly measured) and intolerant behavior. Recent research by Roger Masters suggests that neurotoxins accumulating in humans' brains (such as heavy metals) may affect the likelihood of involvement in antisocial behavior (Masters, 1998). To this point, the results are intriguing, but assessment of brain functioning is most indirect.

Beyond doubt, physiological factors *can* affect political behavior. Nonetheless, biopolitics has been manifestly unsuccessful in interesting others in the discipline in these possible linkages. Why, then, have not more mainstream political scientists followed their biopolitical colleagues into what at first glance seems a promising area of research?

The reasons, we think, are much the same as those mentioned in our discussion of physiological measures of attitudes immediately below. But, if anything, the methodological and logical difficulties are even greater. Conceivably, one could establish that a political figure was ill, or tired, or even intoxicated, when

s/he made a given decision or took a given action – although simply establishing that fact may be no easy matter – but it is quite another matter to demonstrate convincingly that the physiological condition really affected the action or decision. As many who initially opened this avenue of exploration have since reluctantly realized, these are problems that can much more readily be resolved in a laboratory experimental setting than in the study of real life behavior.

Physiological Measures of Political Attitudes and Behavior

John Wahlke has forcefully argued that a serious shortcoming in survey research was that the opinions elicited were not necessarily reliable indicators of actual behavior (Wahlke, 1979). A second major flaw was that survey research relied upon verbal report (i.e. oral or written) and thus provided very little guidance as to the emotional affect or intensity of feeling underlying that response. The latter deficiency, many in biopolitics were convinced, could be corrected if physiological measures were used in conjunction with verbal response (Wahlke & Lodge, 1972). Toward that end, a variety of such metrics were devised and tested – heart rate, eye blink rate, blood pressure, uric acid level, and reaction time (see the summary in Peterson & Somit, 1994).

These lines of exploration seem closely akin to those being pursued in the rest of the discipline. This led to the hope that here, if anywhere, biopolitics would make a significant contribution to political science. Such, however, has not been the case. To date, none of these techniques has been regularly adopted and used by that sizeable group of political scientists for whom survey research is a key research tool.

Closely related was the attempt to develop better physiological measures that might help in understanding actual behavior better. One of these was voice stress analysis, which electronically analyzed the speech of political leaders (Wiegele & Hirsch, 1981). The objective was to identify those parts of a speech where the instruments identified changes in sound production and delivery, changes that might suggest that the subject being studied was experiencing greater stress. Others have explored biochemical correlates of political behavior. Douglas Madsen (1986) assayed levels of serotonin and found that these were correlated with Type-A behavior.

Using yet another approach, Sullivan and Masters have recorded and analyzed non-verbal behavior (and facial expressions in particular) of leaders and then examined the impact of these on ordinary citizens (Sullivan & Masters, 1988). See also Salter, 1996; Mazur & Mueller, 1996). Certain facial behaviors of elites, they found, were associated with citizens' responses: e.g. happy faces engendered more positive emotional responses from viewers. Finally, Kitchin

has used content analysis of presidential speeches to measure hemispheric dominance, based on the assumption that certain word patterns reflect this aspect of brain functioning. Once this assessment was completed, one might then at some future point be able to link hemispheric dominance with decision-making style (Kitchin & Peterson, 1996).

There are doubtless many reasons why other political scientists have not picked up on these lines of inquiry. One, to which we will return, is that so little of the biopolitical literature appears in the regular political science journals that others in the profession may simply be unaware of these developments. A second deterrent, we suggest, is that political scientists contemplating using these techniques must either acquire a whole new body of skills or, alternatively, join a research partnership with someone who already possesses them. The first course requires a substantial investment of time and energy; the second, given the departmentalization of most research in academe, may be neither possible nor practical.

Ideally, of course, such training ought to occur in graduate programs. However, as we observe elsewhere, there is only one doctoral program – at Northern Illinois University – that currently even has a concentration in bio-politics. Thus, gaining mastery of appropriate research technology will, of necessity, be hit-and-miss for graduate students. For those out of graduate school, the process of learning research techniques will have to be by retrofitting one's research capability – hardly the ideal tactic.

A third reason may be the most important. Whatever their merits and promise, these physiologically-based techniques are more complex, and in many instances considerably more expensive to use, than verbal response alone. Furthermore, although they may yield excellent results when used in laboratory-based research involving a modest number of individuals, many of these are essentially impractical for use in surveys involving hundreds – if not thousands – of respondents.

IV. BIOPOLITICS: A BALANCE SHEET[10]

We mentioned earlier that biopolitics can be dated from the mid-1960s. Near the close of the first review of the biopolitical literature, the author rashly expressed the opinion that, to be adjudged successful, biopolitics ". . . will have to yield something substantial in the way of results within the next seven or eight years" (Somit, 1972). With the wisdom of hindsight, it seems painfully obvious that the suggested eight or nine years testing period was much too short for a fair demonstration, let alone a definitive judgement. But now, with biopolitics well into its fourth decade, what can be said?

To begin on a positive note, biopolitics has been reasonably successful in an organizational sense. It enjoys formal recognition from IPSA; it had similar recognition from APSA for almost a decade; it has its own professional association, its own annual meeting, its own journal, and the annually published JAI *Research In Biopolitics* series. Although, as mentioned above, there are some clouds on the organizational horizon, these difficulties are not, at least as yet, insuperable. By this criterion, then, it has done fairly well.

But there is another, and we would have to say substantively more important criterion which must also be applied – one which, as noted above, was understood and accepted from the outset. And that, of course, is the extent to which neo-Darwinian evolutionary concepts have been accepted and utilized by mainstream political science. By this standard, as already adumbrated, biopolitics has not yet had the impact on the profession for which its founders hoped and, in all candor, expected. Unwelcome but near-decisive support for this reluctant judgement was provided by the recently (1997) published *A New Handbook of Political Science*. Some 800 pages in length, and with forty plus American and European contributors, the *Handbook* contains not a single reference to the biopolitical literature, let alone to biopolitics as a subfield within political science.

To be sure, the *Handbook* alone could be readily (and happily) discounted. Unfortunately, as previously observed, much of the other evidence compels the same conclusion. In only one field of the discipline, in fact, do we find any recognition of the possibility that an evolutionary outlook might be helpful. That field, not surprisingly, is international relations, where so much attention needs to be devoted to war and other forms of international conflict. International relations textbooks have increasingly mentioned – but carefully not endorsed – the conviction of some ethologists that there is a relationship between the frequency of warfare and our species' proclivity for aggressive behavior.[11]

That lamentably limited success gives rise to the obvious question: Why has a "more biologically oriented approach" made so little headway in political science? There are, we believe, several reasons.

We earlier remarked that the biopolitical literature contains considerably more than a thousand (now perhaps some 1300) items. Regrettably, very few biopolitically-oriented *articles* have appeared in the leading (or even the so-called "second tier") political science journals. In a way, this problem fed on itself. Among the biopolitical cadre, there early hardened the belief, not altogether unfounded, that these journals were basically inhospitable to biopolitical submissions. Understandably, their authors soon turned to other, more receptive outlets and. naturally, to APLS' own *Politics and the Life Sciences*.

Nor did biopolitical *books* fare much better in the book review pages. To be sure, the two political science journals which have major book review sections

can only review a small fraction of the books they receive and the resulting competition for space is inevitably quite severe. Nonetheless, whether by accident or intent, biopolitical books did no better in this competition than they did in the pages devoted to articles.

The consequences hardly need detailed explication. Many – perhaps most – political scientists rely on the discipline's journals to keep them abreast of the profession and of developments outside their immediate area(s) of specialization. The journals' rare publication of biologically-oriented articles and scant attention to biologically-oriented books has provided their readers scant information about the possible merits of that approach or of what is actually being accomplished by its advocates.

This lack of awareness is no doubt a contributing factor in the inability of biopolitics to "catch on." Even more important, though, is the discipline's commitment, as emphasized above, to the proposition that, for all practical purposes, political behavior is *learned*. Massive investments of research effort, and major reputations, have been made on the basis of that proposition – one whose validity is directly challenged by those who emphasize the importance of genetically transmitted behavioral tendencies. The almost irresistible temptation, then, has been to turn a deaf ear and a blind eye to such a heresy.[12]

There may also be involved here, we think, an even more fundamental opposition. A hundred and fifty years after *The Origin of Species*, almost half of the American population still staunchly believe in creationism, i.e. that the story told in Genesis is literally true; some 70% hold that creationism should be taught alongside Darwinism as equally valid accounts.[13] And, even some of those who accept evolutionary theory in principle insist that it applies only to other species, not to ours. It would be truly remarkable if some sizeable percentage of academics generally, and political scientists, in particular, did not share these views.

One more – and we arc convinced, hardly trivial – deterrent should be mentioned. Biopolitics was never able to develop a truly prestigious institutional base and sponsorship. That is, it was never linked, in the profession's mind, with well-known scholars holding appointments at high-ranking graduate departments. For a new theory or approach, such a linkage is almost an absolute necessity. Thus, to take the post-World War II period, the testamentary examples are group theory (Truman, at Columbia), pluralism (Dahl, at Yale), international *realpolitik* (Morgenthau, at Chicago), behavioralism (the Institute of Social Research and the Survey Research Center at Michigan), and mathematical modeling (Riker, at Rochester).

Unfortunately, such was never the case with biopolitics. None of its initial proponents held an appointment at a "Top Twenty," let alone a "Top Ten"

graduate department. Consequently, biopolitics never enjoyed the attention and respectful hearing almost automatically afforded by a prestigious academic connection. This lack carried with it yet another severe handicap – there was no high ranking department, *whose status carried a reasonable assurance of attractive job offerings*, at which would-be neo-Darwinian political scientists could be trained. Although there was a truly ambitious plan in the early 1970s to create such a department at SUNY's Stony Brook campus (a then-rising star in the academic firmament), that project never came to full fruition. In fact, the only doctoral program in "biology and politics" ever offered was at Northern Illinois University – a good, solid school but hardly a name to conjure with.

V. ISSUES.

A. Can Biology Provide A Basis For A "Scientific" Political Philosophy?

As we all know, central to any political philosophy is a concept of "the nature of human nature," i.e. whether "political man is fundamentally self-centered, altruistic, or a *tabula rasa*; whether human nature is fixed or malleable; whether we are essentially rational or in large measure driven by our emotions. etc. An explicit or implicit view of the nature of political man shapes and colors a political philosophy from its understanding of how political society came into being, to its position on the morally acceptable limits of seeking and exercising political power, and its vision of the ideal – or best possible – state.

From Plato onward, political philosophers have claimed an empirically valid basis for their particular understanding of human nature; from Plato onward they have advanced radically different perceptions of that nature: and from Plato onward, it has remained an issue because, absent conclusive evidence one way or another, the arguments advanced to support a given conception could readily be countered or dismissed by those who believed otherwise. Over the past few decades, however, considerable headway has been made in resolving this long-standing dispute by those working in three fields of biology. One of these fields is human ethology, which seeks to identify those cross-cultural similarities and consistencies of human behavior which, since they occur in vastly disparate social settings and in societies greatly distant from each other, can reasonably be understood as expressions of genetically transmitted behavioral tendencies common to members of our species. A second field is primatology, which seeks to identify both the behaviors manifested by primates generally and those which are observed in individual primate species. The third, of course, is evolutionary theory which attempts to explain how and why these behaviors evolved.

Although the task of explication is by no means yet completed, impressive progress has been made. There is now general agreement that we share with the other social primates a number of behavioral predispositions. Among the most *politically relevant* are the following: We (our species, that is) are inclined toward hierarchical social structures, characterized by dominance at the top and submission below; we are likely to favor and assist, where circumstances require, those to whom we are genetically most closely related, even though that action may entail some personal loss or risk; and our relations with neighboring but non-related groups of fellow speciates are often characterized by aggression, threats, and violence. These familiar proclivities have evolved over many million years of natural selection, neo-Darwinian theory holds, because they tend on balance to maximize the "inclusive fitness" of the individual members of our species.[14] And we share them, to repeat, with the other social primates, especially the chimpanzees to whom we are (in terms of DNA similarity) so closely akin.

But, we hasten to add, evolution has also endowed *Homo sapiens* with one truly remarkable capacity which it shares with no other species, not even the chimpanzees. That is the ability to create belief systems, whether religious or secular, and then, in behalf of (or in conformity with) that system, to act in a manner inconsistent with, or directly contrary to, our innate inclinations.[15] We sometimes do so even when the resulting actions, taken in support of some secular ideology, may cost us our lives. Or, in obedience to some religious conviction, we defy that basic biological (and Biblical) imperative – reproduce yourself. This means that, in many instances, we obey the dictates of our beliefs rather than our genes.

Nonetheless, over the long run, as history overwhelmingly testifies, our evolutionary legacy plays an important, if not necessarily determinative, role in shaping our political behavior and our political and social institutions. Whether we like it or not, the consistency of human behavior and, despite superficial and normally temporary differences, the basic underlying similarity of our political institutions throughout the ages really permits no other conclusion.

Assuming that they are scientifically sound (an assumption some readers may be unwilling to grant), what do these findings imply for political philosophy? From the vantage point of normative political philosophy, the answer may very well be "nothing at all" – a dismissal immediately justified by invoking the "naturalistic fallacy." What has been, what is, even what is likely to be, one may fairly argue, carries neither moral nor logical weight in deciding what *should* be.

On the other hand, a less cavalier and more pragmatic response, while conceding the force of that objection, would also ponder the implications of

these findings for any political philosophy that seeks acceptance and, ultimately, successful implementation. Among the most important of these would surely be: (a) that a predisposition for hierarchical social and political structures, with associated differences of status and power, may explain why the great majority of our political societies have been (and continue to be) authoritarian in nature, with democracies as the infrequent and often short-lived exception; (b) that human behavior is not infinitely malleable; (c) that emotional and genetic factors, as well as "reason," play an important role in shaping political behavior; (d) that although "indoctrinability" sometimes enables (or may seemingly compel) us to act in a manner inconsistent with or even counter to our genetic inclinations, over the long run the latter materially influence our political behavior.

In short, the nature of human nature has been and continues to be a, perhaps *the* pivotal issue in political philosophy. While ethology, primatology, and evolutionary theory have given us a much better understanding of that nature, these findings do not, and probably cannot, provide an *unchallengeable* base on which to build a normative political philosophy. Still, it would be truly rash to ignore the implications of these findings for political philosophy generally and for democratic political philosophy – and democratic governments – especially.

B. Can a "More Biologically Oriented" Political Science Aid in the Formulation of Public Policy?

With regard to *some* types of public policy questions, the answer is "yes." This is surely the case with those policies whose success or failure turns on what David Easton (1975) called the "organic limits" of human behavior.

A not altogether objective observer has written rather unflatteringly about the proclivity of "liberals" to espouse (and if possible) enact policies whose success depends on the assumption that human nature is infinitely malleable so that a new, better, and perhaps perfect nature can be produced by the rearrangement of social institutions (Bork, 1996: 27). If an evolutionary perspective teaches anything, it is that human behavior, just as that of any other species, is not infinitely malleable.

A more biologically oriented perspective, then, can provide a valuable "reality" check in assessing a proposed policy, program, or even a political movement. It would, and should, ask: Are the professed ends and/or the proposed means consistent with what we know about our species' behavioral predispositions? If not, might it not be wise to modify, if not abandon, the

former rather than attempt, by some combination of persuasion and brute force, drastically to alter the latter? The collapse of the Soviet empire is simply another chapter in the long history of failures to recognize these inherent constraints.[16] (So, too, we might add, were the utopian expectations about the brave, new democracies which were envisioned as arising from the ashes of that empire.)

Beyond doubt, an evolutionary perspective can be of inestimable value in applying the touchstone of behavioral reality to public policy proposals. When we turn to other aspects of policy making, though, the extent to which that outlook can contribute in other ways is not altogether clear.

It can very usefully continue, as reported above, to call attention to actual or incipient problems – the environment, AIDS, population, biological terrorism, etc. – which should be addressed but which somehow have yet to receive their proper place on the political agenda. What is not immediately apparent, though, is the degree to which that outlook can of itself further, in substantive policy terms, the resolution of those problems.

A specific example might be helpful here. To take a currently urgent issue, what should done about the way in which humans are polluting their environment? Darwinian theory would seem to suggest that, if a species can be said to have a purpose, it is to perpetuate itself. But even if we take this as our point of departure, difficulties immediately arise. First, does "perpetuate" mean to seek maximum population or, rather, to hold population to what are defined as optimal (itself a matter of sure controversy) limits? The answer(s) offered may dictate quite different policy directions.

Second, the idea that sound policy should be concerned with furthering the welfare of the "species" implies a unity and commonality of interest among the component members of that species which is strikingly absent in *Homo sapiens*. In the majority of political societies, there are often vast differences in economic and political, not to mention religious, ethnic, and racial, interest. Still greater, in some instances, is the economic chasm between the wealthy and the impoverished nations. The former may be quite ready, able and willing to pay the cost of environmental improvement for themselves and for their children; in sharp contrast, for many (perhaps most) third-world nations, environmental protection and improvement could entail serious adverse economic consequences and be literally an unthinkable luxury. Under these circumstances, an evolutionary outlook may well lead to an unequivocal agreement that *something* should be done but to profound disagreement as to what that something should be (and just who should pay for it.) These, however, are precisely the circumstances encountered in many and probably most policy areas and why, in a species so rich in conflicting interests, the dictum "do that which is most likely to benefit the members of the species" is hardly a useful guide to action.

C. A Separate Professional Organization (and journal)?

At the Paris (1975) meeting mentioned above, William Mackenzie urged that political scientists advocating a "more biological approach" should view that approach as "a possible focus of all existing [fields]" rather than as a "new sector" of the discipline. As reported above, that counsel was not heeded; within a decade, those working in biopolitics (Mackenzie also argued forcefully against that term) had established their own professional association *and* their own journal. In principle, Mackenzie may have been right; but, as he himself acknowledged, given the manner in which academic *American* political science was (and is) organized, the strategy he advocated might not have been very effective.

A quarter-century later, several developments have brought that issue again to the fore. One is APLS' recent loss of recognition by the American Political Science Association; a second is that APLS membership is no longer predominantly composed of political scientists; and a third stems from the financial and managerial problems with which that organization is now struggling.

The pivotal question is whether, on balance, a separate professional association and journal will *now* do more to aid/further or to hinder/impede the effort to convince mainstream political scientists that they should adopt a more evolutionary perspective. The second, more pragmatic question is whether the current financial and managerial problems can or cannot be satisfactorily resolved. What the ultimate outcome will be we do not know – but, given the tendencies of organizations to perpetuate themselves, we suspect the decision is more likely to turn on the latter, rather than the former, set of considerations.

VI. A LOOK AHEAD

Aside from the organizational issue, there is another even graver problem for the biopolitical enterprise which looms immediately ahead. Ironically enough, those working in biopolitics may soon be in imminent danger of violating the most important of all evolutionary (not to mention Biblical) commandments – thou shalt reproduce yourself. They are, in fact, on the verge of becoming an endangered species.

Two factors have combined to raise that danger. The first is demographic. With few exceptions, the "founding fathers" have retired or are about to retire. Many of them, to be sure, continue to pursue their research and writing – but they are themselves no longer actively recruiting and training graduate students. That responsibility, as academic practice mandates, has been passed to the next generation of scholars in biology and politics.

But the present academic environment, the second operative factor, is hardly conducive to the successful accomplishment of those tasks. With few exceptions, the great majority of political scientists are opposed to or, at best, disinterested in a more biopolitical approach. None of the top-ranking graduate departments, from whom a doctorate still carries by far the best prospect of future academic employment, provides anything resembling systematic training in the "field"; the only formal program is at a good, but not top-tier institution.

The outlook is further darkened by accumulating, though admittedly anecdotal, evidence that many political science departments discount (or even worse, view negatively) biopolitically oriented articles and books and articles when weighing appointments and promotions.[17] No matter how greatly that approach may appeal to them, doctoral candidates can hardly ignore these environmental realities in choosing their formal fields of specialization. Under such unfavorable circumstances, they may well decide that Darwin, like heaven, can wait.

Nonetheless, granting that the near-term outlook is not altogether encouraging, we think that not very many years from now many more political scientists will have accepted and adopted an evolutionary perspective. Within the profession, there will be an inevitable generational replacement; younger practitioners will have less invested in, and intellectually more willing to move away from, the presently dominant "behavior is learned" paradigm. Paradoxically enough, the major impetus for such a shift may well come from outside, rather than from within, the discipline. Over the past decade or two, neo-Darwinism has been making impressive headway in the other social and behavioral sciences – anthropology, economics, psychology, sociology, even history and jurisprudence.[18] Historically, political science has not hesitated to borrow its concepts from its sister sciences (the wholesale acceptance of "rational choice" theory is but the most recent example) and this is especially likely to be the case with so powerful an explanatory theory as the neo-Darwinian synthesis.[19]

Admittedly, a quarter century ago we were overly optimistic about the likelihood of fundamentally changing the way the profession perceives, studies and explains political behavior. Looking at what is happening in neighboring disciplines, we are increasingly confident – and surely very much hope – that we have not made the same mistake once again.

NOTES

1. Contemporary evolutionary theory, it is important to note here, insists that, for a proper understanding of behavior, we must give proper weight to both learning *and* genetic tendencies.

2. A paradigm shared, at least until very recently, with almost all the other social and behavioral sciences.

3. There is, to be sure, yet another aspect to biopolitics, one concerned with the policy issues that have a significant biological component. Prominent among these issues, to mention some of the best known, are abortion, birth control, drug and alcohol regulation, genetic engineering, water resources, the environment, biological warfare and terrorism, radiation control, risk assessment, and AIDS policy (see, e.g. Cole, 1998; Bonnicksen, 1998; Blank, 1997; Paul, 1998; Caldwell, 1997; J. Schubert, 1997). But, since the biopolitical concern with these issues has been, at best, only peripherally related to political behavior, we will not deal with this area in this essay.

4. Some key works on evolutionary theory include: Darwin, 1859; Dobzhansky, 1951; Fisher, 1930; Huxley, 1942; Mayr, 1970; Simpson, 1944.

5. Among the popularized introductions to evolutionary bases of human behavior, see Ardrey, 1966; Dawkins, 1976; Lorenz, 1966; Morris, 1967; Tiger, 1969; Tiger and Fox, 1971; Wilson, 1978.

6. *Darkness at Noon* and *Nineteen Eighty Four* are, we think it safe to say, the best-known classical fictional portrayals of the process.

7. Our recollection is that most of those present opposed such a move.

8. Somit and Peterson, 1990. For new literature since then, see, for example, the "Recent Books and Articles" section, coming near the end of each issue of the journal, *Politics and the Life Sciences*.

9. Readers who would like to examples of illustrative works from biology and politics in the traditional areas of the discipline might wish to explore the following as a starting point. Of course, the following is hardly an exhaustive listing and many more works could easily have been included. However, this table does serve to provide a sampling of what is available.

Area within Political Science	Examples of Biopolitics Literature
International Relations	Falger, 1997 van der Dennen, 1995 Wiegele, 1979: chapter 5
Methodology	Lodge, 1981 J. Schubert, 1988b Somit and Peterson (1996), especially the chapters by Barner-Barry, J.. Schubert, Mazur and Mueller, Salter Wahlke and Lodge, 1972
Comparative Politics	Flohr, 1994 Mazrui, 1972 Phillips, 1980 Vanhanen, 1984 Watts, 1994
Political Theory	Arnhart, 1998 Corning, 1983 Hartigan, 1988

Political Theory (cont.)	Masters, 1989, 1992 Somit and Losco, 1995
Public Administration	Losco, 1994 White and Losco, 1986
Public Law	Kort, 1994 G. Schubert, 1997
Public Policy	Bonnicksen, 1998 Blank, 1997 Caldwell, 1964, 1997 Carmen, 1986 Caton, 1988 Funke, 1985 Wiegele, 1985
Political Behavior	Davies, 1976 Johnson, 1986 Laponce, 1976 Peterson, 1996 J. Schubert, 1988a Schwartz, Garrison, and Alouf, 1975 Welch, 1976 Willhoite, 1976

10. For a more detailed, but earlier, discussion of this subject, see Somit and Peterson (1998).

11. Of the 31 international politics textbooks published or revised since 1995, almost all briefly mention this possibility. But even this near unanimous inclusion is not quite as encouraging as it might initially appear. With only some exceptions, the references are to the decades-old popular ethological books noted above; the much more recent biopolitical literature is almost totally ignored (Arwine et al., 2000).

12. The field of political socialization, whose basic tenets run counter not only to evolutionary theory but to developmental psychology, is a classic example of this Nelsonian tactic.

13. Dennett, 1995.

14. "Inclusive fitness," a key term in sociobiology, is defined as (Barash, 1982: 392) "the sum of an individual's fitness as measured by reproductive success and that of relatives, with those relatives devalued in proportion to their genetic distance, i.e. as they share fewer genes." In the final analysis, then, individuals ought to behave in such a way as to maximize the number of his or her genes in the next generation. Since relatives share genes, there is some motivation for an individual to also invest energy to enhance the reproductive success of relatives.

15. For lack of a better term, this unique capacity is called "indoctrinability." For greater detail, see Somit and Peterson, 1997.

16. Perhaps, as Lionel Tiger speculated (1975), there may be some genetically transmitted "design defect" which prevents us from comprehending these limitations?

17. Fortunately, this was a problem spared the "founding fathers," almost all of whom had achieved tenure and full professorships when they initially launched the new heresy.

18. See, for example, Somit and Peterson, in press.

19. "The most coherent and all-encompassing theory that scientists have ever had to explain the living world" (Hrdy, 1999: xvii).

REFERENCES

Ardrey, R. (1966). *The Territorial Imperative*. New York: Dell Publishing.

Arnhart, L. (1998). *Darwinian Natural Right*. Albany, NY: SUNY Press.

Arwine, A., Wang, H., Somit, A., & Peterson, S. A. (2000). Impact of biology on political science. Unpublished manuscript, Texas Tech University.

Barash, D. P. (1982). *Sociobiology and Behavior*, 2nd edition. New York: Elsevier.

Blank, R. H. (1997). *The Price of Life*. New York: Columbia University Press.

Bonnicksen, A. L. (1998). Transplanting nuclei between human eggs. P*olitics and the Life Science*, *17*, 3–10.

Booth, A., & Welch, S. (1976). Stress, health, and political participation. Presented at *International Political Science Association*, Edinburgh, Scotland.

Bork, R. M. (1996). *Slouching toward Gomorrah*. New York: Regan Books.

Caldwell, L. K. (1964). Biopolitics: science, ethics, and public policy. *The Yale Review*, *54*, 1–16.

Caldwell, L. K. (1997). *Environmental Policy*. Westport, CT: Greenwood.

Carmen, I. (1986). *Cloning and the Constitution*. Madison: University of Wisconsin Press.

Caton, H. (1988). *Regulating Biomedicine*. Brisbane: Bonhoeffer Institute.

Cole, L. (1998). The poison weapons taboo. *Politics and the Life Sciences*, *17*, 119–132.

Corning, P. (1983). *The Synergism Hypothesis*. New York: McGraw-Hill.

Darwin, C. (1859). *On The Origin of Species*. London: Murray.

Davies, J. C. (1976). Ions of emotion and political behavior: a prototheory. In: A. Somit (Ed.), *Biology and Politics*. The Hague: Mouton.

Dawkins, R. (1989). *The Selfish Gene*, 2nd edition. London: Oxford University Press.

Degler, C. (1991). *In Search of Human Nature*. Oxford: Oxford University Press.

Dobzhansky, T. (1951). *Genetics and the Origin of Species*, 3rd edition. New York: Columbia University Press.

Easton, D. (1976). The relevance of biopolitics to political theory. In: A. Somit (Ed.), *Biology and Politics*. The Hague: Mouton.

Falger, V. (1997). Human nature in modern international relations. In: A. Somit & S. A. Peterson (Eds), *Recent Explorations in Biology and Politics*. Greenwich, CT: JAI Press.

Fisher, R. A. (1930). *The Genetical Theory of Natural Selection*. Oxford: Clarendon Press.

Flohr, H. (1994). Political sociology in biocultural perspective. In: A. Somit & S. A. Peterson (Eds), *Biopolitics and the Mainstream*. Greenwich, CT: JAI Press.

Funke, O. (1985). Biopolitics and public policy. *PS*, *18*, 69–76.

Gilbert, R. E. (1992). *The Mortal Presidency*. New York: Basic Books.

Goodin, R. E., & Klingemann, H.-D. (Eds) (1997). *A New Handbook of Political Science*. Oxford: Oxford University Press.

Hartigan, R. S. (1988). *The Future Remembered: An Essay in Biopolitics*. Notre Dame: Notre Dame University Press.

Hrdy, S. B. (1999). *Mother Nature – A History of Mothers, Infants, and Natural Selection*. New York: Pantheon.

Huxley, J. (1942). *Evolution: The Modern Synthesis*. London: Allen & Unwin.

Johnson, G. R. (1986). Kin selection, socialization, and patriotism. *Politics and the Life Sciences, 4*, 127–140.

Kitchin, W., & Peterson, S. A. (1996). Hemispheric dominance and elite political behavior. In: A. Somit & S. A. Peterson (Eds), *Research in Biopolitics*, Volume 4. Greenwich, CT: JAI Press.

Kort, F. (1994). A biobehavioral approach to law as a component of political processes and institutions. In: A. Somit & S. A. Peterson (Eds), *Biopolitics and the Mainstream*. Greenwich, CT: JAI Press.

Laponce, J. (1976). The left-hander and politics. In: A. Somit (Ed.), *Biology and Politics*. The Hague: Mouton.

Lodge, M. C. (1981). Magnitude Scaling. Sage University Series on *Quantitative Applications in the Social Sciences*, 07–001. Beverly Hills: Sage Publications.

Lorenz, K. (1966). *On Aggression*. New York: Bantam Books.

Losco, J. (1994). Biology and public administration. In: A. Somit & S. A. Peterson (Eds), *Biopolitics and the Mainstream*. Greenwich, CT: JAI Press.

Mackenzie, W. (1976). Biopolitics: a minority viewpoint. In: A. Somit (Ed.), *Biology and Politics*. The Hague: Mouton.

Madsen, D. (1986). Power seekers are different: further bio-chemical evidence. *American Political Science Review, 80*, 261–270.

Marcus, G., Wood, S. L., & Theiss-Morse, E. (1998). Linking neuroscience to political intolerance and political judgment. *Politics and the Life Sciences, 17*, 165–178.

Masters, R. D. (1989). *The Nature of Politics*. New Haven: Yale University Press.

Masters, R. D. (1993). *Beyond Relativism*. Hanover, NH: University Press of New England.

Masters, R. D. (1998). Neurotoxicity and criminal violence. Presented at *Association for Politics and the Life Sciences* meeting, Boston.

Mayr, E. (1970). *Populations, Species, and Evolution*. Cambridge: Harvard University Press.

Mazrui, A. (1972). Political man and the heritage of hair. *British Journal of Political Science*, 2, 1–20.

Mazur, A., & Mueller, U. (1996). Facial dominance. In: A. Somit & S. A. Peterson (Eds), *Research in Biopolitics*, Volume 4. Greenwich, CT: JAI Press.

Morris, D. (1967). *The Naked Ape*. New York: Dell.

Paul, D. B. (1998). *The Politics of Heredity*. Albany: SUNY Press.

Peterson, S. A. (1996). Survey research measures in biopolitics. In: A. Somit & S. A. Peterson (Eds), *Research in Biopolitics*, Volume 4. Greenwich, CT: JAI Press.

Peterson, S. A., & Somit, A. (1994). Research methods derived from the life sciences. In: A. Somit & S. A. Peterson (Eds), *Biopolitics and the Mainstream*. Greenwich, CT: JAI Press.

Phillips, C. S., Jr. (1980). Cultural evolution and political development. Presented at *International Studies Association*, Los Angeles.

Salter, F. (1996). 'Drawn by light': visual recording methods in biopolitics. IIn: A. Somit & S. A. Peterson (Eds), *Research in Biopolitics*, Volume 4. Greenwich, CT: JAI Press.

Schubert, G. (1976). Politics as a life science. In: A. Somit (Ed.), *Biology and Politics*. The Hague: Mouton.

Schubert, G. (1991). *Sexual Politics and Political Feminism*. Greenwich, CT: JAI Press.

Schubert, G. (1997). Oral argument in Supreme Court decision-making. In A. Somit & S. A. Peterson (Eds), *Recent Explorations in Biology and Politics*. Greenwich, CT: JAI Press.

Schubert, J. N. (1988a). Age and active-passive leadership style. *American Political Science Review*, *82*, 763–772.

Schubert, J. N. (1988b). Politics under the microscope. *International Political Science Review*, *9*, 305–325.

Schubert, J. N. (1997). Political culture and public health strategies in state policy response to AIDS. In: A. Somit & S. A. Peterson (Eds), *Recent Explorations in Biology and Politics*. Greenwich, CT: JAI Press.

Schwartz, D. C., Garrison, J., & Alouf, J. (1975). Health, body images, and political socialization. In: D. C. Schwartz & S. K. Schwartz (Eds), *New Directions in Political Socialization*. New York: The Free Press.

Simpson, G. G. (1944). *Tempo and Mode*. New York: Columbia University Press.

Somit, A. (1972). Biopolitics. *British Journal of Political Science*, *2*, 209–238.

Somit, A. (Ed.) (1976). *Biology and Politics*. The Hague: Mouton.

Somit, A., & Losco, J. (Eds) (1995). *Human Nature and Politics*. Greenwich, CT: JAI Press.

Somit, A., & Peterson, S. A. (1990). *Biopolitics and Mainstream Political Science*. DeKalb, IL: Program for Biosocial Research.

Somit, A., & Peterson, S. A. (Eds) (1996). *Research in Biopolitics*, Volume 4. Greenwich, CT: JAI Press.

Somit, A., & Peterson, S. A. (1997). *Darwinism, Dominance, and Democracy*. Westport, CT: Praeger.

Somit, A., & Peterson, S. A. (In press). *Darwinian Theory and the Social and Behavioral Sciences*. Stamford, CT: JAI Press.

Sullivan, D. G., & Masters, R. D. (1988). Happy warriors: Leaders' visual displays,
viewers' emotions, and political support. *American Journal of Political Science*, *80*, 261–270.

Tiger, L. (1969). *Men in Groups*. New York: Vintage Books.

Tiger, L. (1976). Ions of emotion and political theory: notes on a prototheory. In: A. Somit (Ed.), *Biology and Politics*. The Hague: Mouton.

Tiger, L., & Fox, R. (1971). *The Imperial Animal*. New York: Holt, Rinehart & Winston.

van der Dennen, J. M. G. (1995). *The Origin of War*. Groningen: Origin Press.

Vanhanen, T. (1984). *Emergence of Democracy*. Helsinki: Finnish Society of Sciences and Letters.

Wahlke, J. C. (1976). Observations of biopolitical study. In: A. Somit (Ed.), *Biology and Politics*. The Hague: Mouton.

Wahlke, J. C. (1979). Pre-behavioralism in political science. *American Political Science Review*, *73*, 9–32.

Wahlke, J. C., & Lodge, M. (1972). Psychophysiological measures of political attitudes and behavior. *Midwest Journal of Political Science*, *16*, 505–537.

Watts, M. W. (1994). Biopolitical contributions to comparative politics. In: A. Somit & S. A. Peterson (Eds), *Biopolitics and the Mainstream*. Greenwich, CT: JAI Press.

Welch, S. (1976). Crowding and political activity. *Polity*, *9*, 40–62.

White, E. & Losco, J. (Eds) (1986). *Biology and Bureaucracy*. Lanham, MD: University Press of America.

Wiegele, T. C. (1979). *Biopolitics: Search for a More Human Political Science*. Boulder: Westview Press.

Wiegele, T. C. (1985). The biotechnology of sex preselection. *Policy Studies Review*, *4*, 445–460.

Wiegele, T. C., & Hirsch, L. (1981). Methodological aspects of voice stress analysis. In: M. Watts (Ed.), *Biopolitics: Ethological and Physiological Approaches*. San Francisco: Jossey-Bass.

Willhoite, F. Jr. (1976). Primates and political authority. *American Political Science Review*, *70*, 1110–1126.

Wilson, E. O. (1978). *On Human Nature*. Cambridge: Harvard University Press.

IS PSYCHOLOGY UNDERGOING A PARADIGM SHIFT? PAST, PRESENT, AND FUTURE ROLES OF EVOLUTIONARY THEORY

Irwin Silverman and Maryanne L. Fisher

PSYCHOLOGY AS A PROXIMATE SCIENCE

Scientists, in their endeavors, deal with different levels of causation, which can be conceptualized on a continuum of *proximate* to *ultimate*. Proximate causes refer to more immediate variables, ultimate to those more remote. For example, if you were asked why you were reading this chapter, a proximate explanation might be that it was recommended for a course you were teaching or taking; an ultimate reason might be that your species is innately curious.

Levels of causation have much to do with the role of evolutionary theory in psychology. Consider this quote from Einstein and Infeld (1961, p. 31):

> In our endeavor to understand reality we are somewhat like a man trying to understand the mechanism of a closed watch. He sees the face and the moving hands, even hears it ticking, but he has no way of opening the case.

Now envision visiting scientists from some extra-terrestrial domain, with no concept of time as we know it, trying to understand a watch (circa 1960s). Assume further that they are divided into two schools. One, the proximate school, maintains that the proper study of the watch is what it does and how it does it, the here and now of the watch's existence. They focus on

Evolutionary Approaches in the Behavioral Sciences, Volume 8, pages 203–216.

ISBN: 0-7623-0769-2

observable variables associated with the watch, such as the moving hands and the ticking sounds, and develop their theories based on the relationships of these to each other and to external events.

The rival school harbors no objection to the study of proximate causes, but claims that one must also ask the ultimate questions. Why does the watch exist? How was it created and for what purpose? Proximists disparage questions of this kind for various reasons, described below. Ultimists, however, remain convinced that their ways will bring them closer to the essential facts about the watch.

For psychology, the watch is analogous to mind, or brain, or psyche, or whatever concept is used to describe the processes that propel and direct behavior. In this vein, psychologists have been traditionally proximists. Their attempts to understand the processes that propel and direct behavior are largely limited to *ontogenetic* development, events that occur during the lifespan of an organism. *Phylogenetic* development, comprising evolutionary bases of behavior and the genetic predispositions that these have engendered, has a very insignificant role in psychological theorizing.

Consider, for example, the question of why most people enjoy sweet tastes and find bitter unpleasant. The proximate explanation is that stimulation of taste receptors that are sensitive to sweet evoke central nervous system responses that are experienced as pleasurable, while stimulation of taste receptors sensitive to bitter elicit disagreeable sensations. But why should this occur? Why should our tongues and brains be connected in this manner? The ultimate cause is that in our evolutionary history as forest dwellers, individuals with an innate preference for sweet and aversion to bitter tended to eat only ripe fruits, which were rich with sugars, while those without this preference were less discriminating. The former would have been more apt to survive and reproduce, and so the genetic predisposition for these particular tastes was passed on by natural selection (Barash, 1982, p. 146).

Psychologists in the area of perception have been thorough in their studies of neural processes underlying taste and all other modalities. One would be hard pressed, however, to find the evolutionary theory of why sweet is sweet and bitter is bitter. Nor are there any attempts to explain, beyond the occasional superficial conjecture, how and why particular perceptual processes evolved in any sensory domain.

This eschewal of evolutionary theory and ultimate causation pervades every area and conceptual approach within psychology. Ask psychologists why men tend to be more aggressive than women. A physiological psychologist will probably answer that male testosterone levels are higher. A social psychologist might say that males possess stronger tendencies to establish status hierarchies

within their social groups. A developmental psychologist will tell you that males are socialized to be more aggressive by their parents and society. All are valid answers, as far as they go. None, however, deal with ultimate origins. Why is testosterone level linked to aggression? Why do some species, humans among them, form social groups and status levels within these and some not? Why do parents, across cultures and species, tend to treat male offspring differently than females?

The topic of family occupies much of psychology. Clinical, developmental, personality, and social psychologists devote considerable resources to questions about familial processes. Who is attracted to whom? Who marries whom? What determines how they get along? How will they socialize their children? How will the children be affected by their socialization, etc., etc.? Nowhere in the process, however, does the question arise of why people tend to live in families in the first place. Of the million or so identified animal species on the planet, some live in families, others do not. Those that do show a variety of patterns in addition to the so-called nuclear family of mother, father and dependent children which characterizes our species. For example, there are family units comprised solely of a mother and her offspring, or multiple mothers and their offspring, or multiple mothers with fewer fathers. For the evolutionist, the primary questions for the scientific study of families, human or otherwise, are why does family living sometimes occur and sometimes not, and why does it take the various configurations that it does (Trivers, 1985; van den Berghe, 1979).

Human language is another topic that attracts considerable attention in psychology, but again, theories begin and end in ontogeny. Psychologists endeavor to explain the progression from babble and coo to the most complex levels of communication, but none attempt to incorporate their concepts into the ultimate question of why we have language. All species have communication systems of varying nature and complexity, and it seems critical for the study of human language to know how and why our particular version originated.

This pure proximate approach to language was probably basic to psychology's most futile quest, the attempts over decades to teach human language skills to chimpanzees and other infrahuman primates (see Gardner, Gardner & Van Cantfort, 1989). Of course, no one expected the chimps to chat. Perhaps they would learn five symbols, perhaps 50 or 500. They might even learn to put two together, as Washoe appeared to in the well-known watermelon anecdote. But what conceptual goal could be served by such information? Whatever the results of these studies, they were bound from the outset to be uninterpretable in the absence of theory about the ultimate reasons for communication systems and their diversity.

Neural organization, family structure, status hierarchies, socialization practices, language; all are included among the *givens* of traditional psychology. They are regarded as a priori, simply there, like the movements and sounds of the watch to our proximist, inter-planetary scientists.

PSYCHOLOGISTS' TRADITIONAL ARGUMENTS FOR ESCHEWING EVOLUTION AND THEIR REFUTATIONS

The Primacy of Culture Argument

How do traditional psychologists justify their exclusive proximate outlook? The most common argument is that evolutionary theory may be highly relevant to infrahuman species but not to humans, inasmuch as human behavior is largely determined by culture.

The attribution of behavior to culture, however, does not in itself represent a scientific explanation. It relegates the development of the behavior to happenstance. An individual does what he does because he happened to be born in such and such region, or ethnic group, or social class, or whatever. Inuits isolate their aged parents because that is what Inuits do; it is part of their culture. West Indians take their aged parents into their households because that is what West Indians do; it is their culture. The assumption appears to be that somewhere along the way, some influential Inuit happened to treat his aged parents one way and some influential West Indian happened to treat his aged parents in another, and both groups have followed suit ever since.

The theory of culture has no explanatory value unless it includes the ultimate question of how and why cultural traits evolved in the ways that they did (Tooby & Cosmides, 1992). To say simply that something is due to culture is to say that it came about by accident. One is privileged to believe that accidents occur, but the epistemology of science is rooted in the premise that all events have definable causes.

Form and Function

The intrinsic interrelatedness of evolution and culture may be demonstrated in another manner. If asked why human hands evolved, the apparent answer is that the things that hands do, like gripping, carrying, and manipulating, conferred adaptive advantages to their owners. This illustrates the axiom of evolutionary theory that form follows function.

The evolutionary approach to psychology is that this axiom applies as well to the brain. Thus, the cognitive and communicative functions of the brain

involved in the group enterprises called cultures evolved because these enterprises conferred adaptive advantages to those who engaged in them. In contrast, the primacy of culture argument rests on the premise that function follows form; that is, the expanded functions of the human brain simply appeared for some reason and the cultural developments that represent the expression of these functions followed (e.g. Kuper, 1994), a proposition that breaches both the rules of parsimony and logic.

The Teleology Argument

Another frequent defense of the proximist view is that evolutionary explanations tend to be teleological; that is, they embrace the fallacy of explaining causes in terms of consequences. Evolutionary based explanations are often referred to by critics as "just-so stories," signifying that the investigator began with a notion of the apparent purpose of a behavior and then created a suitable story of its origins.

This also is misguided epistemology. Though teleological explanations may sometimes be misleading, as suggested in the frequently cited satirical paper by Gould and Lewontin (1979), they play an indispensable role in the inductive stages of inquiry. Mayr (1983) has shown that most major advances in physiology began with observations about the purpose of a given structure or system for the organism.

Paradoxically, not only has traditional psychology failed to reap the benefits of insightful, generative, just-so stories, but its eschewal of evolutionary theory has rendered it more susceptible to the potential pitfalls of teleological thinking. In the absence of a conceptual basis for formulating explicit assumptions about ultimate origins, psychological theories often harbor implicit and simplistic assumptions that mirror casually observed consequences. For example, because people kill and maim others in warfare, psychologists' assumptions are that the causes of war are to kill and maim others, and it has been explained in terms of the effects of frustration on aggression or displaced death instincts or the like. Because the rapist inflicts harm and suffering upon women, it is assumed his motive is to inflict harm and suffering upon women, and psychological explanations of his motives focus on variables such as anger toward women or impaired capacity for normal sexual function. Because learning often results in concrete, immediate rewards, psychologists assume that animals and humans learn in order to obtain concrete, immediate rewards, and the conditioning paradigm became psychology's prevailing, general theory of behavior.

Furthermore, evolutionary oriented psychologists are as committed to operationalism and empiricism as their traditional counterparts. They do not regard

their theories as truisms, but as conceptual tools, assessed in terms of the scope and predictive value of the testable hypotheses they engender. It will be on these grounds that the contributions of evolutionary theory will be judged.

The Alternative Approaches Argument

Not all traditional psychologists reject the relevance of evolutionary theory to modern human behavior. They maintain, however, that proximate and ultimate causation represent separate approaches, and one can effectively pursue psychological knowledge within one or the other. This is also a flawed notion. First, the dichotomy is not between proximate and ultimate explanations, but between a purist experiential model and a model based on the interaction between evolutionary based innate predispositions and environmental events.

Additionally, explanations that rely exclusively on proximate causation characterize a primitive mode of thinking. If a young child asks where rain comes from, an adult will probably tell him from the clouds. If the child asks how babies are made, the answer will probably be that they grow in mommies' bellies or something similar. Most young children are content with proximate answers of this sort; in fact, they may become confused and upset by more detail. But as they mature, they will require more definitive, ultimate level explanations. Thus, the infusion of evolutionary theory into psychological science may signal its emerging maturity. In fact, no sciences other than the behavioral sciences ignore their ultimate questions. Biologists are deeply concerned with the origins of life, physicists with the origins of matter, and progress on these questions are a mark of the progress of the disciplines.

THE CONTEMPORARY EVOLUTIONARY MOVEMENT IN PSYCHOLOGY

Beginnings

Most writers (e.g. Buss, 1999; Crawford, 1998) attribute the current evolutionary trend in psychology to the neo-Darwinian movement in behavioral biology heralded by Hamilton (1964), Williams (1966), Trivers (1971), Maynard Smith and Price (1973), and others, and reaching an apex with Wilson's (1975) treatise on sociobiology as the grand synthesis of the biological and behavioral sciences.

Though the biological movement was undoubtedly significant, there was a quieter revolution within psychology at about the same time, largely in response to shortcomings of behaviorism as the prevailing general theory in the field.

Breland and Breland (1961) was a classic contribution in this vein, describing the spontaneous substitution of behaviors which had been conditioned by food reinforcers by "instinctive behaviors having to do with the natural food getting behaviors of the particular species," (for example, a chicken conditioned to propel plastic capsules with its beak suddenly tried to peck them open, as if they were seed pods; a raccoon conditioned to place coins in a bank began to rub them together instead, as if washing its food). The authors concluded that, ". . . the behavior of any species cannot be adequately understood, predicted, or controlled without knowledge of its instinctive patterns, evolutionary history, and ecological niche" (pp. 683–684).

Shortly thereafter, Garcia and Koelling (1966) demonstrated that, contrary to behaviorism's core principle of equipotentiality of reinforcers, organisms were programmed in an apparently adaptive manner such that specific behaviors conditioned more or less readily to specific consequences. Attempting to condition taste aversions in rats, they discovered that a small dose of ionizing radiation, sufficient to cause nausea, was highly superior to any other negative reinforcer. Furthermore, counter to the behavioristic principle of contiguity, conditioning could be accomplished in one trial even if nausea was induced several hours after the animal had tasted the food in question (Garcia, Ervin & Koelling, 1966).

A third example was Harlow and Zimmerman's (1959) finding that monkeys placed from birth with inanimate surrogate mothers with either wire or terrycloth surfaces developed attachments to the latter, to whom they could cling, even when the former functioned as their source of food and water. Thereby, the behavioristic explanation of filial attachment as an acquired drive was refuted in favor of the notion that infants possessed an innate, adaptive need for continuing contact with an appropriate maternal figure. Follow-up studies (Harlow, 1971) revealed that optimal social development required that this contact occurred within an innately programmed critical period.

Among other notable psychological contributions of this era relating to evolutionarily adaptive aspects of early attachment were Fantz (1961), who demonstrated an innate preference for newly born human infants to gaze at a schematic of a frontal human face, and Moltz (1960), whose theory of the precise role of genetic and environmental factors in social imprinting represented a prototype for the interactionist perspective of contemporary investigators.

Progress

How salient is the evolutionary movement? How much of an impact has it had on traditional psychology? In the following studies, we searched for answers among the archives and among our colleagues.

Table 1. Frequency of Keywords Across Topics and Time Intervals.

Term	1970–1979	1980–1989	1990–1999	% increase (+) or decrease (−) from 1970s to 1990s
Evolution	1615	3888	6590	+308.0
Conditioning	9948	7001	7381	−34.8
Personality	27837	36308	44227	+58.9
Memory	9998	16235	30558	+205.6
Psychotherapy	9792	17298	24738	+152.6
Socialization	1624	3158	3855	+137.4
Perception	30188	35863	48854	+61.8

Database survey: The PsychINFO database is one of the largest and most comprehensive in psychology. Its sources are journal articles, books, book chapters, technical reports, and dissertations, encompassing all areas within or related to the field.

Table 1 shows the results of a search of specific keywords within PsychINFO, separately for the years 1970–1979, 1980–1989, and 1990–1999. Our primary interest was in changes in the frequencies of publications containing *evolution* as a keyword during these 30 years. Our comparison samples comprised other broad keywords in diverse areas of traditional psychology, selected on the basis of a perusal of chapter headings in current introductory psychology texts.

Evolution showed the largest increase across the 30 years at 308%, significant by Chi-squared analyses from other headings at $p < 0.001$ in all cases. *Memory* showed the second largest gain, which may be a function of advances in brain imaging techniques and/or current emphasis on gerontological issues paralleling increased proportions of the elderly in the population. *Conditioning* was the only keyword showing a decrease, possibly reflecting the beginnings of a "paradigm shift" (Kuhn, 1970) toward evolution and genetics in favor of purely environmental theories.

Introductory Text Survey: Has the increase in evolutionary content in primary sources become manifest in textbooks? We compared samples of traditional introductory psychology texts published during the years 1977–1979, 1987–1989,

and 1997–1999, using ten books in each interval. The source was York University's library and selection was random, with the qualifications that later editions by the same author(s) and texts specifically oriented to or containing one or more chapters devoted to evolution were excluded.[1]

The index of each text was used to locate evolutionary related content, which we measured first in terms of pages. If the content occupied less than a full page, a proportion was calculated by counting lines. Upon discovering that psychology introductory textbooks increased in size over time,[2] page counts were converted to percentages of space within the text.

Based on the decrease for *conditioning* observed in the database survey and its possible implications for a paradigm shift, we did a similar analysis for conditioning related content.

Results are shown in Fig. 1. The anticipated pattern over time of an increase for evolutionary content and a decrease for conditioning content was discernible, but the interaction term for a 2 × 3 ANOVA for content by time interval did not reach significance ($F = 2.11$, $p = 0.13$). A one way ANOVA, however, showed that the trend for evolutionary content approached significance ($F = 2.80$, $p = 0.08$). Given the directions and strength of these trends, it is plausible that the failures to achieve significance were due to small Ns.

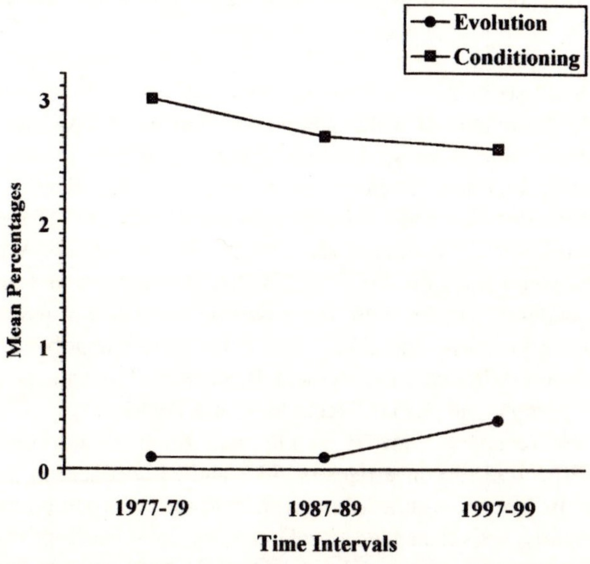

Fig. 1. Mean percentages of space given to evolutionary and conditioning related content in introductory psychology textbooks by time intervals.

There was a significant main effect for content ($F = 77.21$, $p < 0.001$) favoring conditioning. Despite the contrasting trends over time, the percentage of space devoted to conditioning in 1997–1999 was still six times greater than that for evolution. Considering the close numbers between the two in psychology literature keyword frequencies for the 1990s (Table 1), we might conclude that evolutionary theory does not yet seem to be receiving its deserved attention in introductory texts.

We also examined the distribution of topics within evolutionary related content for 1997–1999 texts. This was done by recording brief topical descriptions for relevant references. For each text, if a reference was cited multiple times or if multiple references were used to describe the same specific body of research, the topical description was included only once. These topical descriptions were then subsumed into more encompassing headings and tabulated across texts.

Results showed a diverse distribution. The heading, *learning, language and cognition* received 17 counts; *sexual attraction and mate choice*, 15; *personality and emotion*, 11; *broad theoretical issues*, 5; *family and social processes*, 4; *sleep*, 3; and *general sources*, 2.

Faculty and Graduate Student Survey: Another way to ascertain the beginnings of a paradigm shift, as Kuhn (1970) described the process, is to compare the attitudes of neophyte and established academics. The former, having had less time to develop vested personal interests in traditional paradigms, would be expected to be more open to new movements.

Thus, we conducted a brief e-mail survey of the psychology faculty and graduate student lists at York University. We asked for opinions on three items, regarding: (1) the extent to which Darwinian and neo-Darwinian theory had influenced their own careers, (2) the extent to which it had influenced psychology in general and, (3) the extent to which it will influence psychology in future generations. Responses were given on a scale of: (1) *not at all*, (2) *slightly*, (3) *moderately*, (4) *extensively*, and (5) *very extensively*. We also asked for number of years since the Ph.D. for faculty respondents and year of study for graduate students, and for both, the academic areas within psychology with which they were primarily identified. The latter were categorized in terms of Clinical (including Clinical Developmental), General Experimental (including History and Theory), and Social Psychology and Personality.

Replies were received from 31 faculty and 66 graduate students, which represented respective response rates of 50% and 31% based on published list memberships. It may be assumed, however, that not everyone on the published lists was currently active; hence, actual response rates were probably higher. Two respondents in each group did not answer the questions directly and were excluded.

Table 2. Perceived Influences of Evolutionary Theory By Academic Status.

Question	Position	N	Mean	Standard Deviation
Own scholarly career?	Professor	29	2.93	1.25
	Graduate Student	64	2.27	1.17
Psychology in general?	Professor	29	3.28	0.92
	Graduate Student	64	3.53	0.76
Psychology in future generations?	Professor	29	3.52	0.87
	Graduate Student	64	3.27	0.87

Table 3. Perceived Influences of Evolutionary Theory By Academic Area.

Question	Area	N	Mean	Standard Deviation
Own scholarly career?	Clinical	50	2.32	1.10
	General Experimental	22	2.95	1.46
	Social/ Personality	20	2.30	1.22
Psychology in general?	Clinical	50	3.58	0.70
	General Experimental	22	3.45	0.86
	Social/ Personality	20	3.15	0.99
Psychology in future generations?	Clinical	50	3.34	0.80
	General Experimental	22	3.64	1.05
	Social/ Personality	20	3.00	0.79

Means for faculty vs. graduate student status, by question, are presented across areas in Table 2. Means for area by question, across status, are shown in Table 3. Proportions of graduate students and faculty were approximately the same for two of the three areas, but there were significantly more graduate student respondents in the Clinical area sample. Based on the overall data analyses, however, it did not appear that this difference could have contributed to spurious effects.

Using a $2 \times 3 \times 3$ ANOVA, significant main effects were observed for question ($F = 13.02$, $df = 2$, $p < 0.001$) and area ($F = 4.25$, $df = 2$ $p < 0.05$), and borderline significance was found for status ($F = 3.86$, $df = 1$, $p = 0.05$).

Main effects for question and status were attributable to a significant interaction between them ($F = 3.28$, $df = 2$, $p < 0.05$), with faculty giving more positive responses on the question regarding influences upon their scholarly careers. Congruent to this finding, the correlations of item responses with time in the field yielded one at borderline significance ($r = 0.37$, $p = 0.05$) for faculty on the same question. Thus it appears that for academics at all levels, those with more experience are more open to evolutionary contributions. This may represent an encouraging finding for the field but was inconsistent with the Kuhnian based hypothesis.

There were no interactions at or approaching significance involving the main effect for area; thus, the conclusion was that members of the Social Psychology and Personality area attributed less influence to evolution, across items, than did their counterparts in other areas. This seemed incongruous inasmuch as research on evolution and behavior of the past three decades has intersected more with the topics of social psychology than any other area of the field. As an illustration, Buss' (1999) popular evolutionary based text contains three chapters on sexual and mating strategies, two on family dynamics, two on intergroup interaction, one on conflict between the sexes, and one on status and social dominance. Crawford and Krebs (1998) edited *Handbook of Evolutionary Psychology* includes chapters on social reciprocity, mate choice, culture, morality, family violence, sexual aggression and environmental crises.

On the other hand, given that the traditional perspectives of social psychology would be most compromised by the new evolutionary paradigm, these data do provide indirect support for Kuhn's theory.

Future Prospects

Our findings suggest that the evolutionary movement in psychology is expanding at a rapid rate, though more extensive data will be required to assess whether this truly represents a paradigm in transition. There is another basis, however, on which to expect that the movement will continue to thrive.

The pristine environmentalist paradigm of psychology is largely a social construction, mimicking the historic antipathy of the public toward evolutionary explanations of molar human behavior. There are several reasons for this antipathy. One is the erroneous assumption that genetic causation implies immutability, which is antithetical to peoples' strongly held beliefs that they are capable of and responsible for individual and social change. Another is the specter of Social Darwinism evoked by the "naturalistic fallacy", the mistaken premise that what is biologically natural is also morally righteous. The third, and probably the most salient reason, is that evolutionary theory confronts seminal questions that often reveal the dark side of human nature, exposing the self-serving hypocrisies that characterize our species. For the most part, people prefer to repress their dark sides.

What, then, has prompted current changes in attitudes? One possibility is that breakthroughs in the science and technology of genetics render it difficult to deny the influences of genes in human psychological development. Furthermore, our rapidly increasing capacity to manipulate genetic structures and functions are bound to create a host of individual and societal ethical issues, requiring decisions by an informed populous.

According to some theories, self-deception is ubiquitous in our species because it has been and remains adaptive in a range of situations (see Lockard & Paulhus, 1988). But if self-deception ceases to be feasible, the alternative adaptive strategy may be to learn to live with the realities.

NOTES

1. Four texts (Brown and Cook, 1986; Buss, 1978; Gray, 1991; Hall, 1984) of this nature were encountered, but only Buss was published during the designated year intervals.

2. Page means were: 547 (SD = 88) for 1977–1979, 644 (SD = 76) for 1987-1989, and 709 (SD = 71) for 1997-1999 ($F = 10.80$, $p < 0.001$).

REFERENCES

Barash, D. P. (1982). *Sociobiology and Behavior*. New York: Elsevier.

Breland, K., & Breland, M. (1961). The misbehavior of organisms. *American Psychologist, 16*, 681–684.

Brown, R. M., & Cook, P. P. (1986). *Introductory Psychology*. New York: Holt, Rinehart and Winston.

Buss, A. (1978). *Psychology: Behavior in Perspective*. 2nd ed. New York: John Wiley and Sons.

Buss, D. M. (1999). *Evolutionary Psychology: The New Science of the Mind*. Boston: Allyn and Bacon.

Crawford, C. (1998). The theory of evolution in the study of human behavior: An introduction and overview. In: C. Crawford & D. L. Krebs (Eds), *Handbook of Evolutionary Psychology* (pp. 3–41). Mahwah, NJ: Lawrence Erlbaum.

Crawford, C., & Krebs, D. L. (Eds) (1998). *Handbook of Evolutionary Psychology*. Mahwah, NJ: Lawrence Erlbaum.

Einstein, A., & Infeld, L. (1961). *The Evolution of Physics From Early Concepts to Relativity and Quanta*. New York: Simon and Schuster.

Fantz, R. L. (1961). The origin of form perception. *Scientific American, 204*, 66–72.

Garcia, J., Ervin, F. R., & Koelling, R. A. (1966). Learning with prolonged delay in reinforcement. *Psychonomic Science, 5*, 121–122.

Garcia, J., & Koelling, R. A. (1966). Relation of cue to consequence in avoidance learning. *Psychonomic Science, 4*, 123–124.

Gardner, R. A., Gardner, B. T., & Van Cantfort, T. E. (Eds) (1989). *Teaching Sign Language to Chimpanzees*. New York: State University of New York Press.

Gould, S. J., & Lewontin, R. C. (1979). The spandrels of San Marco and the Panglossian paradigm: A critique of the adaptionist programme. Proceedings of the *Royal Society London, B 205*, 581–598.

Gray, P. (1991). *Psychology*. New York: Worth.

Hall, G. (1983). *Behaviour: An Introduction to Psychology as a Biological Science*. New York: Academic Press.

Hamilton, W. D. (1964). The genetical theory of social behaviour (I). *Journal of Theoretical Biology, 7*, 1–16.

Harlow, H. F. (1971). *Learning to Love*. San Francisco: Albion.

Harlow, H. F., & Zimmermann, R. R. (1959). Affectional responses in the infant monkey. *Science, 130*, 421–432.

Kuhn, T. S. (1970). *The Structure of Scientific Revolutions*. 2nd ed. Chicago: University of Chicago Press.

Kuper, A. (1994). *The Chosen Primate: Human Nature and Cultural Diversity*. Cambridge MA: Harvard University Press.

Lockard, J. S., & Paulus, D. L. (Eds) (1988). *Self-Deception: An Adaptive Mechanism?* Englewood Cliffs NJ: Prentice Hall.

Maynard Smith, J., & Price, G. R. (1973). The logic of animal conflicts. *Nature, 246*, 15–18.

Mayr, E. (1983). How to carry out the adaptionist program? *The American Naturalist, 121*, 324–334.

Moltz, H. (1960). Imprinting: Empirical Basis and Theoretical Significance. *Psychological Bulletin, 57*, 291–314.

Tooby, J., & Cosmides, L. (1992). The psychological foundations of culture. In: J. H. Barkow, L. Cosmides & J. Tooby (Eds), *The Adapted Mind: Evolutionary Psychology and the Generation of Culture* (pp. 19–136). New York: Oxford Press.

Trivers, R. (1985). *Social Evolution*. Menlo Park CA: Benjamin/Cummings.

Trivers, R. L. (1971). The evolution of reciprocal altruism. *Quarterly Review of Biology, 46*, 35–57.

van den Berghe, P. L. (1979). *Human Family Systems: An Evolutionary View*. New York: Elsevier.

Williams, G. C. (1966). *Adaptation and Natural Selection*. Princeton NJ: Princeton University Press.

Wilson, E. O. (1975). *Sociobiology: The New Synthesis*. Cambridge MA: Harvard University Press.

NEO-DARWINIAN THEORY, PSYCHIATRY, AND THE OTHER CLINICAL HUMAN SCIENCES

Russell Gardner, Jr.

HISTORY OF MOVEMENT INCLUDING ITS LEADING FIGURES

Amongst Charles Darwin's writings relevant to evolutionary factors in psychiatry, the most relevant may have been his acclaimed *Expression of Emotions in Man and Animals* published in 1872 that featured some of the first scientific uses of photography. His earlier *Origins of the Species* (1859) had focused on how species changed through natural selection, but the new effort emphasized continuity, on how attributes stayed the same over evolutionary time with many explicit comparisons between humans and non-human animals. For many observations, he leaned on a physician colleague working with the insane and mentally retarded, J. Crichton Browne, a psychiatrist before the specialty was so designated. Disorders illustrated normal communications in exaggeration. This approach resembles a tactic used in recent decades as well. Psychiatric states can be considered naturally occurring communicational processes or states, but wrong for the situation, causing them to be considered pathology. Mania, for example, exemplifies thwarted leadership: the patient behaves as though in control, despite the maladaptive consequences of such behaviors when control lies elsewhere as discussed by Gardner in 1982.

Evolutionary Approaches in the Behavioral Sciences, Volume 8, pages 217–233.

ISBN: 0-7623-0769-2

Contemporaneously with Darwin in 19th century Paris, Paul Pierre Broca practiced neurology before it was a specialty as well as internal medicine and surgery (Schiller, 1992). Curiously, he started with an initial training year in psychiatry. Intensely interested in the brain and its evolution, he also founded French Anthropology. Noting that the dolphin's brain had a well-developed "smell" region, but no olfactory end organ, he coined the term "limbic system" to describe these structures with others in a ring of the cerebrum that bilaterally surrounded the brain stem (limbus or rim). He did not, however, attribute specific function to this system, although famously he did suggest motor speech function for another brain region, an area of left frontal lobe known later as Broca's area. On post-mortem examination of an aphasic patient, he demonstrated pathology in this brain region. His English contemporary, J. Hughlings Jackson, also displayed keen awareness of evolution. In working with seizure patients, he described core behavioral propensities (more primitive, earlier evolved) uncovered by pathology.

Though European psychiatrist Emil Kraepelin was interested in evolution during his early professional years, he later focused not on mechanisms of psychopathology but characteristic symptoms and course of illness. The neurologist who founded psychoanalysis, Sigmund Freud, speculated about phylogenetic history in some formulations, but mostly used machine metaphors popular in the 19th century for explaining the phenomena he heard from the analytic couch (such as hydraulics experienced widely via the steam engine). As summarized by Sulloway, he said nothing of the brain in the twentieth century despite his 19th century training and sophistication, because he felt brain science too distant from his observations. His Swiss psychoanalyst contemporary, Carl Jung, strongly relied on evolutionary thinking, theorizing that ideas, themes and metaphors he called archetypes were inherited. Present Jungians who apply evolutionary thinking to psychiatry and psychology include Englishman Anthony Stevens, and in Texas, David Rosen. Other contemporary psychoanalysts who suggest that evolutionary principles illuminate observations from their clinical work include Melvin Slavin, David Kriegman, Robert Langs and Stanley Palombo.

The European ethologists, Nikos Tinbergen and Konrad Lorenz, working principally after World War II, earned the Nobel Prize for Medicine and Physiology in 1973 (with von Frisch), about a century after Darwin's *Expression of the Emotions.* They relied on detailed behavioral descriptions and compared characteristic patterns across species, presuming inherited mechanisms for behavior; the Nobel Committee apparently made the award because such observations of non-human animals pertained to psychiatric issues. In response, Tinbergen and his wife presented speculations on autism from an ethological

point of view. Presently, John Richer from England carries forward ethological work with autism; Simon Baron-Cohen as well as Christopher and Uta Frith from England have suggested that autistic patients lack a "theory of mind" – meaning they cannot understand that other people think and feel. Two other ethologists from the European tradition established reputations bearing on evolutionary examinations of psychopathology. One, Iraneus Eibl-Eibesfeldt, extensively fostered human ethology by making across-cultural comparisons, and in 1990 published an encyclopedic book with that name. The other, Detlev Ploog, presided over a Max Planck Institute devoted to psychiatric aspects of ethology after first working in the U.S. with Paul D. MacLean.

English ethologist Michael R.A. Chance also began after WWII with the investigation of the behavioral effects of amphetamines, a drug empirically used by Allied pilots to enhance their wakefulness on long bombing runs, but whose overall effects had been little studied. He and colleagues found that rodents given the drug responded differently when grouped; compared to those housed singly, death occurred more often. He concluded that social circumstances must be considered when considering drug effects. Part of his legacy was illustrated in a 1994 collection of papers edited by Cooper and Hendrie. Exemplifying the practicality of this tradition, a Swiss drug company employs ethopharmacologist Keith Dixon.

Chance also contributed observations on the import of tense social rank hierarchies in primates (agonic mode) versus relaxed playful social organization (hedonic mode). His work on social rank hierarchy influenced psychiatrist John S. Price who published the first explicit statement connecting social rank hierarchy with affective disorders in 1967. Chance and Price led the Birmingham Group that met over the years to discuss issues of evolutionary biology, ethology and clinical and political matters. Chance edited *Social Fabrics of the Mind* in 1988 and became the first president of the Across-Species Comparisons and Psychopathology (ASCAP) Society in 1991. Price succeeded him as the second president. With Anthony Stevens, Price co-authored the first psychiatric textbook on evolutionary psychiatry in 1996. Another Birmingham Group participant, John Birtchnell, has published two books on relating, going into depth on this oddly neglected topic. Chance associate, C. U. M. Smith, has published two editions of a widely used text, *Molecular Neurobiology*. This featured evolutionary issues in the formation of the nervous system; he also wrote an editorial on the value of evolutionary biology for psychiatry in the *British Journal of Psychiatry*.

In post-war U.S.A., Paul D. MacLean at Yale University and the NIMH resurrected Broca's limbic system while speculating it involved emotional life. He coined the term "triune brain" suggesting that the core of the brain and

brain stem be termed reptilian because structures there are seen in humans (and other mammalians) are already well developed in reptiles, implying that ancestral forms to both groups of vertebrates had already developed these structures. He suggested the limbic system evolved in early mammalians layered over this core because familial bonding provided adaptive advantage. The neocortex represented the most recent of the triune complex. Though not truly new in that it had already existed in more primitive forms, this most superficial layer hypertrophied greatly in neomammalians surrounding the family-oriented middle structures. It gained especially large size in humans. If flattened, the human cerebral cortex would spread over a four-times greater surface area than would the chimpanzee cortex although 98.4% of the two genomes exhibit identity.

With his NIMH colleagues MacLean observed lizards, finding that death may occur after defeat by a rival. He described a number of behaviors also seen in human psychopathology (and normal people also), such as routinizing behaviors and postures that echo those of another individual, supposing that some of these are core to vertebrate courtship as well as the later development of cordial relationships in species such as humans. In other work he performed studies on the squirrel monkeys noting that simulation of a number of subcortical brain sites produced sexual excitation. MacLean influenced many psychiatrists, notably U.S. psychiatrist Daniel X. Freedman, who as president of the American Psychiatric Association (APA), fostered MacLean's major address to an annual meeting of that group. Child psychiatrist James V. Harris represents a present day author who argues the value of MacLean's ideas for clinicians; he authored a comprehensive two-volume textbook on developmental neuropsychiatry, and has stressed Kropotkin's emphasis on the evolution of cooperative behavior in addition to competitive interactions.

John Paul Scott, a MacLean contemporary, investigated dog behaviors by comparing breeds. He also wrote an evolutionary view of social systems and influenced the psychologist, Daniel L. Friedman, who led an interdisciplinary department featuring behavior, genetics and evolution at the University of Chicago. Friedman's students included current leaders of an integrative approach uniting behavioral and evolutionary genetics. These include California twin researcher, Nancy Segal, and Detroit psychologist, Glenn Weisfeld. Weisfeld has published on the evolutionary aspects of the adolescence phase of development. Another student of Scott, Jaak Panksepp, has extensively studied emotion brain systems in rats, summarising his evidence for seven separable such systems in *Affective Neuroscience*.

In the years that followed MacLean and Scott, primatologists such as Sherman Washburn influenced prominent psychiatrists. For example, David Hamburg,

when chairman of psychiatry at Stanford, provided support to Jane Goodall's work on chimpanzees. She in turn has been greatly admired and often featured at psychiatric meetings. Hamburg noted the relevance of such studies to attachment in line with the impact of John Bowlby whose work on attachment and separation has been enormously influential in American psychiatry and psychoanalysis. Bowlby noted that propensities to attachment are clearly inherited which nudged developmental theorists towards evolutionary biology. Hamburg has wielded enormous influence as head of philanthropic organizations, including the Carnegie Fund. Frans De Waals has contributed to understanding cooperative behaviors in such books as *Peacemaking Among The Chimpanzees*. In related work on bonding viewed from an across-species perspective, Myron Hofer, U.S. psychiatrist and animal researcher, showed that mammalian mothers produce powerful stabilizing influences on their offsprings' autonomic and hormonal systems.

Sociologist Lionel Tiger along with anthropologist Robin Fox published *The Imperial Animal* in 1971 that sold well and had followed the popular works of the self-taught playwright Robert Ardrey in the 1960s that emphasized humans as aggressive creatures for inborn reasons. But putting an unmistakable stamp on the idea that humans were affected by their biology for the public at large, E. O. Wilson in 1975 published *Sociobiology*. He actually spent nearly all that volume on non-human animals and their social organizations. But he projected that all the human social sciences would have their paradigms greatly influenced by evolutionary thinking about behavior. Shortly after, he too was invited to speak at an annual APA meeting. Controversy surrounding his work, however, leveled largely by social scientists propounding a "standard social sciences model" that eschewed variability from inherited factors may have influenced mainstream psychiatric reaction to him as well as to other facets of "selfish gene" theory as described by English biologist Richard Dawkins in a number of books. Prominent critics have included the mollusk expert and prolific essayist Stephen Jay Gould who argues for the avoidance of biological reductionism, i.e. that learned and culturally entrained facets of development and behavior are oversimplified. He and colleagues worry that the rationale underlying eugenics might return so that minority populations could be persecuted under the banner of a "scientific" and falsely legitimated rationale. But while recognizing and respecting such arguments, irreversible recognition has mounted that unbiological culture or behavior cannot exist – all behavior and culture operate through brain actions – contrary to the mind-body split inherited from Descartes and perpetuated since then.

Returning to more specific psychiatric concerns, Michael T. McGuire of UCLA co-edited in 1977 *Ethological Psychiatry: Psychopathology in the*

Context of Evolutionary Biology that brought together many figures working in this arena at that time. McGuire's subsequent work with Michael Raleigh and Arnold Yuwiler connected blood levels of serotonin with alpha status in vervet monkeys and humans. Recently he authored a textbook on Darwinian psychiatry with Alfonso Troisi of Italy. Between these two efforts, Isaac Marks of England published *Fears, Phobias, and Rituals: Panic, Anxiety and Their Disorders* in 1987. This work also described psychiatric symptoms in an evolutionary context. Marks' book has had widespread effect, frequently quoted in discussions of social phobia, for instance.

Randolph Nesse, also a psychiatrist, and Richard Alexander (an entomologist like Wilson) from the Universityof Michigan helped organize the Human Behavior and Evolution Society (HBES) after a series of precursor meetings at Ann Arbor. Nesse authored two well publicized contributions with sociobiologist, George Williams, on the adaptive features of many medical symptoms, e.g. the 1994 book, Why *We Get Sick. The New Science of Darwinian Medicine.* The concerns of HBES group in general, however, do not usually involve clinical issues. Moreover, it has turned away from ethology to focus more on population-focused studies. Martin Daly and Margo Wilson, for instance, examined murder statistics and used the inclusive fitness paradigm resulting from William Hamilton's explanation of altruism determined by shared genes. They suggested that this helped explain why kin kill non-kin (spouses, step-children, in-laws) many times more often than genetically related kin (parents, children, siblings). Canadian Denis de Cantanzaro noted that suicides occurred much more frequently in people without relatives who possess shared genes. Linking sociobiological thought and research to bonding theory, Sarah Blaffer Hrdy wrote in 1999, *Mother Nature: A History of Mothers, Infants, and Natural Selection.* Sociobiologist Linda Mealey has written on sociopathy and published *Sex Differences: Developmental and Evolutionary Strategies* in 2000.

Russell Gardner's paper on the evolutionary origins of mania and depression published in the *Archives of General Psychiatry* in 1982 noted that these so-called mood disorders were also disorders of communication, that is, the patient communicates assumption of too much power in mania and the depressed person powerlessness and low status. In Chance's 1988 edited book, he coined the term psalic, an acronym meaning propensity states antedating language in communication, to highlight the importance of conserved communicational readiness states. These conceptual entities depended on their observation in three settings implying mechanisms in common: (1) normal people, (2) those psychiatrically disturbed, and (3) non-human animals. Thus, alpha psalic can be inferred from its existence in charismatic leaders, mania, and alpha animals. Some other psalics include mating, nurturance-eliciting, nurturance-providing,

and in-group omega, the latter seen in low ranking humans and other animals as well as in depression. Outgroup omega psalic can be inferred in a patient with a persecutory delusion, a communicational state that would not be delusional in a truly persecuted person or animal.

In late 1987, Gardner began publication of *The Across-Species Comparisons and Psychopathology (ASCAP) Newsletter* then issued monthly until January, 2000 (twelve yearly volumes with 146 issues). Multiple contributors wrote material that was distributed internationally. This in turn fostered The ASCAP Society that meets annually, often in conjunction with HBES, but also with the American Psychiatric Association (APA) in New York City and with the World Psychiatric Association (WPA) in Hamburg, Germany; in 1999, the group also called a free-standing two-day meeting to celebrate the work of Paul D. MacLean. Moreover, John Price, when chairman of the section on psychotherapy of the World Psychiatric Association arranged that the newsletter (transformed in 2000 to a quarterly bulletin) become the official news distributing organ of the section, an action approved by the WPA. A number of papers in refereed journals stemmed from discussions in the pages of the newsletter, notably a paper in 1994 by Price et al. of "The social competition hypothesis of depression" in the *British Journal of Psychiatry*, subsequently anthologized by Simon Baron-Cohen. Another ASCAP Society member started the new millennium in the prestigious *Archives of General Psychiatry*, when Randolph Nesse authored the lead article of the January issue, entitled, "Is depression an adaptation?"

ASCAP Society Presidents in order after: (1) Chance and (2) Price include (3) Paul Gilbert, English psychologist, who has published multiple books on evolutionary aspects of depression. (4) Psychiatrist John Pearce, with psychologist Kalman Glantz both of Cambridge, Massachusetts, published in 1989, *Exiles from Eden: Psychotherapy From An Evolutionary Perspective*, a text that argued that many inborn patterns in relating and other characteristic behaviors stem from adaptations that occurred during the long "era of evolutionary adaptation" prior to the recent (10,000 years before present) advent of agriculture and city-focused living. (5) Leon Sloman, child psychiatrist from Toronto, has argued that depression results from involuntary subordination, in concert with formulations of Price, Gardner and Gilbert. (6) Kent G. Bailey, psychologist, published *Human Paleopsychology: Applications to Aggression and Pathological Processes* in 1987, that suggested the concept of phylogenetic regression. (7) Daniel R. Wilson, psychiatrist, founded evolutionary epidemiology, that labels methods using prevalence statistics to determine whether an illness has had adaptive features. (8) Mark Erickson, published in the *American Journal of Psychiatry* an evolutionary perspective on incest taboos in people raised together (contemporary applications of the Westermarck effect). (9) Ivor

Jones, Australian psychiatrist, former Professor and Chairman, Department of Psychiatry at the University of Tasmania, published articles on the description of various psychiatric symptom-behaviors, for example, self-mutilation, juxtaposed to similar behaviors in animals. He has argued that the defeated sugar-glider marsupial is a model for major depression. (10) James Brody, psychologist near Philadelphia has written on chaos theory as a factor in evolutionary views of psychopathology. He has organized an Albert Einstein College of Medicine summer course on Cape Cod over several years that has featured in addition to himself, John Pearce, John Fentress (an ethologist specializing in wolves), Robin Walker (an English veterinarian has observed animal behavioral responses to drugs), Steven Heisel (Boston psychiatrist who has written on obsessive compulsive disorder as connected to epidemic disease), John Price, Russell Gardner and Nancy Segal. (11) Psychologist Thomas Joiner has researched social and interpersonal factors in depression.

EXTENT OF ACCEPTANCE OR REJECTION (AND REASONS)

As detailed in the history section, many figures have pervasively addressed major components of the field. No great hostility has surfaced, but indifference to possible evolutionary implications generally persists. Phenomena that are interesting for psychiatrists such as low brain levels of serotonin in cases of violent suicide and crimes seem less interesting to psychiatrists if they index normal phenomena; thus, that high levels of blood serotonin correlate with high social status seem uninteresting and trivial, although such observations may have implications for illness pathogenesis.

Such neglect and disinterest has a number of likely causes. Medical specialties, of which psychiatry is one, are traditionally highly rigid. For instance, data-based and very clear observations that a spirochete living in acidic gastric fluids causes duodenal ulcers and could be treated with antibiotics nevertheless took many years to become standard practice. Psychiatry presents more problems in that its data-base is often less clear, and the brain more complex and less studied than other organs, at least to date.

Systems of etiopathogenesis parallel to the other specialties have not developed. In general, illnesses are still not considered deviations from normal physiology. Michael McGuire and Russell Gardner have argued independently that evolutionary biology applied to social behavior may help with a new hypothesis-based approach. But only slowly is this penetrating the specialty. Hagop Akiskol, editor-in-chief of the *Journal of Affective Disorders*, argued strongly at a discussant presentation of the WPA psychotherapy section

symposium that evolutionary arguments are under-published, noting that his journal was eager to publish such articles. In response partially to this, January, 2000, Gardner discontinued the monthly 30–35 page *ASCAP Newsletter* with its limited circulation and brain-storming aims in order to concentrate fully on more extensive lay, medical and scientific audiences. At present, he leads the Research Committee of the psychiatric think-tank, the Group for the Advancement of Psychiatry (GAP). Committee members include researchers whose areas of work range from molecular biology to family systems to territoriality and educators in psychiatry and child psychiatry. This group composed a condensed position statement in April, 2000, reproduced later in this chapter, that it has posed to the larger group of the GAP organization.

These efforts address how Psychiatry's problems stem in part from the mystery of how normal body processes, resulting in illness when they go awry, usually operate in the complex human brain. The 1990's "decade of the brain" spurred much research on the organ as well as major progress on the genome project. But only now and very gradually are the secrets revealed of what brain parts and what genetic details operate in such dreaded diseases as Alzheimer's and Huntington's diseases, schizophrenia, and bipolar disorder. In part the problem also derived from the disproportionate power of Freudian psychoanalysis in dominating the field in the U.S.A. after World War II. It was highly theoretical on the one hand but not backed by hypothesis-driven data-collection on the other. Then an extraordinary backlash stemmed from new authorities opposed to its practice. For instance, feeling harmed by his dominating analyst who erred in understanding his illness, Eli Robins led a full and successful revolt from what he perceived as psychoanalysis' antiscientific approach. He chaired the Washington University School department of psychiatry for many years and his colleagues and followers became leaders in other departments in Kansas, Iowa and Minnesota, while publishing widely and achieving considerable success, now dominating the research and clinical paradigm, though psychoanalysis persists. They argued for data-based diagnosis and treatment and for disregarding pathogenetic theory until more is known, though presuming that this would eventually stem from the molecular and pharmacological sciences. Guze summarized this approach in 1992. Moreover, psychopharmacology produced extremely effective drugs. The current fundamental question posed by psychiatrists confronting their patient's illness is often limited to what drug might work for various disorders. Curiously the framers of the APA's diagnostic and statistical manual hoped that these disorders would resemble species in being separate from one another. But they are not; rather many symptoms overlap and the so-called "comorbidities" litter the diagnostic landscape. Psychosis defined by the symptoms of hallucinations and delusions

occurs in depression, mania, schizophrenia, and deliria caused by electrolyte disturbances, head trauma and alcohol withdrawal states.

The paradigm shift occurred in the 1970s and 1980s. Although psychoanalysis has not disappeared, its power in academia has waned. Sympathetic researchers such as Eric Kandel have argued, however, that psychotherapy must leave brain traces, an observation supported by neuroimaging data now accepted throughout the specialty. Kandel departed psychiatry to become a world-class neurobiologist by studying the relatively few neurons of the sea snail and has argued the "fear response" signaled by the withdrawal of the snail's syphon may be homologous to anxiety in humans, and susceptible to molecular analysis.

In summary, fierce opponents to evolutionary factors in psychiatry do not exist amongst its practitioners. But also thinking on how such factors might or do figure in pathogenesis remains absent and hardly influences discussions with patients. How such a perspective helps the clinician with his patient has not penetrated the general awareness of the psychiatric public, nor the public at large. The latter group, however, has become more aware of other facets of their problems and conditions, regularly coming to their physicians with drug company ads or internet information that provide arguments for desired treatments. A combination of powerful drugs that the public knows work effectively, drug company lobbying that patients should request these expensive medications from their doctors, and the health maintenance organizations' push to limit exposure of patients to doctors as needless expense has generated what Karen Dineen Wagner has termed "the ideal of relationshipless psychiatry". The GAP research committee has suggested, contrariwise, that the study of the social brain constitutes a true basic science of psychiatry. Its illnesses are deviations from normal system function just as heart disease represents deviation from the normal physiology of the cardiovascular system. But although medical students, residents and practitioners study the heart's anatomy and physiology very well, counterpart examination of the social brain does not occur at this time.

ORGANIZATIONAL MANIFESTOS

The ASCAP *Newsletter* (now the quarterly ASCAP *Bulletin*), organs of the ASCAP Society, proclaimed the following: it concerns paleobiology, sociophysiology, interpersonal and group relations, and psychopathology. The society primarily involves evolutionary psychiatrists, psychologists and others interested in sociophysiological integration. The mission statement reads as follows:

> The ASCAP Society represents a group of people who view forms of psychopathology in the context of evolutionary biology and who wish to mobilize members and resources of various disciplines so as to enhance the further investigation and study of the conceptual

and research questions involved. This scientific society is concerned with the basic plans of behavior that have evolved over millions of years and that have resulted in psychopathologically related states. We are interested in the integration of various methods of study ranging from cellular processes to individuals in groups.

Turning now to the research committee of the GAP, this specialty think-tank group agreed on the following during its Spring, 2000, meeting:

Members of the Research Committee of the Group for Advancement of Psychiatry (GAP), a specialty think-tank, have addressed psychiatry's need for a unifying scientific foundation. Such a foundation would consider the disorders commonly treated by psychiatrists in terms of the physiological baseline from which they depart, much as heart disease is understood as deviation from normal cardiac function. The relevant physiological focus for psychiatry is the social brain.

The social brain concept focuses on the interaction between brain physiology and the individual's environment. The brain is the organ most influenced on the cellular level by social factors across development; in turn, the expression of brain function determines and structures an individual's personal and social experience. The social brain framework may have greater direct impact on the understanding of some psychiatric disorders than others. However, it helps organize and explain all psychopathology. A single gene-based disorder like Huntingtons disease is expressed to a large extent as social dysfunction. Conversely, traumatic stress has structural impact on the brain as does the socially interactive process of psychotherapy.

Brains, including human brains, derive from ancient adaptations to diverse environments and are themselves repositories of phylogenetic adaptations. In addition, individual experiences shape the brain through epigenesis, i.e. the expression of genes is shaped by environmental influences. Thus, the social brain is also a repository of individual development. On an ongoing basis, the brain is further refined through social interactions; plastic changes continue through life with both physiological and anatomical modifications.

In contrast to the conventional biopsychosocial model, the social brain formulation emphasizes that all psychological and social factors are biological. Conversation, feeling, and thinking can happen only from brain-actions in the involved individuals. Nonbiological and non-social psychiatry cannot exist. Molecular and cellular sciences *offer* fresh and exciting contributions to such a framework but provide limited explanations for the social facets of individual function.

The social brain formulation is consistent with current research and clinical data. Moreover, it ultimately must:

- unify the biological, psychological and social factors in psychiatric illness,
- dissect components of illness into meaningful functional subsets that deviate in definable ways from normal physiology,
- improve diagnostic validity by generating testable clinical formulations from brain based social processes,
- guide psychiatric research and treatment,
- provide an improved language for treating patients as well as educating trainees, patients, their families and the public, and
- account for the role of interpersonal relationships for brain function and health.

In conclusion, the concept of the brain as an organ that manages social life provides significant power for psychiatry's basic science. Burgeoning developments in neural and genetic areas put added demands on the conceptual structures of psychiatry. Findings from such incoming work must be juxtaposed and correlated with the behavioral and experiential facets of psychiatry to give it a complete and rational basis. Psychiatry's full and unified entry into the realm of theory-driven and data-based medical science has arrived. The social brain concept allows psychiatry to utilize pathogenesis in a manner parallel to practice in other specialties.

TRAINING PROGRAMS IF ANY FOR STUDENTS/FACULTY INCLUDING RIGIDITY OF CLINICAL TRAINING

As is true of undergraduate medical education, specialty residency training requires special standards. None that focus extensively on evolutionary factors have yet been put in place. However, special emphasis on such topics may have followed David Hamburg's work at Stanford University. Stanford psychiatrist trained in that era, Brant Wenegrat, was among the earliest psychiatrists who suggested adaptive features in psychiatric symptoms. Other psychiatrists who have been psychiatry chairmen include Cornelis Bakker who has written on territoriality (Bakker & Bakker-Rabdau, 1973) and Russell Gardner (1982). Renewed optimism may be warranted, however, as evolutionist psychiatrist, Daniel R. Wilson, has recently assumed responsibility for a university chairmanship in psychiatry at Creighton University in Omaha, Nebraska.

IMPORTANT BIBLIOGRAPHIC ITEMS

Ardrey, R. (1966). *The Territorial Imperative*. NY: Atheneum.

Ardrey, R. (1976). *The Hunting Hypothesis*. NY: Atheneum.

Bakker, C. B., & Bakker-Rabdau, M. K. (1973). *No Trepassing: Explorations in Human Territoriality*. San Francisco, CA: Chandler & Sharp Publishers, Inc.

Baron-Cohen, S. (Ed.) (1997). *The Maladapted Mind: Classic Readings in Evolutionary Psychopathology*. Hove, East Sussex, UK: Psychology Press.

Birtchnell, J. (1993). *How Humans Relate: A New Interpersonal Theory*. Westport, CT: Praeger.

Birtchnell, J. (1999). *Relating in Psychotherapy: The Application of a New Theory*. Westport, CT: Praeger.

Bowlby, J. (1969). *Attachment and Loss*. New York: Basic Books.

Brown, G., Harris, T. O., & Hepworth, C. (1995). Loss, humiliation and entrapment among women developing depression: a patient and non-patient comparison. *Psychological Medicine*, *25*, 7–21.

Chance, M. R. A. (Ed.) (1988). *Social Fabrics of the Mind*. Hillsdale, USA: Lawrence Erlbaum Associates.

Cooper, S. J., & Hendrie, C. A. (Eds) (1994). *Ethology aced Psychopharmacology*. New York, NY: Wiley.

Daly, M., & Wilson, M. (1988). *Homicide*. NY: Aldine de Gruyter.

Damasio, A. R. (1994). *Descartes' Error: Emotion, Reason, and the Human Brain*. New York, NY: Grosset/Putnam.

Damasio, A. R. (1999). *The Feeling of What Happens*. New York, NY: Harcourt, Brace & Company.

Dawkins, R. (1987). *The Blind Watchmaker: Why the Evidence of Evolution Reveals a Universe Without Design*. NY: WW Norton.

Dawkins, R. (1982). *The Extended Phenotype: The Long Reach of the Gene*. NY: Oxford U Press.

Dawkins, R. (1976, 1989). *The Selfish Gene*. New Edition. NY: Oxford U Press.

De Cantanzaro, D. (1980). Human suicide: a biological perspective. *The Behavioral and Brain Sciences*, *3*, 265–290.

Engel, G. L. (1977). The need for a new medical model: a challenge for biomedicine. *Science*, *196*, 129–136.

Erickson, M. (1993). Rethinking Oedipus: an evolutionary perspective on incest avoidance. *Am JPsychiat*, *150*, 411–416.

Gardner, R. (1995). Sociobiology and its application to psychiatry. In: H. I. Kaplan & B. J. Sadock (Eds), *The Comprehensive Textbook of Psychiatry*, Edition 6. (pp. 365–375).

Gardner, R., & McKinney, W. T. (2000). Ethology and the use of animal models. Peer Reviewed Chapter for Psychiatrie der Gegenwart (English version). Heidelberg, Germany: Springer-Verlag.

Gardner, R. (1996). Psychiatry needs a basic science titled sociophysiology. *Biological Psychiatry*, *39*, 833–834.

Gardner, R. (1997). Sociophysiology as the basic science of psychiatry. *Theoretical Medicine*, *18*, 335–356.

Gardner, R. (1998). The brain and communication are basic for clinical human sciences. *British Journal of Medical Psychology*, *71*, 493–508.

Gardner, R. (1982). Mechanisms in manic-depressive disorder: an evolutionary model. *Arch. Gen. Psychiatry*, *39*, 1436–1441.

Gilbert, P. (1992). *Depression: The Evolution of Powerlessness*. NY: The Guilford Press.

Gilbert, P. (1989). *Human Nature and Suffering*. Hillsdale, N.J., U.S.A: Lawrence Erlbaum Associates.

Glantz, K. (1989). *Pearce JK: Exiles From Eden: Psychotherapy From An Evolutionary Perspective*. NY: Norton.

Gould, S. J. (1991). *Bully for Brontosaurus: Reflections in Natural History*. NY: W.W. Norton.

Gould, S. J. (1977). *Ontogeny and Phylogeny*. Cambridge, MA: Harvard U Press.

Gould, S. J. (1981). *The Mismeasure of Man*. NY: W.W. Norton.

Gould, S. J. (1980). *The Panda's Thumb: More Reflections in Natural History*. NY: W. W. Norton.

Gould, S. J. (1989). *Wonderful Life: The Burgess Shale and the Nature of History*. NY: W.W. Norton.

Guze, S. (1992). *Why Psychiatry is a Branch of Medicine*. NY: Oxford U Press.

Hamburg, D. A. (1963). Emotions in perspective of human evolution. In: P. Knapp (Ed.), *Expression of Emotions in Man*. New York: International Universities Press.

Harris, J. C. (1995). *Developmental Neuropsychiatry*, Volume I: *Fundamentals*. New York, NY: Oxford University Press.

Harris, J. C. (1995). *Developmental Neuropsychiatry*, Volume ff.: *Assessment, Diagnosis and Treatment of the Developmental Disorders*. New York, NY: Oxford University Press.

Henry, J. P., & Stephens, P. M. (1977). *Stress, Health, and the Social Environment: A Sociobiologic Approach to Medicine*. Berlin: Springer-Verlag.

Hofer, M. A. (1981). *The Roots of Human Behavior: An Introduction to the Psychobiology of Early Development*. San Francisco: WH Freeman.

Hrdy, S. B. (1999). *Mother Nature: A History of Mothers, Infants, and Natural Selection*. New York, NY: Pantheon Books.

Jung, C. G. (translators: W. S. Dell, & C. F. Baynes) *Modern Man in Search of Soul*. NY: Harcourt, Brace and Co., 933.

Kandel, E. (1998). A new intellectual framework for psychiatry. *American Journal of Psychiatry, 155*, 457–469.

Kraepelin, E. (1968). (Revised and edited T. Johnstone): *Lectures on Clinical Psychiatry*. Facsimile of the 1904 edition. NY: Hafner Publishing,

Langs, R. (1996). *The Evolution of the Emotion Processing Mind*. Madison, CT: International Universities Press.

MacLean, P. D. (1990). *The Triune Brain in Evolution: Role of Paleocerebral Function*. New York, NY: Plenum.

Marks, I. M. (1987). *Fears, Phobias, and Rituals: Panic, Anxiety and Their Disorders*. New York: Oxford University Press.

McGuire, M. T., & Fairbanks, L. A. (1977). *Ethological Psychiatry: Psychopathology in the Context of Evolutionary Biology*. NY: Grune & Stratton.

McKinney, W. T. (1988). *Models of Mental Disorders: A New Comparative Psychiatry*. New York, NY: Plenum Book Company.

McKinney, W. T., Gardner, R., Barlow, G. W., & McGuire, M. T. (1994). Conceptual basis of animal models in psychiatry: a conference summary. *Ethology & Sociobiology, 15*, 369–383.

McGuire, M. T. & Troisi, A. (1998). *Darwinian Psychiatry*. New York: Oxford University Press.

Mealey, L. (2000). *Sex Differences – Developmental and Evolutionary Strategies*. New York: Academic Press.

Nesse, R., & Williams, G. C. (1994). *Why We Get Sick: The New Science of Darwinian Medicine*. New York: Vintage Books.

Nesse, R. (2000). Is depression an adaptation? *Archives of General Psychiatry, 57*, 14–20.

Palombo, S. R. (1999). *The Emergent Ego: Complexity and Coevolution in the Psychoanalytic Process*. Madison, CT: International Universities Press.

Panksepp, J. (1998). *Affective Neuroscience: The Foundations of Human and Animal Emotions*. New York, NY: Oxford University Press.

Price, J. S. (1967). Hypothesis: The dominance hierarchy and the evolution of mental illness. *Lancet*, 2, 243–246.

Price, J., Sloman, L., Gardner, R. Jr., Gilbert, P., & Rohde, P. (1994). The social competition hypothesis of depression. *British Journal of Psychiatry, 164*, 309–315.

Richer, J. M. (1979). Human ethology and psychiatry. In: H. Van Praag et al. (Eds), *Handbook of Biological Psychiatry*. New York: Plenum Press.

Schiller, F. (1992). *Paul Broca: Explorer of the Brain*. NY: Oxford U Press.

Scott, J. P., & Fuller, J. L. (1965). *Dog Behavior: The Genetic Basis*. Chicago: U Chicago Press.

Scott, J. P. (1989). *The Evolution of Social Systems*. NY: Gordon and Breach Science Publishers.

Segal, N. L. (1999). *Entwined Lives: Twins and What They Tell Us About Human Behavior*. New York, NY: Dutton.

Shair, H. N., Barr, G. A., & Hofer, M. A. (Eds) (1991). *Developmental Psychobiology: New Methods and Changing Concepts*. New York, NY: Oxford University Press.

Slavin, M. O., & Kriegman, D. (1992). *The Adaptive Design of the Human Psyche: Psychoanalysis, Evolutionary Biology, and the Therapeutic Process*. NY: The Guilford Press.

Sloman, L., Price, J. S., Gilbert, P., & Gardner, R. (1994). Adaptive function of depression: psychotherapeutic implications. *American Journal of Psychotherapy*, *48*, 401–416.

Smith, C. U. M. (1993). Evolutionary biology and psychiatry. *British Journal of Psychiatry*, *162*, 149–153.

Smith, C. U. M. (1996). *Elements of Molecular Neurobiology*, 2nd Ed. New York: Wiley.

Smith, W. J. (1977). *The Behavior of Communicating: An Ethological Approach*. Cambridge, MA: Harvard University Press.

Stevens, A., & Price, J. S. (1996). *Evolutionary Psychiatry: A New Beginning*. London: Routledge.

Sulloway, F. J. (1979, 1992). *Freud. Biologist of the Mind*. Cambridge, MA: Harvard U Press.

Taylor, J. (Ed.) (1958). *Selected Writings of John Hughlings Jackson*, Volume 1: *On epilepsy and epileptiform convulsions*. London: Staples Press.

Taylor, J. (Ed.) (1958). *Selected Writings of John Hughlings Jackson*, Volume 2: *Evolution and dissociation of the nervous system. Speech*. Various papers, addresses and lectures. London: Staples Press.

Tiger, L., & Fox, R. (1971, 1989). *The Imperial Animal*. NY: Henry Holt.

Tiger, L. (1969, 1970). *Men in Groups*. NY: Vintage Books.

Tiger, L. (1992). *The Pursuit of Pleasure*. Boston: Little, Brown.

Weisfeld, G. (1999). *Evolutionary Principles of Human Adolescence (Lives in Context)*. Westview Press.

Wenegrat, B. (1984). *Sociobiology and Mental Disorder: A New View*. California: Addison-Wesley.

Williams, G., & Nesse, R. (1991). The dawn of Darwinian medicine. *Quarterly Review of Biology*, *66*, 1–22.

Wilson, D. R. (1993). Evolutionary epidemiology: Darwinian theory in the service of medicine and psychiatry. *Acta Biotheoretica*, *41*, 205–218.

CURRENT TRENDS AND POSSIBLE FUTURE PROSPECTS

A number of cross-currents suggest that evolutionary sciences will impact psychiatry in coming years. After the twentieth century battles between analysts and their challengers, a vacuum remains that may to be filled in with scientific data stemming from brain and genome research. Unfortunately at the time of this writing, normal social behavior – the third component of the needed genome-neural-behavioral analysis – does not receive the same systematic scrutiny although think-tank groups and a burgeoning literature links behavior to mechanisms at these levels. Impeding this are the tremendous strides in psychopharmacology that the pharmaceutical houses have achieved because these perpetrate simple-minded twisted-molecule models of illness. Also careless across species comparisons can offend sensitive public groups. Thus an NIMH leader compared inner city violence to actions in primate groupings and minority leaders protested that he was labeling them mere animals. This cost

him his position. Moreover, in the U.S., religious groupings often insist on literal interpretations of the Bible that eschew evolutionary theory "which has not been proven," in their eyes. Still the strong and obvious connections between humans and other animals are obvious to any observer and pet psychology represents a popular pastime for many people. Most powerful in persuasion, however, are such best-selling and very well written works as those composed by Antonio Damasio. He has been featured in the New York Times Magazine Section as are the brain-images of thought patterns compiled by his wife, the neuroradiologist, Hannah Damasio. These well-received works indicate greater acceptance of such matters by lay, medical and scientific publics.

An example of evolutionary factors that contribute to patterns of pathogenetic data was illustrated in the 2000 Adolph Meyer Memorial Lecture at the annual meeting of the American Psychiatric Association in Chicago. Schizophrenia researcher, Daniel Weinberger, brought together many threads of his investigations on the pathogenesis of the illness. One showed that patients with this illness inherit a vulnerability that exists in relatives as well. Thus lateral frontal lobe areas with lesser synaptic densities exhibit more intense neuroimaging when a challenging working memory task is presented them. This in turn stems from a variant of a dopamine-processing enzyme that resembles that in chimpanzees more than the usual variant in humans. Thus, the enzyme works in both species, in patient as well as in normal groupings, but a single amino acid difference that seemingly has been adaptive in humans separates the groups. Family members of the patients often do not have the illness despite having the same chimpanzee-like variant, as many factors work together to produce the disease. But this kind of research that teases apart the roles of varied factors in the production of complex human disease illustrates how evolutionary thinking and awareness will likely figure in future psychiatric work.

An image popular for the past few decades in U.S. psychiatry was named the "biopsychosocial model." George Engel urged psychiatrists and other physicians also to consider factors from all three domains when formulating the illnesses of patients. The term has helped guide education of medical students and the examination of practitioners when sitting for board certification, but a criticism has held the model to be a fragmenting concept. Illness does not result from the breakdown of a multilevel system. The term allows practitioners who would prefer to avoid the social and psychological facets of illnesses to "specialize" in only the biological (fostered by the managed care industry as it considers this less expensive treatment). In fact, of course, the social and psychological are biological too, in that brain activity occurs in any social or psychological behavior. Integrating terms are required instead such as a sociophysiological basic science of psychiatry or the social brain image.

Of course the fixture lies in the younger members of the profession. Aaron T. Beck, a founder of the idea that psychotherapies should be assessed and measurements made of patients so treated, has persistently wondered about the mechanisms of psychiatric illness. He has strongly backed the ASCAP Society's conferring the Aaron T. Beck ASCAP Award for deserving essays that are then presented at the annual meeting. The five winners to this point include:

(1) 1995: Nicholas Allen, University of Melbourne, Melbourne, Australia with "Towards a computational theory of depression".
(2) 1996: Souhir Ben Hamida, Northwestern University, Chicago, USA with "Human mate preferences: implications for the gender difference in unipolar depression."
(3) 1997: Edward Hagen, University of California, Santa Barbara, USA with "Delusional and somatoform disorders as possible examples of intraspecific exploitative mimicry in humans."
(4) 1998: Bruce Ellis, Vanderbilt University, Nashville, USA with "Psychosocial antecedents of variation in girls' pubertal timing: maternal depression, stepfather presence, and marital and family stress."
(5) 1999: No award was given in 1999 as no submitted essay met the judges' standards.
(6) 2000: Jennifer Connolly, Griffith University, Brisbane, Australia, with "Children's' preferences for waist-to-hip ratios: a developmental study."

Dr. Beck took the initiative to meet personally with the first three winners to encourage their further work in these areas. In such efforts lies hope for the future.

In summary, evolutionary factors in psychiatry undoubtedly exist and the groundwork for a more satisfactory basic science for psychiatry has already been laid. Powerful forces de-emphasize such thinking about these ills, but research models, think-tank efforts, sympathetic journal editors, and the public's hunger for such explanations may hasten the wide-spread use of such explanations in the clinician's office.

DARWIN AND SOCIOLOGY: OIL AND WATER*

Allan Mazur

INTRODUCTION

Most sociologists ignore Darwin. There are two good reasons. Ideologically, natural selection is antithetical to sociology's support for the oppressed. Empirically, natural selection has provided a poor description of human behavior. Darwinism will not convert sociology to the faith until it provides a cogent description of modern behavior that is morally appealing.

IDEOLOGY

Sociology in the late 19th and early 20th centuries was explicitly evolutionary in outlook, teaching that mankind had progressed from a hunting/gathering state of "savagery," through the "barbarism" of agrarian society, finally arriving at the stage of "civilization" enjoyed by rich industrial nations. The world of 1900 contained representatives of all three stages – compare South Africa, China and Britain. Natural selection – "survival of the fittest" – was in vogue as an explanation for the superiority of white male elites and for the subordination of women and inferior races.

An exemplar of Social Darwinism during the 1920s was Professor Charles Ellwood, president of the American Sociological Society (a name later changed because of its unfortunate acronym to the American Sociological

*I appreciate the advice of Alan Booth.

Evolutionary Approaches in the Behavioral Sciences, Volume 8, pages 235–245.

ISBN: 0-7623-0769-2

Association). Ellwood was a progressive social reformer, yet some of his theoretical views, expressed in a popular undergraduate text, are today appalling:

> (T)he African environment of the ancestors of the present negroes in the United States deeply stamped itself upon the innate traits and tendencies of the race. For example, the tropical environment is generally unfavorable to severe bodily labor. Persons who work hard in the tropics are, in other words, apt to be eliminated by natural selection. On the other hand, nature furnishes a bountiful supply of food without much labor. Hence, the tropical environment of the negro failed to develop in him an energetic nature, but favored the survival of those naturally shiftless and lazy. Again, the extremely high death rate in Africa necessitated a correspondingly high birth rate in order that any race living there might survive; hence, nature fixed in the negro strong sexual propensities in order to secure such a high birth rate (1919: 248–249).

By the 1930s sociology and kindred disciplines were taking a more egalitarian posture. Jews entering these fields in disproportionate number disdained Social Darwinism with its implied anti-Semitism. Revulsion to Nazi racism soon stifled whatever remained of that thinking. By the 1950s sociology was cleansed of biology. Emile Durheim's (1938) earlier advice, that social facts should be explained only by other social facts, had become the ruling dictum.

By the 1960s, little was being said about disadvantaged peoples that sounded pejorative. Rather it was elites that were criticized and opposed for their subjugation of lower classes and racial minorities. During that turbulent decade, sociologists joined the battle against oppression – sometimes literally. By the 1970s, as women and gays increasingly entered the field, their rights also became primary foci.

Darwinian notions have nothing to offer post-World War II sociologists, whose ideological focus remains the alleviation of oppression. Natural selection will never have sociological appeal unless it is framed in a way that fosters the interests of the underdog.

But ideology is not the only reason that sociologists disdain Darwinism. Equally important, natural selection has failed as a description of human behavior. Consider its history since World War II.

ETHOLOGY AND ITS PROGENY

Ethology arose in Europe as a reaction to the "behaviorist" school of psychology, popular in the U.S. after World War II. Led by B. F. Skinner of Harvard University, behaviorists studied how an animal, usually a laboratory rat or pigeon, learns to obtain food rewards or to avoid electric shocks. The behaviorists' ultimate goal was to explain all of human behavior, even language acquisition, in terms of simple learning principles derived from their animal

studies. Their accomplishments were impressive, both in predicting behavior under various reward contingencies, and in training animals to do complex tasks that were unusual for the species, like getting a pigeon to shoot baskets.

Konrad Lorenz and Niko Tinbergen, the most famous ethologists, objected that such contrived laboratory tasks are irrelevant to understanding animals. They were interested in *natural* behavior, which they regarded as much a characteristic of a species as its anatomy. Each species' most important behaviors are instinctual, or "wired in," occurring automatically without learning. Shaped by natural selection, instincts are "adaptive" because they neatly fit the animal to its environment. A favorite technique of the ethologists was to rear young animals in isolation from adults, and then demonstrate that they have the normal behaviors of their species. (Sometimes untutored young produce only rudimentary behaviors, showing that instincts frequently do require some learning to develop fully.)

Ethology provided a valuable corrective to Skinnerian behaviorism, but being based on stereotypic instincts, it had little to say about humans whose behavior is plastic. By 1972, when Nobel Prizes were awarded to Lorenz and Tinbergen, ethology was losing favor because of extravagant claims by Lorenz (1966) and others – more literary or philosophical claims than scientific ones – about the instinctively aggressive or territorial nature of humans as the root of warfare. The derivative image of man as a killer ape was the basis for the opening sequence of Stanley Kubrick's classic film, *2001: A Space Oddysey*. (Mel Brooks's wonderful parody in the opening sequence of *History of the World: Part II* is more sociologically innocuous.)

With the appearance in 1975 of *Sociobiology: The New Synthesis*, by Harvard professor E. O. Wilson, the rubric of selectionist explanation changed from ethology to sociobiology. Wilson proposed that all animal and human social behavior be explained by evolutionary theory, and that ethics could be derived to boot. A pre-publication controversy over whether or not human behavior can be explained genetically (with racist and sexist implications), pitting Wilson against other prominent professors, was reported in *The New York Times*, guaranteeing *Sociobiology* widespread attention (Mazur, 1981).

Wilson, an entomologist, was influenced by new selectionist thinking about social insects. Theorists pondered why beehives and ant colonies contain only one breeding female, the queen, while the numerous female workers are sterile. The problem they posed is now well known: Why does natural selection preserve workers that do not reproduce their own kind? Possibly it was the queens that were selected for, those producing sterile workers having an advantage over queens that do not. But no, the preferred solution was "kin selection". Among social insects, workers are so closely related to the queen's other offspring that

they pass more copies of "their genes" to the next generation by helping their kin than by having their own offspring.

How do ethology and sociobiology differ? As evolutionary theorists, the ethologists had been casual and nonmathematical; sociobiology is couched in the sophisticated mathematics of kin selection and game theory. In practice, the ethologists had been less interested in the *ultimate* explanation of a behavior – why did it evolve? – than in *proximate* mechanisms, the instinctual releasers and responses that directly induce animals to feed their young or build their nests; sociobiology does the opposite, focusing on ultimate causation in the distant past, ignoring proximate mechanisms. Birds and fish are the paradigm animals for ethology, social insects for sociobiology. Another hallmark of sociobiology is anthropomorphism, speaking of "selfish genes" as if genes had motives, or of insect "altruism", a specialized term in genetics, as if it had the common meaning of *regard for the welfare of others*, which in this context it does not.

The central tenet of sociobiology is that individuals behave so as to maximize their inclusive fitness. ("Inclusive fitness" is one's genetic representation in the next generation, calculated as a weighted sum of the number of one's own children, plus the children of one's kin. Genetically speaking, leaving four nephews in the next generation is as good as leaving two children.) Behaviors that maximize inclusive fitness (and are genetically determined) are adaptive, meaning that they are preferentially passed to the next generation and eventually become characteristic of the species.

A clever theorist can make up a plausible reason why any given behavior (real or imagined) is adaptive and therefore evolved via natural selection. Once I published as satire an evolutionary explanation of suicide – a maladaptive behavior if ever there was one – by arguing that killing oneself is adaptive since it frees up resources for the use of relatives, thus enhancing one's inclusive fitness. Within the year I saw the same argument in a serious journal article (Mazur, 1979; deCatanzaro, 1980). As in Rudyard Kipling's *Just So Stories*, the most outlandish explanations can be proposed for why something – anything – evolved. With no serious way to test speculations about events in the distant past, such claims are impervious to empirical refutation. Strict environmentalists also use "just so" stories, an examplar being Jared Diamond's *Guns, Germs, and Steel* (1997), which explains that white Europeans came to dominate the world largely because Eurasia, but not Africa or the Americas, has a predominately east-west axis. (Probably if he had been writing in the year 1500, Diamond could equally well have explained why China had the most glorious civilization on earth.) This is not necessarily a problem for sociologists who readily accept untestable explanations such as are found in functionalism, symbolic interaction, and postmodernism.

Sociobiology introduced two other problems that did seriously annoy sociologists.

First is its focus on *ultimate* causes of human nature, ignoring those *proximate* mechanisms through which behavior actually operates. A well known example is the selectionist theory of sex differences in mating strategy. Males are capable of fathering offspring with no more investment of resources than an ejaculation. Mammalian females must invest a prolonged period of pregnancy and nursing. Therefore males maximize their fitness by indiscriminately spreading their seed among many females, whereas females maximize their fitness by mating selectively, devoting their limited pregnancies to the finest sires and, if feasible, withholding sexual favors until they receive from the male a commitment for child support. (This reasoning seems to many sociologists a justification for the double standard, though Darwinians deny that implication.) The theory speaks of evolution that occurred long ago, ignoring those proximate influences – our psychology and culture – that are the explanatory currency of the social sciences. Thus it cannot explain or even query why one culture is polygamous, another monogamous; why marriages in some societies are arranged by parents and in others by romantic attraction; why divorce and birth rates are sometimes high, sometimes low.

The second serious annoyance to sociologists is that some of sociobiology's central claims defy common observation. An obvious example is the assertion that we act to maximize our inclusive fitness. Most sociologists do nothing of the kind. Instead we limit our children to two or less, even have ourselves sterilized, and some of us "waste" our resources by adopting unrelated infants. In another contrary example, Robert Trivers (1974) argues that offspring may increase their personal genetic fitness by continuing to demand the services of the parents when raising a second brood would be more profitable for the parents. Therefore, according to Trivers, parents break away from children (to raise others) before children want to break away from parents. Perhaps Trivers was never father to a teenager. My own children, now in their thirties, have been trying to break the bonds of dependency for two decades while I've resisted the whole time. E. O. Wilson (1975) likens human homosexuals to sterile insect workers that maximize their inclusive fitness not by having their own offspring but by nurturing the fitness of siblings. This implies that homosexuals devote twice the resources to nieces and nephews that heterosexuals devote to their own children, an implausible assertion to anyone who has put their kids through college. Sociobiology may sound convincing for insects but not for humans.

These annoyances gave rise to a third generation of natural selectionists, calling themselves "evolutionary psychologists." The new leaders are scholars of human rather than animal behavior, hence their interest in a reformulation

that avoids absurd statements about humans. Their most important innovation has been to re-introduce proximate causes, not ethology's fixed instincts – good for fish and birds but not for humans – but the quintessential human organ, a thinking, talking brain. When sociologists of the 1950s spoke of brains, which was not often, they thought of John Locke's blank slate, upon which a particular culture and language were written during a period of socialization. Today's evolutionary psychologists have an entirely different concept, the "adapted mind," with specialized mechanisms for parenting, emotional communication, kinship, mate choice, sex, aggression, danger avoidance, mate guarding, child care, and so on (Barkow, Cosmides & Tooby, 1992: 99). Most notable is the Chomskyan mechanism for language, pre-wired though requiring a period of learning to accommodate any particular language.

Evolutionary psychologists tell the same "just so" stories as sociobiologists but emphasize that it is our minds that have evolved, not our disembodied behaviors: Evolution produced the human mind, and the mind produces human behavior. This is an ingenious corrective to sociobiology's exclusive focus on ultimate causes. With a language-speaking mind as a proximate mechanism, it is easy to incorporate learning, socialization, and cultural differences – all notions that are important to social science but excluded from insect-level sociobiology.

According to evolutionary psychology, our minds became shaped by natural selection to a "human environment of evolutionary adaptedness (EEA) – i.e. the Pleistocene environment in which the overwhelming majority of human evolution occurred" (Symons, 1992). Often this is envisioned as an African savannah, in any case very different from Chicago or New York. Humans left the EEA only 10,000 years ago, when agriculture and settled communities began, so there has been too little time for our minds to re-adapt to these newer conditions. As a result, we cope with industrial society using Stone Age brains. Behaviors that were adaptive in the EEA may not be adaptive – even maladaptive – in Chicago. For example, we possess a taste mechanism that favors sweetness, an adaptation that improved our ancestors' reproductive fitness because sugar-producing fruit is most nutritious when its sugar content is highest, but which now leads us into unhealthy eating habits.

The EEA is reminiscent of the EOS (environment of original sin), the Garden of Eden where simpler creatures were transformed into fully human beings. As scientific concepts they have similar credibility. There is little reason to assume that human capabilities developed as an integrated package during a discreet time period beginning some five million years ago, when the human lineage split from apes, and ending with the Pleistocene. Brain evolution is highly conservative, with newer structures added upon or modifying ancient structures.

Our preference for sweets evolved long before there were humans, as did unequal investment by the sexes in progeny. Even human speech has important features that emerged before apes split from monkeys (Werker & Vouloumanos, 2000). Almost certainly the human mind developed piecemeal, a mixture of some mental mechanisms that are ancient, others that are new.

Having introduced the adapted mind, evolutionary psychology is able to eliminate the untenable tenet that we act to maximize our inclusive fitness. Instead our actions are explained by the calculations of these minds, which balance, say, a diabetic's primitive urge to eat sweets with her physician's advice to abstain. Married couples, whom sociobiologists required to have as many children as possible, are allowed by evolutionary psychologists to weigh their reproductive urges against the costs of a college education or the demands of a two-career family. The male mind still desires sex with many women, but the penis may be held in check by moral strictures or caution about venereal disease or fear of discovery by a wife. The female brain still seeks long-term commitment from a dominant male, but actual behavior may depend as well on availability of birth control pills and an active bar scene.

Having solved some of sociobiology's problems, evolutionary psychology introduces a few of its own. Beside the problematic concept of an EEA is the dubiously elaborated mind, far overreaching what is known from research on the brain (which is not the same as the mind in any case). Also, evolutionary psychologists focus too narrowly on humans, ignoring our cousins, the nonhuman primates, which offer instructive models of behavior. Consider, for example, the theory of male and female mating strategies, which should apply to apes as well as humans. Male gorillas and orangutans seek multiple mates while females are usually limited to a single male, consistent with the theory. But gibbons and siamangs are monogamous. Among chimpanzees, both sexes mate with multiple partners (Goodall, 1986). Among bonobos (pygmy chimpanzees), females freely seek sexual relations with males and other females (de Waal, 1995). Thus the hypothesized sex difference is not apparent in most apes nor in those most closely related to humans, the chimpanzees, casting doubt on the theory's underlying reasoning.

Despite its shortcomings, evolutionary psychology – by introducing a sophisticated concept of the mind – has taken a long step toward reconciling natural selection with sociology. This version accommodates psychology, socialization and culture, and it eliminates absurdities in sociobiology's description of contemporary human behavior. But the ideological divide between sociology and natural selection remains wide. Social Darwinism in its various incarnations has too long a history of justifying why some people are inferior to others, or why some groups oppress others. This cauldron was stirred most recently by

the evolutionary theory of rape promoted by Thornhill and Palmer (2000; also see Shields & Shields, 1983), who claim that forced intercourse is a successful reproductive strategy for males, and that all men are potential rapists. (There is an analogous argument for the evolution of cuckolding in women.) Thornhill and Palmer condemn rape as an immoral action but at the same time suggest a behavioral apology for the rapist: we (men) are all sinners, there but for God go I. Such theoretical rationalizations for oppression blow away any appeal that evolutionary psychology might have held for sociologists.

Rape theory, by the way, suffers from the fallacy of misplaced specificity, drawing too narrow a conclusion from a broad argument. Hunger is as natural a motive for eating as libido is for coitus. Both appetites increase with the duration of time since the last consummation and both are stimulated by the presence of attractive objects of consumption. Normally, both appetites are satisfied without coercion, but either may be satisfied forcibly, by rape or by robbery of food (or of money to buy food). If one accepts the Thornhill-Palmer argument that men have psychological traits designed for the specific purpose of rape, one might as well accept that men have psychological traits designed specifically for food robbery. Indeed, the argument can be made for robbing any specific food, my favorite being cheese Danish, which I have stolen on numerous occasions (Mazur, 1992).

RECONCILIATION?

Hardly any biological scientist today doubts that evolution by natural selection is the true key to understanding the diversity of species and their adaptations to specific environments. Yet sociology is an ostrich, likely to keep its head in the sand until Darwinism is reformulated to support the perspective of the disadvantaged. It is not easy to start with "survival of the fittest" and end with a sympathetic portrayal of those who are the weakest. But it is worth a try.

One plausible path begins with the diverse meanings given to "natural selection." Some theorists regard the genetic variants that are selected during the creation of new species as normal variations within existing populations, others believe these variants to be mutants that appear only rarely. Some regard evolution as gradual and continuous, others argue that long periods of stasis are occasionally "punctuated" by short episodes of speciation (Somit & Peterson, 1989).

In all these varieties of natural selection the organism itself is passive. For Darwin, the environment – whether the broad physical environment or the limited environment of sexual competition – was analogous to a human breeder of fancy pigeons who selects the characteristics of one generation that are to

be passed on, at the same time weeding out undesired characteristics. From the organism's own perspective, it is all a matter of luck or of fate whether the environment/breeder chooses it to be fruitful or to become extinct.

Less often discussed is *active* natural selection where the organism chooses its own environment rather than being captive to it. A rare variant may discover a better life if it moves away from the usual niche of its species, perhaps slightly higher up the mountain, or into a drier zone, or into previously unused vegetation. If two of these rare variants move into the new niche at the same time, they are more likely to locate one another for reproduction than if they had stayed among their normal cousins. Their offspring might displace other occupants of the new niche, eventually becoming a unique species in its special environment. The result in this case is not due to the environment selecting variants, but to variants selecting their own environment.

Obviously the human mind is capable of selecting, even constructing, new environments, of building igloos on ice, planting and irrigating fields, assembling stations in space. Material habitats aside, cultures incorporating language and rules (norms) are our most impressive surroundings, the hallmark of human society. At least since we have had the capability of language, humans have actively selected our material and cultural environments, more than environments have selected humans.

Which humans had the power to actively construct an adaptive environment for themselves? In virtually all sexually dimorphic species, males are the stronger sex and can physically coerce females. In species with dominance hierarchies, the leading males have access to valued prerogatives that are not available to subordinate males. Physical size and strength are important contributors to dominant position, allowing maturing males to move upward in the hierarchy while aging males with ebbing strength and health move downward. Also, political alliances are well developed in monkeys and apes, with top positions in a hierarchy often held by a coalition of mutually supportive males (de Waal, 1989). For these reasons, leading males have inordinate power in primate groups.

As language and culture developed, leading men realized that they could make rules for their group, including rules to prolong their own privilege as they aged, and to protect the privileges of their children and friends. The most obvious prerogatives to be maintained were material wealth, authority, and sexual access to attractive women. The realization that these perquisites could be preserved better by cultural rules than brute strength is as inevitable as the discovery that cooked meat is better than raw, or the recognition of a spirit world in dreams of acquaintances recently dead. These simple ideas occur nearly universally in language-using groups and are passed down the generations, attaining the authority of tradition.

One of the earliest insights of anthropology is that alliances are maintained by the exchange of food, gifts or favors (Mauss, 1925). As primitive cultures developed lore on the collection or manufacture of material items, these must have become currencies for exchange to establish solidarity among leading men. Captives would serve the same purpose, with females more valued than males as slaves because they could more easily be physically controlled, and attractive women would be particularly desired for sexual pleasure. Perhaps women indigenous to the group, especially younger or lower status women, also were seen as a medium for exchange among leading men.

In any case, early human groups must have regularly incorporated rules that reinforced the privileges of leading males, ensuring them wealth, authority, and sexual access to women, and extending these privileges to their male children. Captive women or those of low status often became chattel. If not literally property, women still must have held an inferior position to men, serving at their pleasure. Thus, the selection by dominating males of an environment enforcing sexual inequality was an inevitable consequence of the emergence of language-based culture, requiring no new mating strategy beyond that common to the higher primates. (Similar ideas, independently derived, appear in Lough, [1999]).

In this active version of natural selection, the oppression of women derives from the superior physical power men could wield to influence primitive cultures, not in behavioral genes or the biologically adapted mind. Sexist traditions, once formed, passed from generation to generation, at least until the present time, when in some societies they are changing, if not yet completely.

Can this "just so" story reconcile sociology with Darwin? Perhaps it is more palatable to sociologists than to evolutionary psychologists.

REFERENCES

Barkow, J., Cosmides, L., & Tooby, J. (Eds) (1992). *The Adapted Mind*. New York: Oxford University Press.

DeCatanzaro, D. (1980). Human Suicide: A Biological Perspective. *Behavioral and Brain Sciences*, *3*, 265–290.

De Waal, F. (1989). *Chimpanzee Politics*. Baltimore: Johns Hopkins University Press.

De Waal, F. (1995). Bonobo Sex and Society. *Scientific American*, (March), 82–88.

Diamond, J. (1997). *Guns, Germs, and Steel*. New York: W.W. Norton.

Durkheim, E. (1938). *The Rules of Sociological Method*. Chicago: University of Chicago Press.

Ellwood, Charles. (1919). Sociology and Modern Social Problems, new edition. New York: American Book Company.

Goodall, J. (1986). *The Chimpanzees of Gombe*. Cambridge MA: Harvard University Press.

Lorenz, K. (1966). *On Aggression*. New York: Harcourt Brace Jovanovich.

Lough, T. (1999). Energy, Agriculture, Patriarchy and Ecocide. *Human Ecology Review*, *6* (winter), 100–111.

Mauss, M. (1925, 1967). *The Gift: Forms and Functions of Exchange in Archiac Societies*. New York: Norton.

Mazur, A. (1979). Sociobiological Theory of Suicide. *Sociological Symposium*, *27* (summer), 75–78.

Mazur, A. (1981). Media Coverage and Public Opinion on Scientific Controversies. *Journal of Communication*, *31*, 106–115.

Mazur, A. (1992). The Evolutionary Psychology of Rape and Food Robbery. *Behavioral and Brain Sciences*, *12*, 397.

Shields, W., & Shields, L. (1983). Forcible Rape: An Evolutionary Perspective. *Ethology and Sociobiology*, *4*, 115–136.

Somit, A., & Peterson, S. (Eds) (1989). *The Dynamics of Evolution*. Ithaca NY: Cornell University Press.

Symons, D. (1992). On the Use and Misuse of Darwinism in the Study of Human Behavior. In: J. Barkow et al. (Eds), *The Adapted Mind* (pp. 137–159). New York: Oxford University Press.

Thornhill, R., & Palmer, C. (2000). *A Natural History of Rape: The Biological Basis of Sexual Coercion*. Cambridge MA: MIT Press.

Trivers, R. (1974). Parent-offspring Conflict. *American Zoologist*, *14*, 249–264.

Werker, J., & Vouloumanos, A. (2000). Who's Got Rhythm? *Science*, *288* (14 April), 280–281.

Wilson, E. (1975). *Sociobiology: The New Synthesis*. Cambridge MA: Harvard University Press.

'BIOPOLITICS? NEVER HEARD OF IT': A REPORT FROM AUSTRALIA

Hiram Caton[1]

When the editors of this volume proposed an overview of biopolitical research in Australia, I replied that there is nothing to report. Intrigued, they invited me to explain why an available niche isn't occupied. Since I had given the matter some thought, I accepted on the understanding that I might broaden the scope to all research involving synthesis between social and biological sciences. I felt that to ask why a development did not occur in only one discipline provided no point of comparison. To put the same question about several fields gives you something to work with. Besides, the answer to why no biopolitics in Australia is short and sweet: it just isn't the sort of thing we do.

To indicate my observer bias, I'll describe my context. My entry point was the Australian National University in the nation's capital, Canberra. The time was 1971. I arrived as a Research Fellow in the Research School of Social Sciences, one of six research schools at the ANU (there are now eight). There is a school dedicated to the biological sciences and another to medicine. Access to their staff was easy, so I was soon talking with them about behavior. The younger scientists were keen, the senior scientists less so. Eventually I met a zoologist willing to tutor me in ethology. We formed an enduring friendship; I shall have more to say about him. Another colleague at the ANU was Derek Freeman, who would subsequently fire a shot heard around the world. I shall have more to say about him as well. In 1976 I accepted a chair at Griffith University, then a new establishment committed to the interdisciplinary modality. The flux of a new university presents many opportunities, but biopolitics proved not to be among them. In the course of three decades, I made many

Evolutionary Approaches in the Behavioral Sciences, Volume 8, pages 247–269.
2001 by Elsevier Science Ltd.
ISBN: 0-7623-0769-2

tests of the receptivity of the Australia academic milieu to biobehavioral research. It was not very receptive.

To draw on the opinion of colleagues, I devised a survey and circulated it to thirty-one persons who are in relevant fields, or who I judged to be informed on-lookers in and out of universities. In the cover letter, I stated that one diagnosis I ruled out is that the climate of opinion was forbiddingly hostile to the emergence of biobehavioral research. The responses were unstructured and respondents were invited to reply at any length, on or off the record. I received eighteen responses. I responded to my own survey and opted to keep some of my responses off the record. So from time to time I shall cite my responses without attribution. Further detail about the survey is relegated to Appendix I.

A second source of information was university websites (Appendix II). Many detail the entire curriculum, together with the research interests of staff and their contact addresses, current grants, and departmental research foci. This information was an important check on the accuracy of respondent generalisations as well as on my own impressions. The websites also revealed some trends not mentioned by respondents. I browsed the websites of nine universities, primarily to ascertain curriculum and research patterns. A further source of information was government websites. Since all but two universities (out of 42) are funded by the Commonwealth government, policies impact across the entire system in one fell swoop. Government sites provide detailed information about policy at all levels of education, about research initiatives in which government has an interest, about economic issues, and about policy on immigration, multiculturalism and ethnic affairs. These sources may accessed by readers who desire further detail about the policies I discuss.

Overview: A Thirty Year Perspective. Australian scientists put in strong performances in neuroscience, genetics, molecular bioscience, immunology, birth technology, animal behavior, and environmental science, and this list isn't exhaustive. Excellence in evolutionary biology is more a matter of individual achievement. There have been ebbs and flows of quality and changes of fashion; I don't pretend to know them. Suffice it that molecular bioscience is today high profile because of its promise of large financial return. Research funding carrots and sticks encourage scientists to orient on commercially promising research and joint ventures with industry partners.

The humanities have undergone considerable change over this period. The Foucault fashion came, conquered, and withered, leaving behind the lisping idioms of social construction. Feminism was a champ in the 1980s, but in recent years student interest seems to have dwindled. It remains prominent in curricula, but it is not the sacred cause that it was. Marxism, traditionally a force in

Australian academic and cultural life, is scarcely detectable in curricula or in public discourse. The working class has been displaced from its pre-eminent position in victim jeremiads by the disadvantaged, minorities and indigenous people. Labor politicians invoke the once talismanic phrase 'working class' only on ceremonial occasions. Multiculturalism and Aboriginal studies, which came on stream in the mid-1980s, continue to attract students and to enjoy a major presence in the public space. Funding for projects in these areas is generous.

The social sciences have also undergone a major 'structural adjustment'. Illustrative is the transformation of the Research School of Social Sciences. The presiding agenda on my arrival was political pluralism. The key departments – demography, sociology, political science, economics – selected their research to align with concerns of policy makers. Pluralism was, in retrospect, an early version of multiculturalism. In 1971, Australians considered themselves to be whites of British heritage. 'God Save the Queen' sent a tingle down the spine and the 'Royals' enjoyed high public esteem. But there was a large influx of immigrants from the Mediterranean basin in the decades after the war, and this created adjustment problems for new-comers as well as for old Australians. Also in the ambience of pluralism was coping with the consequences of rejecting the White Australia immigration policy (1966) and of granting Aborigines the right to vote (1967). Non-white immigration was miniscule in 1971, but Vietnamese, Chinese, Philippinos, and East Timorese became a larger stream from about 1975 owing to turmoil that made them refugees and asylum seekers. The remodeled RSSS tracked changes in government policy. The Eurocentric focus has been replaced by a regional focus in economics, demography, and security. The political philosophy of pluralism has been displaced by the human rights focus encompassing regional democratization processes (including, currently, Australia's role in Indonesia) and structural readjustment. The 'hard' theory is rational choice-based reconciliation of consumerism with public choice.

To round out the overview, let me touch briefly on change in two disciplines, anthropology and government. The end of colonialism was also the end of anthropology as it had been. Colonial administrators regarded anthropologists as sources of information about colonial peoples and indigenous populations. Few anthropologists went into the field without government sanction and some were commissioned to conduct government research. Anthropologists neither required the approval of their research subjects nor sought it. As whites among people of color, they enjoyed the deference that subjects give their masters. They had 'rank to burn' and could 'order people about' as the young Margaret Mead boasted of her government-conferred status in Samoa. Anthropology was shaped by the colonial influence.[2] Research subjects were conceived to be

'primitives' or 'lower races' subordinate to the superior intellectual and material firepower of whites. Just as the colonial power possessed territories, so each anthropologist possessed a tribe. It was his or her turf and poaching was discouraged. It was thus that anthropology developed into an idiosyncratic field featuring a gallery of personal investigations of the exotic customs of inaccessible savages. Decolonialization changed all this. As Europeans withdrew, migrations of indigenous peoples to Europe and North America increased. Primitives mutated to full humanhood by taking Ph.D.s and joining the ranks of anthropologists. The penetration of communication and economic ventures into formerly virgin lands eliminated the isolated tribe, which might now have more members in cities overseas than in the original habitat. Anthropology restructured by taking its cue from the new relation between European powers and developing nations – assisting in social, economic, environmental and infrastructure development. The leading edge of anthropology in Australia today is the applied regional study of social change and land management. Derek Freeman's old department in the Research School of Asian and Pacific Studies is engaged in such applied research in some thirty localities in the Asia-Pacific; ditto the traditional leader in anthropology, the University of Melbourne. Research outputs feed into the World Bank, IMF and national economic assessment.

Government. The core curriculum was administration, party politics, welfare and labor issues, political theory based on the Greats or on Marxism, and political histories of Australia and the U.K. Part of this core has been displaced and part of it has been reconfigured. Administration has been reconfigured to track government reform of the public service (beginning about 1985). Reform has brought the public service in line with corporate practice of financial accountability, contract conditions for senior appointments, and client-centered service delivery. Following in the wake of this change has been the introduction of subjects in public policy. The collapse of communism took the mickey out of political theory because nothing fundamental seemed any longer at stake. If it has a replacement, it is probably a combination of teaching on globalization, regionalism, and special issues, such as environment and development, human rights, and indigenous peoples (see Appendix II).

The Survey Bias. The survey questions implied that there was something odd or dysfunctional about the failure of biobehavioral research to develop in Australia. Nearly all respondents accepted this perspective and replied with a barrage of comment about the parochialism of our universities. But two respondents bucked the bias. One, a marine ecologist, deplored the harassment of his

policy advice by environmental activists, but he has no time for sociobiology (which he assumed provides the basis of biopolitics). The second respondent is quite certain that Australia is better off without biopolitics or any other hybrid. His response is the longest submitted. It is indeed his just published book, *Science, Myth or Magic?* The respondent is retired zoologist S. A. Barnett, my erstwhile tutor in ethology and long-time friend. For nearly four decades he has, as science publicist and author, discouraged the birth of the dreaded hybrid. Since he is the most knowledgeable and persistent exponent of negative views that appear frequently in public discussion, a summary of his position is warranted.

The first thing to note is the solidity of Barnett's credentials. He is an expert on a major pest, *Rattus norvegicus*. His first assignment as a young zoologist was a posting to Malta, during the war, to terminate bubonic plague. He has consulted to governments about control of rat populations. Thanks to this background he knows learning theory as well as the genetics and developmental biology of the most common experimental animal. When therefore he took on Konrad Lorenz, as he did decades ago, his criticism was unchallengeable by anyone who lacked scientific credibility matching his own. This made him a formidable presence to any social scientist who might stray into human ethology or sociobiology: Barnett could arrive on your patch at anytime (as a grant referee, for example) to deliver disabling blows.

The second point about Barnett is his style of self-presentation. He is the impartial, disinterested scientist examining evidence and theory on their merits. He disavows preconceptions and has no axes to grind. He is a confessed advocate only for scientific rationality against magic, superstition . . . and bogus science. He eschews any appearance of aligning with a school of thought or a partisan fight. Thus he does not identify with the Left Darwinians (Stephen Rose, S. J. Gould, Richard Lewontin, etc.) or mingle in brawls. He does not engage in vindictive blaming of the kind described by Andrew Brown in *The Darwin Wars*, although he does indulge a penchant for biting sarcasm. Yet by a remarkable pre-established harmony, his *betes noirs* are also the *betes noirs* of Left Darwinians, it is sometimes difficult to detect substantive differences in their criticisms, and he cites these authors freely.

Barnett does not claim that his position is value free. Underlying it all is his commitment to freedom, dignity, and autonomy. These undoubted human goods require commitment going beyond evidence because so many humans are deprived of them by other humans. The faith element is the conviction that if we just keep at it, an inclusive common culture, protective of the rights of each, will eventually develop. Opposition to biosocial science follows from this fundamental value position. Because it purports to discover patterns of natural

necessity amidst the choices we make, biosocial science draws limits to the horizon of hope. Thus in his current book Barnett chastises biologically inspired stereotypes of human nature as prone to selfishness, violence, mendacity, opportunistic sex, competitiveness, and exploitation. Those who accept such views of human nature are likely to abandon faith in the eventual triumph of goodness – a society in which these unlovely traits have vanished. The trouble for Barnett's position is not only that these unlovely traits are empirically quite pronounced, they also follow from Darwinian theory. Why not accept the observed facts together with the theory and abandon utopia? Must he not reject the root of sociobiology, Darwinism, if he is to cling to the hopeful vision? In his response to the survey, Barnett noted that his zoology colleagues accepted sociobiology, and he pointed out that Macfarlane Burnet, an Australian Nobelist in immunology, had taken his affirmation of it to 'absurd' lengths. Might then the most *noir* of Barnett's *betes noirs*, E. O. Wilson, be right after all?

Barnett does not follow his logic that far. But the other dissenting respondent does. T. A. Underwood, Fellow of the Australian Academy of Science, is a marine ecologist at the University of Sydney. He dismisses sociobiology as 'metaphysical mumbo-jumbo' (Barnett's view as well) but he grants that scientists who accept natural selection must either accept sociobiology or 'be hypocritical'. This is the bridge that Barnett does not cross: he clings to Darwin as talismanic (non-ocular) evidence that evolutionary thought is consistent with the highest ideals. Since Underwood is possibly the only leading Australian biologist to repudiate natural selection openly, as he did in a public lecture at the Australian Museum, let's hear what he has to say.

> I can only offer a personal view. I have always found any discussion about models or theories of natural selection to be outside science. The only science that I can defend (i.e. as a principle of demarcation) is that amenable to falsificationist test. None of the usual speculative babble unleashed by natural selectionists fits the relevant criteria. So, I really have no interest in its wider flights of fancy – particularly those that seek to interpret human behaviour in terms of genetic history. The recent issue about rape is a good example. So what if there is some genetic residue "encouraging" rape? Why should that be thought relevant to issues of social behaviour, cognizance of law, ethical dealing with other people and all the other social checks on outrageous behaviour? All the sociobiologists have done is unleash an excuse for individuals to fail to take personal responsibility for their actions. Speculation based on untestable theories is to me not different from any other sorts of speculation or mysticism. Special creation, natural selection, the person in the moon, my ancestors' strong belief in the once-and-future-King-Arthur (nynsyu marolb myghtern Arthur), invaders from outer space all have, to very different degrees, the same characteristics from the point of view of an experimental ecologist. I strongly support the late R.H. Peters' views on evolutionary theories and tautologies. So, I guess I am not helping much. There are, of course, many biologists who believe in natural selection as a useful theory and many of them would therefore have to support notions of sociobiology (or be hypocritical – which, of course, never happens, does it?).

Underwood's position is supported by Ullica Segerstråle's just published *Defenders of the Truth: the battle for science in the sociobiology debate and beyond.* This study, like Brown's *Darwin Wars*, probes the personalities, motivations, and brawlings of scientists who claim to defend the truth and nothing but the truth, but who yet take opposite positions. Brown awards the palm to the Left Darwinians. Not so Segerstrale. She shows that the sociobiological explanation of animal social behavior is now orthodoxy in zoology and related sciences. Barnett and Left Darwinians appear to have only two options. The first is to retreat into what Vincent Sarich calls 'social creationism'. Between Homo erectus and Homo sapiens a line is drawn and one declares: 'Miracle happens here'. The miracle is that Homo sapiens escapes the dominion of natural selection thanks to the evolution of mind, language, or culture. This encourages Left scientists to attack as fraudulent or misconceived *every* purported demonstration of evolved traits. The other option is to throw out the theory of natural selection as untestable mysticism. Underwood's authority for this radical rupture is ecologist R. H. Peter's *A Critique of Ecology*. This is deeply shocking to Ultra-darwinians, who tend to declare Darwinian theory to be beyond dispute ('incontrovertible'). But the falsificationist regards this as mystagogery, since all hypotheses are mortal, their infant mortality rate is high, and they all eventually perish.

The way out of this tangle, I shall subsequently suggest, is to disengage empirical evidence for biologically-based traits from the 'ultimate causation' that is thought to ground the traits. Imprinting, to take an example from ethology that Barnett accepts, is clearly adaptive. It falls into a continuum of attachment that is well confirmed in primate and human infant bonding to mother. The phenomenon is independent of the ultimate adaptive causes that are in disrepute as 'just so stories' (even though adaptiveness in this case is obvious intuitively and experimentally).

Australian Parocialism. The temper of the responses is captured in comments by a leading criminologist. He wrote: 'I actually don't know the first thing about your topic. Biopolitical science topics have never appeared in any of the [National Health and Medical Research Council], [Australian Research Council] or Criminology Research Council grant applications that I have reviewed – which probably just reflects the narrow law/sociology base of criminology and related areas in Australia'. This respondent plays a leading role in the high priority National Crime Prevention program, having chaired the team that produced the report, *Pathways to Prevention: Developmental and Early Intervention Approaches to Crime in Australia.* The Report's extensive literature review found evidence of many biological factors associated with crime, viz,

failure of early social bonding, low intelligence, impulsivity, birth injury, low birth weight, and lack of empathy. The Report's description of familial and local community aspects of high risk juveniles is recognizable as the habitat of *r* reproductive strategy. Yet this respondent, situated at the center of a high priority initiative, disclaims knowing 'the first thing' about biopolitics.

A respondent who publishes on Australian culture provided an detailed analysis of its 'shallowness'. Science education at all levels, he says, is indifferent or hostile to historical and philosophical depth. The HPS curriculum available to most tertiary students is a token acknowledgement of the need for depth, but it is not a remedy. HPS curricula are largely dominated by the social constructionist approach pioneered by the UK's Open University. By interpreting science as an expression of social institutions, constructionism cuts off access to the philosophical and scientific problems that scientific theories solve or try to solve. From this vantage point, science may be made to appear to be a capitalist instrument of domination (a theme especially marked in feminist writing) or as a humanised benign agency consistent with social progress. When the later approach prevails, it is inconceivable that 'real' science might threaten the equality premise assumed by social progress agendas. The upshot is that few among us are prepared to give sustained attention to Darwin's *really* dangerous idea – that biological variation imposes a multitude of inequalities that have consequences for social agendas.

This respondent suggested that Australian cultural life is less than the sum of its parts. There are, he believes, elite circles where thinking outside the box is normal. The circles are not only academic; they include activists, the professions, politicians, even journalists. But they are private enclaves and they do not communicate with one another or even know about the existence of other circles. Consequently, this privately sequestered thought has no public impact. The contrasting conditions, for this respondent, are the more open elite cultures of European and North American cities and universities.

Another humanities respondent seems to concur. He states however that broad-minded elites exist only in a few metropolitan areas, and only in recent years have they expanded their conversational agendas beyond merely topical questions. He points to a worrying reverse process resulting from recent structural changes in Australian universities: 'Academics no longer talk to one another. Research commitments so dominate their attention that they haven't the leisure to sustain on-going collegial conversation. That's the good news. The bad news is that serious discussion is negatively perceived because it drags people into paradigm conflicts and stressful cognitive dissonance. Can an evolutionist *enjoy* a conversation with a creationist? Academic life today is honeycombed by standoffs of this kind'.

Another respondent suggested that Australia is heavily influenced by American culture, owing to the penetration of American media. However, Australia cannot replicate the diversity of ideas in America because it lacks the population base. The result is that Australia absorbs the worst of dumbed-down American culture stereotyped thinking without the redeeming feature of diversity.

A research biologist lamented the quality of biology education. He estimates that few biologists understand evolution and natural selection, even fewer understand sociobiology, and biological sciences departments are reluctant to accept advanced courses in evolutionary theory into curricula already packed with disciplinary detail. He and another respondent state that molecular biologists are particularly at fault. (The other respondent disparages them as 'molecular jocks whose only interest outside the lab is football pools'). Biological scientists generally do not think outside the boxes defined as acceptable research topics in the peer review system of grant bodies. He writes:

> while research into hierarchies among chickens is acceptable, similar research into humans has the potential to earn one the annoyance of colleagues, especially the religious . . . who are hostile to discussions of Darwinism and human behavior. One would also face a very uncertain funding future. This would depend in no small part on how the research was sold to the wider public – e.g. [Richard] Lewontin vs [J. Phillipe] Rushton. If it were like the former, no worries; if it were like the latter, good night Vienna.
>
> As Wilson notes in *Consilience*, there are three requirements for a good scientist: obsession, knowledge and courage. Anyone doing research on race differences, for instance, would find it stressful in most departments and would need to tenured and very strong-willed. Academics here lack the courage of their American counterparts; some sort of mysterious cultural difference or entrepreneurism. Australia is a small poor country prone to political fashion.

A respondent conducting research using an ethology model for a political topic reported encountering (in the late-eighties) stubborn resistance in repeated tests of peer review. He writes: 'Although my application was given top rating by four overseas referees, the fifth, a local social psychologist, killed it. He claimed crowd behavior as his discipline's turf and wouldn't countenance poaching. I eventually obtained funding from an overseas foundation'.

If this is the situation, how does a behavior geneticist in the Hans Eysenck tradition, where human variation is fundamental, obtain funding? The answer: he doesn't. This respondent heads the Australian Twin Register. His is funded to study the genetics of cancer, a priority item in current medical funding. He piggy-backs behavior genetics research – his actual interest – on data obtained under medical grants.

There were two New Zealand respondents. Robert Blank (University of Canterbury, Christchurch) is one of many American political scientists in New

Zealand. He writes that the field is 'a very closed, conservative discipline dominated by rational choice and environmental paradigms . . . the fact that many political scientists personally are intimidated by the sciences reinforced opposition to recognition of biopolitics'. Biopolitics is perceived to be fringe, and one that has been made 'politically incorrect among the more radical elements in the discipline'. Blank states that there is no evidence of flux in these attitudes. James Flynn concurs. Flynn (philosophy, University of Otago) is the discoverer of the 'Flynn effect' and author of *Race, IQ, and Jensen*. Although he is a critic of *The Bell Curve*, he is also a critic of S.J. Gould's *Mismeasure of Man* and he supports biologically-based studies of human nature. Flynn looks for but doesn't find much in the way of interdisciplinary understanding. He advises that 'few academics want to learn anything new after they leave graduate school. And even today, only a few will be confronted with genetic explanation in graduate school'.

At the time that I circulated the survey, the *Sydney Morning Herald* front-paged a 'universities in crisis' story. Since this is a standard topic that grabs attention from time to time, one takes such reports with a pinch of salt. Nevertheless, the story, 'University Leaders Lash the System' (SMH Feb 14), helps contextualize the survey responses. It pinpoints the stress arising from the circumstance that universities are mid-way between their former wholly public status and the new privatized status. The new status requires that universities become entrepreneurial. The old status limits managerial flexibility by imposing a broad agenda of requirements. The outcome, according to the story, is the worst of both worlds-the negatives of state control and the negatives of competition. Thus, there has been a 50% increase of funded enrolments since 1990, but not a doubling of funded teaching positions. To make ends meet, universities have increased the staff/student ratio by about 40% since 1990. But this forces a reduction in the number of contact hours, which in turn forces curricular diluation. Science departments have experienced staff cuts between 7–25% in the last two years, and science deans are reported to be 'putting their heads in the oven'. The salary scale for academics has declined appreciably relative to the public service. Teaching staff are demoralized. Those who can seek employment overseas or migrate to industry; those near retirement parachute out. The remainder grind their teeth. Student response mirrors this pattern. Thirty-three percent do not complete their course despite improvements in university facilities and services. This gross and costly inefficiency the Education Minister blames on the decision fifteen years ago to smarten up the nation's competitive capacity by greatly increasing the number of persons qualified with tertiary degrees. The Minister doesn't say so, but the decision was based on the belief that graduates who elected not to attend university were

deterred by price; and that this could be changed by turning up the funding spigot. No consideration was given to possibility that students elect not to attend university because they self-assess as not being cut to the cloth of scholars. Heavy promotion of university study increased enrolments, only to have students negatively self-assess after a year or two. The outcome was that average student ability shifted downward from a standard deviation above the mean to the mean. 'The predicament,' one respondent wrote, 'would not have arisen had universities not responded to increased enrolments by lowering standards. These days a student must work hard actually to fail. When I said something like that publicly a few years ago, the roof caved in. My remarks made the front page and the Education Minister told Parliament that my views were a baseless aspersion on teachers and students. "Lowered standards" is a tabooed phrase. What a circus!'

Political Correctness. This label was devised by the New Left decades ago as the mocking appellation for the attitudes of sodden Old Left bosses. For the Old Left, all social analysis derived ultimately from class struggle, but the New Left were introducing gender, youth, and popular culture as quasi-independent elements of analysis. This the Old Left deemed to be 'incorrect'. 'Political correctness', then, is a new label for the old tension between establishment and innovators, dogma and heresy. In view of the flux in and diversity of the university environment, I designed questions to encourage respondents to pinpoint just where allegiance to old paradigms or sentiment discouraged the uptake of biosocial research.

Respondents grasped this topic with gusto. James Flynn thought that the IQ controversy in the United States had negative effects among us. 'With half their minds', he writes, 'they suspect that there ARE genetically-based racial differences that would deal a devastating blow to say black pride in their racial group. Even my position, mainly agnostic with a slight lean to black = white, does not offer the absolute reassurance they want. After all, I am willing to look at the evidence which just might go against "us"'. Flynn reports that he discerns little willingness among those espousing old paradigms to examine evidence. Only two other respondents addressed the local impact of the IQ wars in the United States. They agreed that it was negative but not highly visible because IQ is not an item on the Australian social agenda.

An American bioanthropologist who took up an appointment in a human biology department states that his work is pinched between attitudes of biologists, who think that the evolutionary perspective is 'too soft, and the arts and humanities thinking we are racist, sexist, genetic-determinist louts'. He reports that on his arrival in 1995, anthropology graduate students confided to him that

their supervisors 'warned them not to get involved with evolutionary perspectives because of the political dangers to their careers'. It doesn't get more blunt than that.

An evolutionary psychologist who returned to her old university in the United Stated after her position was made redundant began her response with an outburst: 'Oh jeez! I'm not sure what I can do that won't put me in even further hot water'. This respondent's department has long funded at least one staff member in animal psychology. The official reason for terminating her appointment was a funding short-fall, but the document rationalizing the decision made clear that teaching ethology, evolution and genetics was not considered to be an important or desirable aim of the department. Her classes were acknowledged to be well attended and her research output excellent. The department has since replaced her with a cognitive psychologist whose orientation is primate and human infant cognition. The department also supports two senior appointments in neuro-psychology. Paradigm conflicts would thus seem to be only marginally a problem. When the respondent submitted her answers, political correctness was not her primary concern. 'The problem lies in the fact that a hierarchy is built into the academic system that rewards inertia . . . full professors have so much more power and so many more perquisites than younger investigators that old, sometimes worn-out, traditional research topics get priority over newer. The "old boy network" is firmly entrenched'. The traditional research topics that command resources are topics in applied psychology that promise immediate solutions to perceived practical problems, viz, controlling drug use in prison. Her research grant applications in the field of criminology are returned with comments indicative that referees are unfamiliar with her problem area. Her students also do not receive the marks they deserve. One of them won the $1000 Aaron Beck student prize awarded by the ASCAP organization in Houston, but the winning submission attracted only a 'B' as an honors thesis.

Another respondent, an anthropologist, was definitely smacked by political correctness. He resigned his position after a protracted struggle. 'Am I bitter? You bet. I shoulda sued the bastards'. Trouble began when he introduced elements of evolutionary psychology into his first year teaching. The trouble wasn't with students. They responded so positively that he introduced a second-third year subject devoted entirely to the cross-cultural explanation of social behavior. He writes: 'Very tame stuff, really, as I included much reasoned criticism where I thought it warranted, and I tiptoed ever so softly through topics that I knew would attract political attention'. But then came the fatal conceit: enrolments for the course shot up; which meant enrolments elsewhere shot down. Enter the green-eyed monster weaving webs of intrigue: 'I was

accused of being a sexist, racist and phrenologist. What I believe to have been an organised campaign to oust me ground along, mostly behind the scenes, but resulted in a series of bogus charges being laid against me, most covertly through a letter writing campaign to the Vice-Chancellor. This brought about a series of highly publicized investigations by the V–C . . .' The documents of this case show the familiar patterns of academic intrigue. Its driver was a respected arts faculty professor who wanted the respondent out so that his unemployed anthropologist wife might be appointed in his place. And this eventually happened. There is fortunately a happy ending. The respondent accepted a golden handshake and expanded his consultancy practice in Pacific region agriculture. He is 'making a lot more money' now than as an academic.

Another respondent reports being frustrated in his attempt to secure (in the mid-eighties) a joint appointment in a new academic activity that embraced behavioral science. The respondent's appointment in a humanities faculty made it awkward for him to teach undergraduate subjects in biopolitics and to supervise graduate work in this field, although he was at the time doing both. His dean agreed that relocation was appropriate and supported the initiative. The initial proposal of joint appointments in humanities and behavioral sciences did not win favor because, he was advised, every dollar was needed to support the teaching program. The proposal was scaled down to an adjunct appointment to supervise graduate students. Again resourcing was tabled as a 'problem', since the new faculty would be required to fund those students. After months of negotiation, a carefully scoped proposition for adjunct appointment was put to a faculty board vote. It lost 11 to nil. The respondent writes: 'At no point in the discussions was it acknowledged that human ethology might have something to offer their program. Social psychologists were wary of any hint that biology might influence behavior. In your list of negative influences, I guess this classifies as the inertia of antiquated paradigms'.

Derek Freeman and Anthropology. Freeman's *Margaret Mead and Samoa: The Making and Unmaking of an Anthropological Myth* (1983) scored a king hit on a highly politicized discipline. Freeman turned to ethology and psychiatry when, about 1960, he realized that social anthropology had no theoretical base for interpreting certain peculiarities of Iban (Borneo) culture. This was regarded by his colleagues and department head as most eccentric and he was strongly discouraged. But he persisted and widened his reading to include neurobiology and, when it came on stream, sociobiology. After he was named department head, he could commit resources. Adam Kendon, a specialist in nonverbal behavior, was employed and set up a laboratory. From his strong institutional position in the Research School of Asian and Pacific Studies, one might have

expected him to build networks into anthropology departments in Sydney and Melbourne. However, on his retirement (1982), there was nothing of the kind. Kendon moved on after a few years and his lab was closed. The dream run of *Margaret Mead and Samoa* did not have a positive effect among Australian academics. The only acknowledgement of his book by organized Australian anthropology was a special issue of *Canberra Anthropology*, in two volumes, featuring a suite of criticisms and Freeman's rejoinders. There were no invitations to give public lectures. The 1988 prize-winning documentary by Sydney film producer Frank Heimans, *Margaret Mead and Samoa*, is not used in undergraduate anthropology instruction; it is dismissed as 'a valueless promotion for Freeman'. An anthropology respondent informed me that he has 'few academic friends'. He attributed this to the 'degraded state of anthropology in this country'. He noted that in internet communications, Freeman is subjected to 'bull-baiting. Australians have primitive pleasures, and bull-baiting is one of them'. Freeman's writings are not assigned reading, the respondent explains, because they 'uphold the distinction between fact and fiction. This distinction is one most Australian anthropologists spend their lives denying'. This is a harsh judgment, but it has a certain credibility. Anthropologists frequently provide expert information to courts adjudicating Aboriginal land rights claims. Pastoralists and mining industries against whom these claims are made complain that many depend on bogus and counterfeit evidence invented by lawyers, anthropologists, and Aboriginal leaders. They sometimes see land rights legislation as a government sanctioned rort of producers by bludgers. Another respondent, also an anthropologist, underscored this view. In a letter to a leading newspaper, he disclosed facts about mendacious anthropological testimony to land claims courts. The university department he identified declared that it would sue for defamation, but never carried through the threat.

The Media. Does media coverage of medical and scientific findings about behavior support or discourage confidence in biobehavioral research? My own perception was that media coverage has changed so much of late that a comparison over three decades is pointless. It was also my perception that the coverage cuts both ways, but that on balance it is positive. The leading newspapers have upgraded their biological and medical coverage because of its greatly increased scope and because of its broad interface with lifestyle issues. The online editions provide links to additional coverage. Animals, whose only appearance on television used to be as pets or farm animals, are now stars of numerous television features, from Attenborough-type documentaries to warm-and-cuddly veterinary sitcoms. The beauty, adaptedness, and social qualities of animals are frequently exposed. It is no longer insulting to compare humans with animals

– a major attitudinal change. Australia's leading producer of science shows, the ABC (Australian Broadcasting Corporation) Science Unit, interprets its brief to include encouraging public awareness of and confidence in science. The Science Unit is very aware of developments in Darwinian social science and gives them considerable exposure. The Unit covers the spectrum of opinion, from Barnett to Dawkins.

Respondent comment on the media was sparse, but what little there was is consistent with my assessment. One said that journalists – like students – are very interested in her work on the psychology of sex differences. Another respondent underscored the importance of lifestyle implications of applied medical research as synergistic with behavioral research. He writes: 'Many have been sensitized to biological parameters of behavior by 'getting into' fitness. Sports fans (9 out of 10 of us) know that East German women smashed Olympic swimming records powered by steroids. The latest lifestyle fashion is testosterone supplements to ratchet up manliness. A thousand cuts like this elevate the credibility of behavioral biology and undermine cultural determinist twaddle'. – These observations are impressionistic. I had hoped to find a content analysis study to reduce reliance on impressions, but time did not permit me to make the search.

Multiculturalism and Ethnic Politics (see Appendix III). That multiculturalism and ethnicity are of high importance in Australian culture and in day-to-day politics is implicit in what I have written. I might add that it is important in New Zealand, where the indigenous minority is the Maori. It is a common view that the center of gravity of the national self-image, sentiments, and expectations has shifted over the past decade. A one sentence description might be that Australia has aligned with the attitudes and expectations projected by the 'global culture', meaning, the media and two political portals, the White House and 10 Downing Street. We are no longer suspicious of Asians or patronizing toward Aborigines. The nation is responsive to the need to implement sustainable future environmental policies. It accepts that it may no longer retreat behind protective tariffs and must be able to compete in world markets. It emphatically embraces the IT revolution and the homogenization of attitudes and practices that go with it. These and other particulars that might be mentioned have made the Australian culture more cosmopolitan and livable. However, these contemporary attitudes are in unresolved tension with their opposites. Asian immigration illustrates. Australia, like many other nations, has tightened its vigilance on illegal entry and reduced its annual quota of immigrants. For Asians fearful of persecution or simply aspiring to a better life, Australia looks like paradise worth risking life to reach. The number of Asians in this category is likely to be millions.

But these 'boat people' must cross a rough sea. There is bipartisan agreement, with 80% public support, that Australia should not open its doors to them. Illegal arrivals with prima facie claims to refugee status are placed in remote internment camps where the desert heat is punitive. They are allowed to languish for years (to 'send a message' the Immigration Department once admitted) or else are promptly shipped out when the crisis that brought them here passes (the Kosovars). There is support from the manufacturing sector to increase the inflow of migrants, but manufacturers are opposed by those whose jobs such a policy would threaten. There is a large Asian presence, mostly of Chinese extraction, in our universities. Many enrol in IT or business courses. These students are often integrated into an existing family business registered in Australia. That is why so often these students drive flashy cars. These things are noticed by students who use public transport. Thus, while there is no official discrimination, and personal discrimination seems to be low level, Australia is likely to maintain the inflow of Asians at a modest proportion of allocated places. It would not be surprising if tensions increased as the success of Asian businesses becomes more conspicuous.

Aborigines present a different picture. The local history, briefly, is perception of Aborigines as the most primitive of all the world's 'lower races'[3]. Primitives were generally thought to be 'dying races', and this was the local perception of Aborigines. It was difficult to integrate them into the urban habitat and job routines: absenteeism, alcoholism, domestic and community violence, squalor, indifference to self-care, indifference to elementary education were the obstacles. These same traits also expressed in life on their reservations. (Aboriginals were valued as stockman, trackers, sportsmen, and soldiers). This perception was, from the earliest years of settlement, shadowed by the perception of the Aborigine as victim of brutal colonial repression, mendacity, and betrayal. From the mid-seventies, the shadow overtook the majority view and finally displaced it as a shameful, iniquitous past. Landmarks of this change were the *Native Title Act 1993*, which gave indigenous peoples limited use rights to about 40% of the total Australian land mass; and the 'stolen children' investigation and report (1997) by the Australian Human Rights and Equal Opportunity Commission. 'Stolen children' are indigenous children who were taken into protective custody and fostered in white families or orphanages between about 1900 and 1970, when the practice was discontinued. The AHREOC report, titled *Bringing Them Home*, styled this fostering 'genocide', and an apology by the Prime Minister on behalf of the Australian people was demanded. The 'stolen children' question was and remains emotional and contentious. Although indigenous peoples make up only 2% of the population, issues pertaining to

them take up about 90% of our moral space. Many Australians thought that the genocide allegation was over the top because fostering children seemed the reverse of killing. Prime Minister John Howard rejected the allegation and refused to apologize despite a media campaign to manufacture his consent. He successfully excluded the issue from the agenda of the last general election, insisting instead that the issues were economic. He was returned to office. In a separate but related vote, the 1999 referendum on the republic was soundly defeated, even though royalist sentiment has vanished. Ethnic and Aborigine groups strongly supported the republic because they viewed it as transitional from colonial to modern attitudes. The Prime Minister opposed the republic and carried 55% of the electorate with him (with 96% of eligible voters casting a ballot). These two ballots are crude measures of attitudes on multiculturalism and ethnicity. But had the vote gone the other way – particularly the republic referendum – it would have been hailed by as endorsement of multiculturalism and reconciliation.

Where might biosocial science be inserted into our ethnic policy? One insertion point is crime prevention. The objective is prevention and foci are domestic violence, drug abuse, youth suicide, and juvenile crime. Aboriginal households and juveniles are so deeply implicated that justice administration agencies are integral to Aboriginal community structure (see Appendix III). The criminologist respondent cited above says that biological variables are to some extent taken into consideration by his group's model (eg, age, gender, and race), but, he adds, 'the Leftist sociologists hate it'. Let us recall the comment of the respondent who imagined research on hierarchies: '[Its viability] would depend in no small part on how the research was sold to the wider public – e.g. [Richard] Lewontin vs. [J. Phillipe] Rushton. If it were like the former, no worries; if it were like the latter, good night Vienna'. As best I can see, Rushton's three race model and its associated personality traits fit the facts of the our Caucasian, Asian, Aborigine ethnic trichotomy. But I share the respondent's judgment that the Rushton model has no chance of emerging in our ethnic affairs public space.

Conclusion. I began by stating that the survey excluded as an answer the judgment that Darwinian social science could not emerge because of a hostile climate of opinion. I excluded that answer because it is unhelpful as a tool for detecting opportunity differences in a diverse landscape. What does this survey permit us to say, now that we have a more detailed picture?

First I would underscore the importance of government initiatives in molding elite opinion. To the surprise of many, the Labor government of the last decade slaughtered some of its sacred cows and adopted reforms originated by the

Thatcher government (but of course it didn't admit that provenance). Policy change was driven by a deep view of economic realities. This view did not emerge from the Right elite (if such a thing exists here), but from a very small group of academics and senior public servants in Treasury. It is possible that future policy changes might directly or indirectly affect the acceptability of one or another initiative in biosocial science. To illustrate – while the Australian polity is stable, it is affected by regional turbulence. (The Labor government's reforms were prompted by recognition of the implications for Australia of Asian economic dynamism). Africa could be our next regional focus. Many African countries are in dire straits and the momentum is downward despite all efforts of the World Bank and IMF. The structural readjustment models used by these agencies are based on the newly industrialized nations, which happen to be Asian. People on the ground in Africa insist that the models do not work and call for Africa-specific models. The African crisis could reach the point that 'recolonialization' may viewed as the only feasible remedy in the short term. Papua New Guinea, with which Australia has a special relationship, is in a predicament very like Africa: capital has taken flight, expatriates essential to infrastructure have departed, property and life are insecure, civil disorder is chronic, corruption is the norm, hunger affects many. Meltdown in Africa or PNG or both would probably have a major impact on Australian policy and public opinion, by challenging our commitment to humanitarian relief. Aid that we give is miniscule relative to need. Everyone knows that. But if the suffering magnified several times above the present level, something would have to give – and not only in Australia.[4]

The least hospitable disciplinary locations for the emergence of biosocial science appear to be anthropology, sociology, and government. As for government, three other persons, in addition to myself, have fallen on stony ground in their efforts to find a niche for biopolitical teaching and research. None of those three succeeded in obtaining a regular academic appointment, despite outstanding publication records. I am unable even to locate two of them. The third, Frank Salter, left Australia upon completing his Ph.D. to join the Max Planck Institute for Human Ethology. His grounding-breaking study, *Emotions in Command,* published by the Oxford University Press, would, in a different environment, be a sure ticket to a career track. But Salter sees no future in his native land because 'I am as much repulsed by the departments most likely to employ me as they are repulsed by me'. He is unwilling to teach materials that he regards as false, as he would be required to do in any government department. As for myself, I abandoned efforts to teach biobehavioral materials to buy peace with colleagues. My sole involvement in that area is an HPS course on Darwin and Darwinism – hardly biopolitics.

The most promising discipline is psychology, especially now that the University of Sydney and the ANU have opened their high profile joint initiative, The Center for the Mind. The aim of the Center is to foster synthesis efforts across the spectrum from philosophy and cognitive psychology to neuroscience and psychiatry. There are many opportunities here.

The internal dynamic of knowledge growth points to the molecular biosciences as the likely star performer in the near term. State and commonwealth governments identify this field as a major economic growth factor and have allocated substantial funds to help lift it to international commercial competitiveness. No respondent identified this field as holding important implications for biosocial science. This I found surprising. Even if these scientists are 'molecular jocks' (a wisecrack, nothing more), the institutional vision for this field's future is well in hand where it counts: at the highest levels. I have mentioned state and commonwealth government awareness. As part of the research for this paper, I spoke with senior scientists who willingly discussed their business plan. The plan's premise is recognition that molecular bioscience will impact heavily on the economy, consumer habits, lifestyle, and health care. It is believed that this technology is the first step in the control of the microevolution of many organisms, including possibly ourselves. It is the field's stock in trade that minute changes in nucleotide sequences can have dramatic effects on specific outputs of organisms or fragments of organisms. A fresh view of evolution is implicated here and it is a theory that has a multitude of consumer outcomes. The plan as described to me recognizes the need to put in place a continuing community education effort, supported by serious money, to apprise the public about what is going on. In a nutshell: the media are already awash in gene stories; public fascination with them appears to be on the increase, and concrete economic goals are the drivers. Continued denial of somatic influence on human behavior, it seems to me, has no future.

For these reasons I think that biosocial science has a future in Australia. To help promote it, I suggest observance of a few maxims of personal conduct that I distilled from my experience.

Eschew arrogance and exaggerated certainty. These blemishes are prominent among some of Darwinism's best known advocates. They boast of evolution's 'incontrovertible' truths and congratulate themselves on the triumph of science. This swagger smacks of adolescent showboating. It puts off those trying to make up their minds and it encourages antagonists to adopt the guerilla war mode where sabotage and assassination are honorable. Be civil, remember that you are mortal, and always have at the ready a couple of stories about biological humbug to conciliate your Other.

Tailor your research program to feed into recognized areas of applied research. Win collegial confidence by explaining how your work complements

theirs. Be open, and acknowledge conflicting evidence. Who knows, they might admit you to their circle.

Play the game. Don't pour energy into symbolic disputation. If you talk to the press, be careful about what you say. A single ill-considered statement can put you on the back foot. Never talk off the record because some journalists do not respect confidence.

Be aware that students and the wider public are potentially strong supports. Students respond positively to Darwinian themes when they are not presented in a threatening way.

Downplay ultimate causation and play up proximate causation. Ultimate causation is the mother's milk of evolutionary explanation, but to people not acquainted with Darwinian thinking (99.9% of the public), it often seems bizarre. The response to the Thornhill and Palmer *Natural History of Rape* is an example. In coverage I read, a journalist was aghast at the idea that scorpionfly mating anatomy could have any bearing on human rape. Natural to us, freaky to Other.

Let me finish by addressing the challenge I set for myself when I began research for this essay. The challenge was: if I could look forward to a couple of decades of university service, where would I attempt to insert biopolitics? I would beat a path to public policy in the molecular biosciences field. Public policy is legit; molecular bioscience is the coming man. Thus one could talk about multiple levels of biological causation invulnerable to attacks on the legitimacy of one's discourse. That's a new window of opportunity.

NOTES

1. Hiram Caton was formerly professor of politics and history at Griffith University and is presently a Fellow in the Key Center of Law, Ethics, Justice, and Governance at Griffith University. Email address: h.caton@gu.edu.au

2. The literature on anthropology's reinvention of itself is extensive. The anthropologist most closely identified with it is George E. Marcus, who has written or edited numerous books. The reinvention, which extended over decades, took place under the auspices of 'postmodern reflexivity' and philosophical relativism. This was the prevailing mood when Derek Freeman's *Margaret Mead* and *Samoa* appeared. The intensity of the polemic against Freeman's espousal of an evolution-based anthropology is due in part to awareness that anthropologists had collaborated with colonial and neo-colonial governments in the governance of 'primitives'.

3. The anthropological depiction of this 'savage state' is due mainly to Baldwin Spencer and F.J. Gillen, who wrote widely-read studies of the Arunda tribe and the tribes of central Australia, circa 1900. A comprehensive over-view was published in 1927 by G. H. Pitt-Rivers as *The Clash of Cultures and the Contact of Races*. The long subtitle of this book tells of the 'laws of racial adaptability' in relation to depopulation and 'the government of subject races'. Historical investigations prompted by

the reconciliation movement brought to light many examples of brutal Darwinian thinking, eg, pastoralist H. K. Rusden explained in 1876 that 'survival of the fittest means that might is right. And we thus invoke and remorselessly fulfil the inexorable law of natural selection when exterminating the inferior Australian and Maori races . . . and we appropriate their patrimony cooly'. The existence of this body of opinion is not widely known, but should biobehavioral science achieve prominence in undergraduate study, it is likely to be given wider currency.

4. After this paragraph was written, the Clinton administration declared that African AIDS is a threat to the security of the United States. This is most extraordinary. Never before has a disease been declared a security risk, and the political events forecast as the direct threat-wars and extensive disruption in Africa-do not now and will not in the foreseeable future threaten US security. This bogus threat could be the surrogate of a quite real moral threat to the integrity of 'new world order'-namely, the specter of a humanitarian catastrophe in Africa. AIDS is the scapegoat that shifts blame away from the rich north (observe that HIV is said to have originated in Africa). The security threat lays the groundwork for massive intervention by peace-keeping forces and many thousands of infrastructure cadres from rich nations to assist combating the epidemic.

APPENDIX 1

The Survey Instrument

There were eighteen respondents from a sample of thirty-one. There were also five secondary respondents, as I call persons with whom I discussed specific questions, but who did not write a response. The sample was age and gender skewed. Most fall into the 40–55 year old bracket; only one respondent was female. The survey was not circulated to anyone in the media, although a journalist was among the secondary respondents. I received no response from neurobiologists. That field is not considered in my essay. The cover letter did not define 'biopolitcs'. In lieu of a definition, I attached a list of relevant books that had received attention in the Australian media.

Digest of factors material to the non-occurrence of biopolitical research, 1970-present, on which you are invited to comment.

- *The disciplines that might have provided leadership were not favorably disposed to research paradigms incorporating concepts of biological influence on human behavior. (Anthropology / government / economics / sociology / psychology / social psychology)*
- *The existing research models used in the social and behavioral sciences were felt to be adequate*
- *The controversies about biobehavioral research had a negative impact (e.g. the sociobiology debate, the Samoa controversy, the IQ controversy)*

- *Absence of interest from science disciplines (e.g. zoology, genetics, neurobiology)*
- *Specific interests, beliefs, or paradigms were perceived to be threatened by biosocial research (e.g. indigenous peoples, gays, feminists)*
- *Institutional inertia, plus restricted research funds, discourage new initiatives*
- *Negative influence of paradigms and influences: liberal, marxist, feminist, gay, postmodern, environmental, animal liberation*
- *Negative perception of the known biobehavioral paradigms: sociobiology, ethology, evolutionary psychology, behavior genetics.*
- *Science journalism helped propagate and maintain negative perceptions of biobehavioral research*
- *In the absence of legitimation of a research area, research grant applications have little chance of success*
- *Professional associations that might have helped to initiate biosocial research did not do so (e.g. Australian Association for Human Biology, Australian Institute of Biology)*

Please comment on changes, favorable or unfavorable, in the climate of opinion over the indicated period of time.

APPENDIX II

Websites

The entry point for all Commonwealth Government websites is http://www.fed.gov.au/; for all universities http://www.detya.gov.au/highered/ausunis/htm; for the Department of Immigration and Multicultural Affairs http://www.immi.gov.au/; for Aboriginal and Torres Straits Islander Affairs, http:/www.atsic.gov.au; for Human Rights and Equal Opportunity, http://www.hreoc.gov.au; for ABC Science Unit, http://www.abc.net.au/science/.

The full report, Pathways to Prevention, is online at http://ncp.gov.au/ncp/division_3/acrobat/EI_report.pdf

APPENDIX III

Australian Demography

The population is 18.5 million, a fivefold increase over the course of a century. The current rate of increase is 1.2% per annum. 3.9 million Australians were born overseas; 3.8 million parents were born overseas. There are 92 religious

denominations and 282 languages, including 170 Aboriginal languages. By race Australia is 91% Caucasian, 6% Asian, 2% Aboriginal and Torres Strait Islander, and 1% Polynesian and other. There has been a 30% increase in the size of the Aboriginal population since 1991, when Australians were given the option of self-identifying as Aboriginal. The Aboriginal population is estimated to have been 300,000–750,000 at the time of contact.

The current migration quota is 76,000. Most of these places are allocated for family reunions and skills. Applications for migration visas exceed 1 million a year.

The peak body for indigenous peoples is the Aboriginal and Torres Strait Islander Commission. It promotes commercial ventures, land rights, culture and heritage issues, housing, legal aid and human services; its operating budget is about $1 billion per annum. Data on key social indicators show the incomplete adaptation between European and Aboriginal cultures. Aboriginal juveniles in police custody comprise 40% of the total; unemployment is 38% vs. 10.5% among non-Aborigines, and so on. This non-alignment is the source of ever-new frictions, e.g. about deaths in custody, mandatory sentencing of juvenile offenders, and land rights.

A PARADIGM IGNORED: BIOPOLITICS AND POLITICAL SCIENCE IN POLAND

Jerzy J. Wiatr

Reviewing the political science literature in Poland one comes easily to the conclusion that biopolitics is the most totally ignored approach, contrasting sharply with other new approaches, such as game theory, rational choice analysis, or mathematical models, not to mention the more traditional behavioral approaches such as survey research or content analysis. There are no references to the main writings on biopolitics in the most widely used academic textbooks in political science (Bodnar, 1984; Ryszka, 1984; Antoszewski, Herbut, 1996). The political science journals do not publish papers based on the biopolitical approach. Even in the otherwise very good and innovative collection of studies in political psychology (Skarzynska, 1999) there are no references to biopolitics and no attempt to point to the role of biological factors in shaping political behavior. Sociologists have been somehow more interested in sociobiology and Edward O. Wilson's *On Human Nature* has been published in Poland with an interesting introduction by Barbara Szacka (Wilson, 1988). The only textbooks offering some information on the development of biopolitics abroad are those which deal with political sociology (Wiatr, 1977, 1999), but even political sociologists do not conduct empirical research on biological factors of politics. When age or gender is studied they are treated as sociological rather than biological categories. Practically, biopolitics is an ignored approach in Poland.

Poland is not alone in this respect. I failed to find any reference to biopolitics in the current political science literature of those East and Central European

Evolutionary Approaches in the Behavioral Sciences, Volume 8, pages 271–275.

ISBN: 0-7623-0769-2

countries that I am able to read (that means this which is published in Slavic languages). But even in Western Europe the situation is only somehow better. The publications quoted in the special issue of the *International Political Science Review* on "Main Currents in Biopolitics" come almost exclusively from the United States, with only a few from France, Germany and the United Kingdom. Ignoring biopolitics seems to be a more common weakness in political science, not limited to any single country. In their paper presented at the World Congress of Political Science in Quebec, August 2000, the founders of biopolitics lamented the decreasing impact of biopolitics on the younger generation of American political scientists, largely due to the unfriendly attitude toward this approach in most of the political science departments (Somit & Peterson, 2000). Still, the situation in this respect seems to be better in the USA than in the other countries.

It is, nonetheless, intriguing why this particular approach encounters such strong (even if silent) opposition in political science, particularly outside the United States. Poland may be an interesting case in this respect, even if it is not an isolated case.

There are, in my opinion, four main reasons why Polish political scientists have ignored biopolitics.

The first is the simplest. Biopolitics is a new trend in political science and as such needs time to be generally accepted. Even in the United States, where it was born, biopolitics is absent in many political science departments for the very simple reason that there are yet too few political scientists familiar with and interested in this particular approach. Even more, the younger generation of American political scientists has not been persuaded to accept this approach. If such is the situation in the USA, it should not be difficult to understand why it will take time before biopolitics becomes popular everywhere. In smaller countries, like Poland, accepting scholarly innovations may take even longer.

The second reason is the structure of university education. Unlike the American students, Poles start their university education within a selected section of the university, for instance in political science, and their curriculum is built in the way which reflects the almost total separation between social and humanistic studies on the one hand, and sciences on the other. Unless a student of political science takes a special effort, he or she gets no training in natural sciences, including biology. The lack of academic background in biology is one of the obstacles to undertaking research in biopolitics and may also explain the lack of interest in this field of political science.

The third reason is ideological. Marxism that for such a long time was the dominant current in political science in Poland rejects the idea of biological conditioning of political (as well as social) behavior. The idea of a total trans-

formation of society required the creation of a wholly *new man*, and therefore had to reject the idea of constant biologically conditioned characteristics of the human nature (Berry, 1986). Seeing man as *the social being*, Marx in a radical way rejected the view that the biological nature of humans determines their social behavior believing instead in the historical character of social life (Masters, 1977: 90). Because of this, Marxist social theory has taken a distinctly anti-biological orientation.

In later years, Marxists as well as other leftist social thinkers rejected strongly the ideological foundations of such biologically oriented theories as Konrad Lorenz theory of the inherited "aggression instinct", in which they saw the ideological rationalization of Lorenz's sympathy for the Nazi regime; in 1971 I had the opportunity to participate in an international symposium at which leftist scholars from twelve countries presented their criticism of Lorenz as well as of other biologically oriented theories of war (Hollitscher, 1973)

Marxism, however, is not the only reason why there have been ideological barriers to the acceptance of the biological interpretation of politics; Political scientists take their theoretical orientation mostly from sociology, which is not only the older but also the more theoretically oriented social science. Sociological theory from its beginning emphasized the *social* character of human nature. Social learning rather than biological inheritance was seen as the source of human behavior. In early twenties, the debate between the *nativists* and the *environmentalists* contributed to the dominance of the latter among sociologists. In Poland the debate got a particularly sharp ideological coloring. The growth of racist ideologies, particularly of Nazism, which took place in the period between the two world wars, provoked Polish scholars to formulating theories that constituted a radical rejection not only of the racist ideas but of the biological interpretation of human nature and of society in general. The most characteristic in this respect was the book written by Stanislaw Ossowski between 1933 and 1938 under the title *Social Bonds and the Heritage of Blood*. It was published in 1939 but all copies perished during the bombing of Warsaw in September 1939. In 1948 the new, unchanged, edition was published becoming one of the most influential books in Polish post-war social sciences. The author, who after the second world war became the leading authority in Polish sociology, argued against the view which stressed the importance of "inborn dispositions" and emphasized the importance of the "social heritage". In a categorical way Ossowski argued, that " natural genealogy directly determines neither the paths of cultural inheritance nor the paths of biological heritage" (Ossowski, 1966: 252). Furthermore, he considered the theories that put emphasis on biological conditioning of human nature as nothing more than a form of social myths. "Today – he wrote – the very rapidity of social,

economic and cultural changes paralyses the Darwinian mechanism of natural selection. Man himself may replace Nature in improving the human race" (Ossowski, 1966: 262).

Ossowski was not alone among the most influential Polish sociologists in his rejection of the biological interpretation of human nature. Other prominent sociologists of the pre-war period – Florian Znaniecki, Tadeusz Szczurkiewicz, Jan Szczepanski – devoted considerable part of their writings to the critique of race theories. It is easy to see the connection between these academic interests and the political situation of Poland on the eve of the Nazi conquest (Wiatr, 1997).

This intellectual tradition constitutes the fourth reason for the lack of progress in biopolitics in Polish political science. Polish political scientists tend to turn to sociology when it comes to formulating the theoretical foundations of their discipline and the way in which Polish sociology has been shaped, makes a sympathetic approach to the study of biological aspects of political behavior more difficult.

This being the case, it is important to undertake a more critical review of the theoretical foundations of our understanding of society. During more than sixty years that passed since the writing of Stanislaw Ossowski's monumental study, biology made a great progress. New areas of research have been penetrated and mechanism of biological heritage is now much better known and understood than in the years when "nativists" and "environmentalists" conducted their intellectual battles. If biopolitics is to become an important new approach to political science it has to go beyond empirical case studies of some forms of political behavior and should rethink the very concept of the biological component of human nature. In this sense, the future of biopolitics is linked to the fundamental discussion on the character of social science in general.

REFERENCES

Antoszewski, A., & Herbut, R. (Eds) (1996). *Leksykon politologii*. Wroclaw: Atla 2.

Berry, C. J. (1986). *Human Nature*. Atlantic Highlands NJ: Humanities Press International.

Bodnar, Ar. (Ed.) (1984). *Nauka o polityce*. Warsaw. Panstwowe Wydawnictwo Naukowe.

Hollitscher, W. (Ed.) (1973). *Aggressionstrieb and Krieg*. Stuttgart: Deutsche Verlags-Anstalt.

Masters, R. D. (1977). Human Nature, Nature, and Political Thought. In: J. R. Pennock & J. W. Chapman (Eds), *Human Nature in Politics* (pp. 69–110). New York: New York University Press.

Ossowski, S. (1966). fez spoleczna i dziedzictwo krwi, Warszawa: Panstwowe Wydawnictwo Naukowe (Stanislaw Ossowski, Dziela Vol. II; quotations are from Ossowski's own English summary).

Ryszka, F. (1984). *Nauka o polityce: rozwazania metodologiczne*. Warsaw: Panstwowe Wydawmictwo Naukowe.

Somit, A. & Peterson, S. A. (2000). Biopolitics in the Year 2000. paper prepared for delivery at *International Political Science Association Congress*, Quebec City, August 2000.

Skarzynska, K. (Ed.) (1999). *Psychologia polityczna*. Poznan: Wydawnictwo Zysk i S-ka.

Wiatr, J. J. (1977). *Socjologia stosunkow politycznych*. Warszawa: Panstwowe Wydawnictwo Naukowe.

Wiatr, J. J. (1997). Spor o role czynnika rasowego z perspektywy polwiecza. In: S. Kozyr-Kowalski, A. Przestalski & J. Wtodarek (Eds), *Krytyka rozumu socjologicznegv*. Poznan: Wydawnictwo Zysk i S-ka, 499–504.

Wiatr, J. J. (1999). *Socjologia polityki*. Warszawa: Wydawnictwo Naukowc Scholar.

Wilson, E. O. (1988). *O naturze ludzkiej*. Warszawa: Panstwowy Instytut Wydawniczy.

THE RECEPTION OF SOCIOBIOLOGY IN GERMAN PSYCHOLOGY AND ANTHROPOLOGY

Harald A. Euler and Eckart Voland

DARK SHADOWS FROM THE RECENT PAST

To understand past German reserve about sociobiology, it may help to look back to the 19th century. Darwin's work was translated and popularized by Ernst Heinrich Haeckel, a towering academic figure of his time, still known today for his biogenetic law that ontogeny partially recapitulates phylogeny, a renowned marine biologist, a gifted illustrator, but also Wilhelminian imperialist, a nationalistic political activist, an ideologist for racism, and a self-declared social scientist (Gasman, 1971). When Haeckel popularized Darwin's theory, key concepts underwent a change in meaning: Darwin's "struggle for existence" became a "fight for existence" (*Kampf ums Dasein*), and Spencer's "survival of the fittest" became "survival of the strongest" (*Überleben des Stärkeren*). Haeckel disseminated a personal brand of Social Darwinism, interwoven with his romantic idealism, scientific positivism and materialism. Under the hands of Haeckel, Darwinism was transformed beyond recognition to a total explanation of the world. In 1899 his "World Riddles" (*Die Welträtsel*) was published with a paperback version for the price of 1 Mark. It was a bestseller of the turn of the century, the most popular book after the bible, with half a million copies sold in the German speaking countries and translated into 25 languages. Especially intellectual circles outside academia read and discussed Haeckel enthusiastically.

Evolutionary Approaches in the Behavioral Sciences, Volume 8, pages 277–286.

ISBN: 0-7623-0769-2

After the turn of the century there lived in Vienna a not-so-gifted young and poor postcard painter who either read Haeckel's "World Riddles" or was influenced by writings of obscure world explainers (Hamann, 1998) who themselves had absorbed Haeckel's grandiose theories. The later published work and utterances from this postcard painter, *Mein Kampf* and his secret table conversations, contain many similarities, even the same expressions and half-sentences found in Haeckels "Riddles", such that Haeckel can be considered a scientific forerunner of Nazi ideology (Gasman, 1971). So when Germans hear the name of Darwin, they tend to associate immediately Haeckelian Social Darwinism of the crudest kind, and the worst memories of the German past are bound to be activated: Darwin, evil doctrine, led already once to hell on earth. However, as will be shown later, this reaction does not characterize any longer the current student generation in Germany.

German psychologists first encountered Neodarwinian thought in magazines and popular science journals which reported about the sociobiology debate after publication of Wilson's "Sociobiology" (1975) and Dawkins' "The Selfish Gene" (1976, German version 1978), and in TV nature documentaries. As long as evolutionary analyses were applied to animal behavior only, German psychology did not feel touched. "Animal psychology" (*Tierpsychologie)* is not the "real" psychology, but generally considered the domain of ethologists from an altogether different faculty. Germans still define psychology anthropocentrically. For most of them, psychology is the science of man as an autonomous subject, and human cognition is targeted in a Continental rationalistic tradition (e.g. Zimmermann, 2000). Animal behavior, however, belongs to a different drawer, because Cartesian dualism is entrenched deeply.

THE PSYCHOLOGY OF EMOTION AS A PRIMER FOR THE THEORY OF EVOLUTION

In the late 70s, Marxist ideology was already waning but still widespread among German students, especially those from the social sciences, and these students were apparently more doggedly determined than those in the U.S. Sociology was the leading discipline and was conceded prerogative of interpretation for all aspects of society. Sociobiology was, therefore, met with fundamental opposition in Germany. But psychology was already primed for evolutionary thinking by evolutionary epistemology (e.g. Lorenz, 1973; Vollmer, 1975) and especially by the renaissance of emotion as a topic. The publications of leading emotion theorists, notably Ekman (e.g. Ekman, Friesen & Ellsworth, 1972, German version 1974), Izard (1977, German version 1981) and Plutchik (1980), who all considered evolution as a constituent of human emotions, were read and well

received. Unfortunately, the most influential of those three was probably Plutchik, who had ignored the Neodarwinian developments. His popular structural model of the emotions was rather scholastic and counter-evolutionary, because this model of bipolar, complementary emotions was constructed by principles of aesthetics and closure and did not reflect the open, historical principle of natural design. Fortunately, the scientific journalist Dieter E. Zimmer (1981) published a good popular introduction to the new emotion psychology which was serially reprinted in the prestigious German weekly DIE ZEIT. The handbook of emotion by Euler and Mandl (1983) finally contained a contribution about the sociobiology of emotions (Schneider, 1983). The seed for evolutionary approaches within German psychology was laid, but it still took more than a decade to germinate as can be seen by a survey of German psychology textbooks between 1975 and 2000.

EVOLUTIONARY ASPECTS IN PSYCHOLOGY TEXTBOOKS 1975–2000

Cornwell, Palmer, and Davis (2000) reviewed 229 English-language general introductory textbooks published between 1975 and 2000 for their treatment of evolutionary approaches. They registered the amount of text devoted to sociobiology or evolutionary psychology as well as tone and accuracy of treatment. During the 25-year period, amount of coverage and number of positive citations

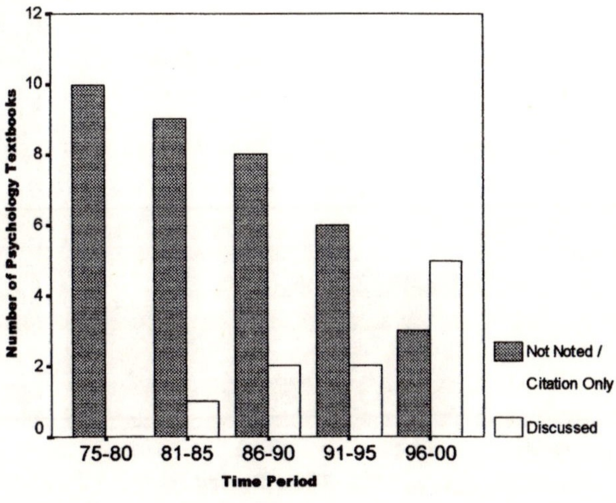

Fig. 1. Number of Psychology Textbooks which either ignore/just cite Sociobiology/Evolutionary Psychology, or discuss it, from year 1975 to 2000.

increased and number of negative citations decreased. Coverage showed an upward trend in the 90s with a dramatic upswing within the last 5 years.

The same analysis was repeated in Germany. A total of 46 psychology textbooks, excluding translations of American or English textbooks, was looked at. Our sample is smaller than the one by Cornwell et al. (2000) because the book market is smaller, and the academic textbook culture is considerably less established in Germany than in the U.S. Figure 1 shows that up to about 1995, the majority of textbooks ignore or just cite the topic of evolution, with textbooks that ignore evolution being the large majority in that category. Only in the last five years can we see more textbooks covering evolution than textbooks ignoring it. The amount of text (number of paragraphs) discussing either sociobiology or evolutionary psychology is shown in Fig. 2. Here we see what appears to be the general trend. In the wake of the early sociobiology debate in the U.S., there is some coverage in the early 80s, but no increase within the next ten years. The last 5-year-period of the century, however, reveals

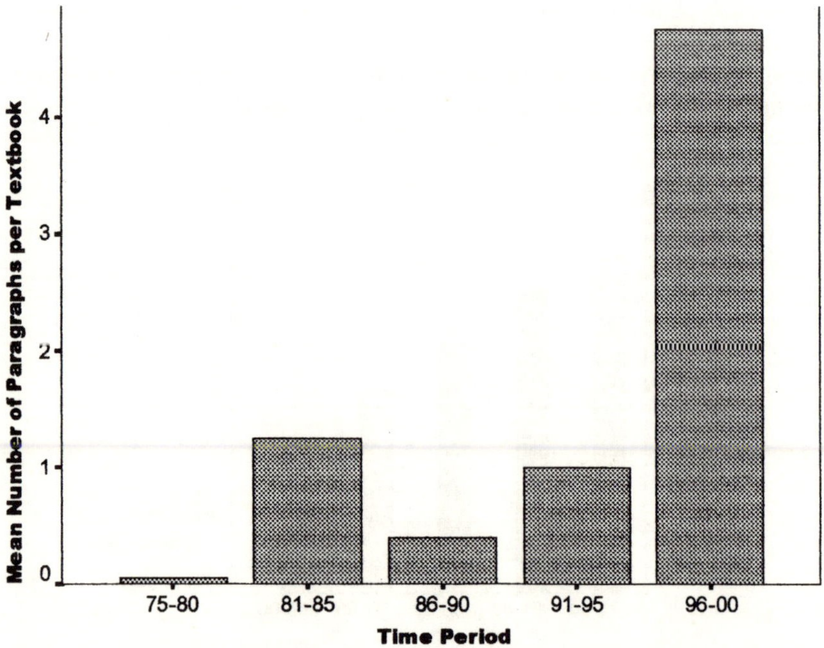

Fig. 2. Mean Number of Paragraphs per Textbook devoted to Sociobiology or Evolutionary Psychology, from year 1975 to 2000.

a clear upswing. Personal observations from the classrooms confirm that in the last years the attitudes towards evolutionary ideas have changed. The recent upswing gives hope in the light of German faculty demographics.

FACULTY DEMOGRAPHICS IN GERMANY

In the aftermath of the Sputnik shock the German educational system was revised, and higher education expanded in the 60s and 70s with a series of newly founded universities and many academic posts being established. At the end of the 70s, there were many young tenured professors at German universities, who were to remain there for quite some time. These young professors had written the Standard Social Science Model (SSSM, Tooby & Cosmides, 1992) on their banner, which had enabled them to get the cherished university posts. The SSSM could utilize the expansion of universities for its establishment. Yet, the same has happened in the U.S. to the theory of evolution over a 100 years ago. In the last quarter of the 19th century many new universities were founded there. Between 1883 and 1900, 41 psychological laboratories were set up (Benjamin, 2000). New faculty positions were available at that time for young evolutionary pychologists, like Stanley Hall and Mark Baldwin (Boakes, 1984). Experimental psychology in Germany had also grown at the end of the 19th century, but evolutionary ideas did not enter these laboratories. Why did the expansion of the university system at that time facilitate evolutionary thinking in the U.S., but not in Germany? Haeckel's version of Darwinism might be one reason, but another one might be equally important and still prevails today. The Wundtian tradition of psychological methods prescibes the experiment as the gold standard, because this tradition stemmed from the physiological study of sensory processes. The comparative method, however, which is so conducive to evolutionary insights and in which Darwin excelled, was considered impure and problematic, and it still plays at best a peripheral role in German psychology curricula today.

The theory of evolution is treated regularly and fairly in German high schools, unattacked by religious fundamentalists. Evolutionary ideas pop up on TV nature documentaries and in popular scientific journals. Most students who nowadays enroll in university psychology courses are familiar with the theory of evolution, and many are open-minded and interested in evolutionary analyses of human behavior. They like applications to everyday life phenomena more than the standard psychology staple, and many of them delve deeper into the subject matter despite little or no support and guidance from their psychology faculty. As Max Planck (1949) has noted, "a new scientific truth usually does not gain acceptance by way of opponents becoming convinced who then declare

themselves enlightened, but rather because its opponents gradually die out, and a new generation grows up that is familiar with the truth from the beginning (p. 15, transl. by authors)." The average age of faculty members is currently very high in Germany. The next expansion of the German university system, after long years of stagnation, seems to be around the corner, which gives hope for a solid establishment of evolutionary thinking in German psychology.

THE SPECIAL SITUATION OF GERMAN ANTHROPOLOGY

The case of German anthropology (physical anthropology), however, appears less favorable than psychology. The Social Darwinistic trauma is a burden for anthropology as it is for other behavioral sciences. But German post-war physical anthropology has a typical structural problem, which hampers or even paralyses any kind of innovative orientation: It is a discipline with relatively few professors, but unfortunately some of them harbored or even still harbor defiantly conservative attitudes towards the national-socialistic period. German anthropology never completely liberated itself from its Third Reich entanglement. Too many of the faculty members after the war had begun their career before the end of the war, and a politically desireable reappraising clearance might have endangered the discipline altogether. The "Ehrhardt Case" (Vogel, 1984) is one in point. The conservatism of many post-war German anthropologists has blocked any innovative openness and a sustained modernization of this discipline. Thus German anthropology became progressively confronted with the problem of its own legitimization. With the purpose and benefit of a hopelessly dusty anthropology being questioned, vacant faculty positions in anthropology were rarely filled with anthropologists, and the field entered a self-made vicious circle. In the early 1970s, there existed in West Germany about 14 academic departments where anthropology was represented by a full professor (a so-called C-4 professor). Currently (September, 2000), only one such department is left in all of Germany, both West and East. Anthropology now is being represented by associate professors (C-3 professors) who administer, so to speak, historical remains in departments with their main focus on zoology or human genetics. In the course of this shrinking process of an already small discipline, the critical threshold to keep a field innovative and progressive was reached. The import of sociobiological thought with young faculty members, which we can hope for in psychology, is not seen for anthropology. A renovation of the field through young professors has not taken place. The reason why sociobiology did not enter anthropological seminars is thus not mainly due to a Social Darwinistic disgust reflex, but perhaps more to a

historically caused lack of ressources for a thematic modern orientation of the field in general.

Additionally, there are theoretical obstacles which impeded a diffusion of sociobiology into German anthropology. Classical anthropology is rooted in typological thinking, as characterized, for example, by the research tradition of race systematics, whereas population thinking, characteristic of sociobiology, is hardly established. Also, group selection has more affinities with 'volkish' thought than individual selection or genetic selfishness. But physical anthropology is basically a biological discipline and thus originally well prepared for sociobiological thought, more so than psychology. Evolutionary and comparative thinking belong to the traditional tool kit of anthropology, and sociobiology would not have needed to tear down anti-naturalistic sentiments, as in psychology. For sociobiology not having fallen on fertile grounds in German anthropology, there must be other reasons than the intellectual unwieldiness of the theory of the selfish gene. We point again to the stagnation of the field, and we can add another relevant observation: The most prominent critics of human sociobiology in Germany, as, for example, Altner, Hemminger, and Koslowski, are not anthropologists. Anthropology was too weak to even refute sociobiology by its own efforts.

The picture of a neophobe discipline sketched here is, of course, a crude generalization. There was a German anthropologist under whose guidance a fruitful reception of sociobiological theory could flourish. Christian Vogel (1933–1994), director of the Institute for Anthropology at the University of Göttingen, headed a long-term project on the eco-ethology of Indian langurs when the discussion about infanticide, which was so important for the sociobiological paradigm change, took its start. First reports about infanticidal male langurs were quite anecdotal, appeared hardly reliable, and did not seem to stand up to a critical analysis. So Christian Vogel (1979) felt encouraged to write a pointedly critical and sceptical article on this subject. The whole empirical basis seemed to rest on just three cases reported by one and the same person, too small for a hypothesis which at that time appeared bizarre. Christian Vogel felt well equipped to discuss this deviant hypothesis critically, when in August 1982 he was about to leave for a symposium at Ithaca, NY, just before the International Primatology Congress. But immediately before flight departure he received a telegram from his then doctoral student Volker Sommer from India, with the succinct but consequential message that Volker Sommer had not only seen a case of infanticide, but even photographed it! The slides, he was told, were on the way to Ithaca. Unquestionably, the prepared talk had not only to be rewritten, but a complete scientific argument had to be turned on its head in the shortest time.

With this paradigm change Christian Vogel managed to consolidate a productive enclave within an otherwise uninterested anthropological surrounding (Voland, 1999). Illness and the death of Christian Vogel led to the end of the work group at Göttingen and, as it looks, nearly to the end of sociobiological research within academic anthropology in Germany. The previous years, however, from about 1979 to 1992, yielded a rich harvest from sociobiologically inspired work (Vogel, 2000).

However, and this is the good news, the Max-Planck Society recently founded, against the trend described above, the "Institute for Evolutionary Anthropology" in Leipzig. Because none of its directors stems from traditional German anthropology, many observers of this discipline see chances here to give anthropology in Germany a completely new profile. It has to be seen which role sociobiology will play in the research work of this institute.

Even before the foundation of this institute, there was a place for anthropological work in the Max Planck Society, namely the "Arbeitsgruppe Humanethologie" (Work Group for Human Ethology) at the "Max Planck Institut für Verhaltensphysiologie" in Seewiesen. This group was headed by Irenäus Eibl-Eibesfeldt, a student of Konrad Lorenz. Lorenz himself had rejected sociobiology curtly and uncompromisingly until his death in 1989, although he had been astute enough to recognize certain explanatory deficits of the principle of group selection (Bischof, 1991). Eibl-Eibesfeldt was more open for a discussion of sociobiological ideas, and as early as 1978 a quite extensive presentation of sociobiology can be found in the 5th edition of his "Grundriss der vergleichenden Verhaltensforschung". The basic tenor there was and still is today that sociobiology furthers the understanding of animal behavior: "There is no controversy between ethologists and sociobiologists" (Eibl-Eibelsfeldt, 1995, p. 108). A direct extension of sociobiology into the human realm, however, could not be justified. This scepticism of Eibl-Eibesfeldt is grounded in his extensive research in human ethology which seems to teach again and again that "in humans group selection is the rule and kin selection the exception" (Eibl-Eibesfeldt, 1978, p. 422). This assessment, arrived at early and continuing throughout his work, leaves an oddly undecided situation: "Both individual and group selection operate in humans" (Eibl-Eibesfeldt, 1995, p. 108). The principle of selection at the species level was discarded as the ultimate explanation of human social behavior without, however, ascribing to the "selfish gene" the explanatory power which it has in animal and human sociobiology.

Sociobiology thus encountered three hurdles in German anthropology: (1) A scepticism towards "biologistic" concepts founded in the experience with Social Darwinism, (2) the great influence of classical ethology in German

speaking countries, with its staunch adherence to the concept of group selection, and (3) a rigidly paralyzed discipline unable to innovate itself as the biggest hurdle.

REFERENCES

Benjamin, L. T. Jr. (2000). The psychological laboratory at the turn of the 20th century. *American Psychologist, 55*, 318–321.

Bischof, N. (1991). *Gescheiter als alle die Laffen – Ein Psychogramm von Konrad Lorenz* [Brighter than all these fops. A psychogram of Konrad Lorenz]. Hamburg, Germany: Rasch & Röhring.

Boakes, R. (1984). *From Darwin to Behaviourism: Psychology and the Minds of Animals*. Cambridge, England: Cambridge University Press.

Cornwell, R. E., Palmer, C. T., & Davis, H. P. (2000, June). Sociobiology and evolutionary psychology: A 25 year retrospective on change and treatment in psychology . Paper presented at *The Annual Meeting of the Human Behavior and Evolution Society*, Amherst, MA.

Dawkins, R. (1976) *The Selfish Gene*. Oxford: Oxford University Press.

Eibl-Eibesfeldt, I. (1978). *Grundriss der vergleichenden Verhaltensforschung* (5th ed.) [Outline of comparative behavioral science]. München, Germany: Piper.

Eibl-Eibesfeldt, I. (1995). Stellungnahme zu Wuketits' Manuskript: Is ethology obsolete? On some current trends in the behavioral science [Comment on Wuketits' manuscript: *Is ethology obsolete? . . .*]. *Biologisches Zentralblatt, 114*, 108–109.

Ekman, P., Friesen, W. V., & Ellsworth, P. (1972). *Emotion in the Human Face: Guidelines for Reasearch and Integration of Findings*. New York, NY: Pergamon.

Euler, H. A., & Mandl, H. (Eds) (1983). *Emotionspsychologie* [Psychology of emotion]. München, Germany: Urban & Schwarzenberg.

Gasman, D. (1971). *The Scientific Origins of National Socialisms*. London, England: Macdonald.

Hamann, B. (1998). *Hitlers Wien. Lehrjahre eines Diktators* [Hitler's Vienna. A dictator's years of apprenticeship]. München, Germany: Piper.

Izard, C. E. (1977). *Human Emotions*. New York, NY: Plenum Press.

Lorenz, K. (1973). *Die Rückseite des Spiegels*. Versuch einer Naturgeschichte menschlichen Erkennens [The back side of the mirror. A natural history of human knowledge]. München, Germany: Piper.

Planck, M. (1948). Wissenschaftliche Selbstbiographie [scientific autobiography]. Leipzig, Germany: Verlag Johann Ambrosius Barth.

Plutchik, R. (1980). *Emotion: A Psychoevolutionary Synthesis*. New York, NY: Harper & Row.

Schneider, K. (1983). Psychobiologische und soziobiologische Ansätze [Psychobiological and sociobiological approaches]. In: H. A. Euler & H. Mandl (Eds), *Emotionspsychologie* (pp. 37–45). München, Germany: Urban & Schwarzenberg.

Tooby, J., & Cosmides, L. (1992). The psychological foundations of culture. In: J. H. Barkow, L. Cosmides, & J. Tooby (Eds), *The Adapted Mind* (pp. 19–136). New York, NY: Oxford University Press.

Vogel, C. (1979). Der Hanuman-Langur (*Presbytis entellus*), ein Parade-Exempel für die theoretischen Konzepte der "Soziobiologie"? [Is the Hanuman langur (*Presbytis entellus*) an outstanding example for the theoretical concepts of "sociobiology"?]. *Verhandlungen der Deutschen Zoologischen Gesellschaft, 1979*, 73–89.

Vogel, C. (1984). Ethische Überlegungen zur Anthropologie und Ethologie [Ethical reflections on anthropology and ethology]. In: M.-P.-Gesellschaft (Ed.), *Verantwortung und Ethik in der Wissenschaft* (pp. 115–136). München, Germany: Max-Planck-Gesellschaft.

Vogel, C. (2000). *Anthropologische Spuren – Zur Natur des Menschen*, V. Sommer (Ed.). [Anthropological traces – On human nature]. Stuttgart, Germany: Hirzel.

Voland, E. (1999). Christian Vogel 16. September 1933 – 2. Dezember 1994. In: *Jahrbuch der Akademie der Wissenschaft zu Göttingen 1997* (pp. 316–321). Göttingen, Germany: Vandenhoeck & Ruprecht.

Vollmer, G. (1975). *Evolutionäre Erkenntnistheorie: angeborene Erkenntnisstrukturen im Kontext von Biologie, Psychologie, Linguistik, Philosophie und Wissenschaftstheorie* [Evolutionary epistemology: Inborn knowledge structures in the context of biology, psychology, linguistics, philosophy, and theory of science]. Stuttgart, Germany: Hirzel.

Wilson, E. O. (1975). *Sociobiology*. Cambridge, MA: Harvard University Press.

Zimmer, D. E. (1980). *Die Vernunft der Gefühle* [The reason of feelings]. München, Germany: Piper.

Zimmermann, T. (2000, March). Einseitiger Blick auf den Menschen [One-sided view at man]. *Psychologie Heute*, *27*, 16.

PERSONAL OBSERVATIONS ON THE SOCIOBIOLOGY CONTROVERSY (1975–2000)[1]

Edward O. Wilson

At the time of the publication of *Sociobiology: The New Synthesis* in the summer of 1975 and onwards, I had no doubt about its intellectual soundness. It fitted a lot of previously discordant information together and was consilient with the rest of biology. It soon reoriented thinking about the study of animal behavior and in 1989 was voted by members of the Animal Behaviour Society the most influential book on the subject of all time. I was pretty sure it would also succeed for the same reason in human behavior, because the core of the opposition that erupted – in anglophone countries at least – was political, uninformed, and frivolous. I felt that if the several Harvard professors who led the attack in their Science for the People group could do no better than their public announcements, ultimate success was assured. I did worry a lot, however, about the damage to my personal reputation caused by the ferocity and the dishonesty of the attacks.

In fact, the opposition to human sociobiology, or evolutionary psychology as it is now often called, was never as intense as the media made out at the time. Several years ago I assayed the 200 or so books published on the subject from 1975 to 1995. They run 20 to 1 predominantly in favor of human sociobiology. In the past five years there has been a veritable flood of new articles and journals, even more predominantly in favor. The most prominent holdout by

1 Text of a talk given at the joint meeting of the European Sociobiological Society and Association for Politics and the Life Sciences, 31 August 2000.

Evolutionary Approaches in the Behavioral Sciences, Volume 8, pages 287–290.
2001 by Elsevier Science Ltd.
ISBN: 0-7623-0769-2

far has been the prodigiously prolific author Stephen Jay Gould, but his creditability as an evolutionary biologist is nowadays diminishing rapidly, at least among evolutionary biologists. Also, the Marxism to which he is devoted has retreated, as one wag has put it, a bit unfairly, to Departments of English where it hopes to die a comfortable death.

Over the years ideological critics such as Gould and his close colleague Richard C. Lewontin have drawn on a half dozen more or less legitimate concerns to erect a barricade against sociobiology, or to mix metaphors, to strangle it in its crib. These are the following:

- The distaste for the notion of human beings as an animals species, however exalted.
- A fear of genetic determinism and the seeming challenge to free will, even though, despite the claims of the critics, sociobiology was always based on interactionist models with no serious challenge to the concept of free will.
- There was also an understandable fear of the support that genetic interpretations might give to sexism, racism, and mandatory eugenics. This concern has faded almost to zero.
- Finally, and far from least, the biological view of human instinct implicit in sociobiology happens to be profoundly at variance with the Marxist view of the human mind and of historical materialism.

The core of the human sociobiology program today is the consilience of neuroscience, human genetics, and cognitive psychology to give account of the physical basis of mind, combined with the theory of evolution by natural selection to trace the origin of the biological basis of social behavior.

This is a research program. It constitutes a discipline, and works through testable hypotheses. It is not a top-down doctrine, as portrayed by its chief critics. It could in some way be foundationally wrong, but with each passing year that outcome seems less and less likely.

What is the opposing view, as propounded by the leading critics, namely Gould and Lewontin in this country and Steven Rose in England? They have never crystallized it except as an implied default position with reference to sociobiology. Discredit sociobiology with an air of moral and philosophical superiority, the approach has been, and their ideas (whatever they are) will ooze into the vacuum as a kind of ideational ectoplasm. In *Defenders of the Truth* (Oxford University Press, 2000) Ullica Segerstråle has made a heroic effort to explain what Lewontin and Gould in particular meant to say, and to give them some semblance of intellectual parity. I've puzzled over it a bit too, and here briefly is what appears to be what they intended.

They wanted to take science entirely out of human social behavior and social institutions. They proposed what might be called Marxian holism. They conceived of a higher epistemology than conventional science, especially for the human condition, one compounded from scientific truth and ideological truth. Ideology can then be given freer reign, and its experts would be empowered to pass judgment on ordinary scientific approaches to human social behavior and institutions. If this interpretation is correct it helps to explain why Lewontin, with Rose and Leon Kamin, have written that the purpose of critical science is to promote socialism, which, they say, is the only just form of government. It also explains why Lewontin has claimed that nothing written by Marx, Lenin, and Mao can be refuted by science. The last previous person who recommended such an advance in scientific thinking, to the best of my recollection, was Trofim Denisovich Lysenko in 1948. This holistic interpretation may also explain why Gould believes the human mind arose by accident, not evolution by natural selection, and that science should leave spiritual and moral leadership to religion-presumably to include the purer forms of the correct political ideology.

So where is this controversy today? Perhaps the issue can be best judged by the outcome. I'll now close with a series of observations in absolute terms. I've passed them informally by several knowledgeable colleagues, who have not found reason to consider them less than absolute. Nevertheless, I would of course welcome exceptions and corrections.

- There are now hundreds, possibly thousands of natural and social scientists working in human sociobiology (a.k.a. evolutionary psychology); many are in professional organizations such as the Human Behavior and Evolution Society and the European Sociobiological Society. Virtually all are pro-sociobiology. Only a handful of scientists, perhaps a half dozen, remain actively contra-sociobiology.
- There are four refereed technical journals devoted to human sociobiology.
- None of pro-sociobiologists professes a political agenda in their research and writing. All the half dozen or so contras do profess such an agenda.
- The private political views of the pros range from left to right. Those of the contras, which are very public, are all on the far left.
- All younger people entering the arena, and they number in the hundreds, are pro-sociobiology. All of the contras, in contrast, are aging – the youngest I believe is Gould (b. 1941) – and no young scientists in the field are joining them. They are not handing the torch to the next generation. They are passing it around within a small circle of senior citizens.

- The contras have contributed not a single original idea to the understanding of human mental and social development, not even one testable hypothesis, in defense of their position.
- More than 90% of media reports on the subject, including those in *Nature* and *Science*, are pro. The contras are usually featured only when the focus of the article is on them.
- The controversy is kept alive solely as an occasional media story featuring the surviving contras.

POSTLUDE

Albert Somit and Steven A. Peterson

Many readers who believe as we do, that a "more evolutionary approach" would significantly enhance our ability to explain (or even predict) human behavior, may well be both troubled and disappointed by many of the foregoing pages. Although it has established a foothold in practically all of the social and behavioral sciences, neo-Darwinian evolutionary theory still remains very much a minority perspective. No less perturbing, in many parts of the world, as reported above for Australia, Germany, and Central Europe, it has almost literally made no headway at all. Other work suggests that we can add France to those countries where evolutionary theory has had difficulty gaining a hearing (Jaisson, 2000). In addition, there are very few adherents in South America, Asia, and Africa.

No doubt, the first thought (or hope), which comes to mind, is that our contributors are being overly pessimistic. Possible – but not very likely. We say this for two reasons. One is that all of them are convinced Darwinians and the enthusiasts are notoriously prone to exaggerate, rather than underestimate, their successes. The other is that in several instances we have discussed these assessments with other knowledgeable scholars of the same disciplines and, in most instances, the verdict has been corroborated.

Were those of us who had hoped for a more sweeping acceptance simply expecting too much, too soon? That, too, is possible. As our contributors repeatedly emphasize, a biobehavioral perspective requires drastic changes in the long-dominant Standard Social Science Model (SSSM) that holds that human behavior is learned and that genetic influences can and should be ignored. So radical a change, surely, will take time. As Pinker notes (1997: 45):

Evolutionary Approaches in the Behavioral Sciences, Volume 8, pages 291–293.
2001 by Elsevier Science Ltd.
ISBN: 0-7623-0769-2

> The SSSM not only has become an intellectual orthodoxy, but also has acquired a moral authority. When sociobiologists first began to challenge it, they met with a ferocity that is unusually even by the standards of academic invective.

True enough, if not altogether reassuring. The attempt to introduce neo-Darwinism in the social and behavioral sciences now dates back at least four decades (see, for instance, Degler, 1991) and there are still very few disciplines with programs offering graduate programs in that approach. Given these offsetting considerations, the question of whether biobehavioralism has made "satisfactory" progress over a near half-century may be another case of the classic "is the glass half-full or half-empty" issue.

In a baker's dozen chapters, our chapter authors have described the historical and, for want of a better term, what we will call paradigmatic obstacles that have so far effectively impeded the acceptance of an evolutionary outlook in their respective fields (and geographic areas). A few of these factors may be unique to a given discipline or region but, as the reader has undoubtedly noted, most of them tend to cut across practically all of the social and behavioral sciences. The regularity with which they recur testifies, we think it fair to say, to the essentially common intellectual outlook of the great majority of practitioners in this family of sciences.

The obvious question, then, is what basic changes in this pervasive mindset will be required for evolutionary thinking to become widely accepted and, even better, widely applied? We can identify at least three:

1. That evolution does occur – and that the so-called neo-Darwinian synthesis, with its emphasis on natural selection as the driving force, satisfactorily accounts for that process. As the media constantly remind us, perhaps half of the American people still believe in some form of creationism; it is not totally unthinkable that a significant number of social and behavioral scientists feel likewise – as would their students.
2. That human behavior, just as that of other species, is significantly influenced by our genetic legacy. Many of those who accept evolutionary thinking in principle steadfastly deny that it also applies to *Homo sapiens*. This is hardly surprising, since the acceptance of this proposition could have profound and far-reaching personal religious and philosophical consequences. As Edward Gibbon remarked, there are many (1953: 8) ". . . who are afraid to deduce, even from principles they acknowledge, conclusions they know to be logically exact."
3. That our evolutionary history and genetic legacy have not only played a role in shaping our social and political behavior and institutions in the past but may also impose constraints on the extent and rapidity to which those behaviors and institutions can be changed.

As of this writing, we have probably reached, or are close to, majority acceptance of the first proposition. But we have not yet approached the threshold of the second (although Edward Wilson's essay suggests that movement in this direction is occurring) – and are likely an even greater distance from the third. In the long run, no doubt, these changes will come about; still, as Keynes' aphorism reminds us, long-run occurrences are not always comforting. Understandably, many of those who launched and have led the biobehavioral movement may wonder if they, like Moses, will see from afar, but not live to enter if possible the Promised Land.

REFERENCES

Degler, C. (1991). *In Search of Human Nature*. Oxford: Oxford University Press.

Gibbon, E. (1953). Essay on the study of literature. In: D. A. Saunders (Ed), *The Portable Gibbon*. New York: Viking.

Jaisson, P. (2000). Perception of sociobiology in Lamarck's country. Presented at *Association for Politics and the Life Sciences* meeting, Washington, D. C.

Pinker, S. (1997). *How the Mind Works*. New York: Norton.

ABOUT THE CONTRIBUTORS

Hiram Caton. Was Professor of Politics and History at Griffith University (Australia) prior to retirement. Now a Fellow of the Key Center for Ethics, Law, Justice, and Governance at Griffith University. Principal publications listed at amazon.com.

Lee Cronk. Associate Professor of Anthropology and a member of the Center for Human Evolutionary Studies at Rutgers University. Received his Ph.D. from Northwestern University. Has conducted fieldwork among the Mukogodo of Kenya and on the Honduran island of Utila.

Carl Degler. Received his Ph.D. in 1952 in History. Has taught at Vassar College and served as Byrne Professor of American History at Stanford University. Has authored books on Southern history, race relations, women's history, and evolutionary theory (*In Search of Human Nature*, 1991).

Arthur T. Denzau. His research has focused on public sector resource allocation. While teaching History of Economic Thought, he became interested in Evolutionary Epistemology, this leading him to explore biological ideas in the field of economics.

Harald A. Euler. Received his B.A. in Psychology at the University of Bonn in 1965 and as a Fulbright Scholar his Ph.D. at Washington State University in 1972. Since 1974 he is Professor of Psychology at the University of Kassel. His current research interests include the evolutionary psychology of family relations, human sperm competition, and the therapy of stuttering.

Vincent S. E. Falger. Senior Lecturer in International Relations at the University of Utrecht. Research interests cover evolutionary approaches to human nature, international politics, in-group-out-group phenomena, and the abuse of biological theories for political purposes. He is active in the European Sociobiological Society and the Association for Politics and the Life Sciences.

Maryanne Fisher. M. Sc. From McMaster University in Hamilton, Ontario. Recipient of Natural Sciences and Engineering Research Council of Canada Fellowship. Pursuing Ph.D.. at York University. Has presented and published research on evolutionary psychology.

Russell Gardner. Has served as Professor and Chair of Psychiatry and Behavuioral Sciences at the University of North Dakota; has served as Harry K. Davis Professor at the University of Texas at Galveston. Initiated the *Across-Species Comparisons and Psychopathology (ASCAP) Newsletter*. Currently serves as President of the Neuropsychiatry and Social Brain Institute at the University of Wisconsin at Madison.

Margaret Gruter. Doctor of Jurisprudence from the University of Heidelberg and J.S.M. from Stanford University Law School. President and Executive Director, Gruter Institute for Law and Behavioral Research. Has taught at Stanford University and Heidelberg University. Her research interest focuses on the interaction between law and biologically based human behavior.

Jean Lachapelle. In the Department of Humanities at Champlain College in St-Lambert, Quebec. Graduated from the University of Guelph with a Ph.D.. in 1999. His dissertation was on "Cultural Evolution and the Nature of Evolutionary Explanations."

Joseph Losco. Ph.D.. from Temple University. Teaches Political Science at Ball State University. Has published in political theory, biopolitics, and public policy.

Allan Mazur. Professor of Public Affairs in the Maxwell School at Syracuse University. B.S. in Physics (Illinois Institute of Technology), M.S. in Engineering (UCLA), and Ph.D.. is Sociology (Johns Hopkins). Fellow of the American Association for the Advancement of Science. Has authored many works in sociology, including research on evolutionary theory and its relationship to social phenomena.

Monika Gruter Morhenn. Received a J.D. from Georgetown University. Associate in law firm of Wilson Sonsini Goodrich & Rosato. Secretary, Board of Directors, Gruter Institute for Law and Behavioral Research. Has published in the area of evolution and the law.

Steven A. Peterson. Professor of Politics and Public Affairs and Director of the School of Public Affairs at Penn State Harrisburg. Active in Research Committee # 12 (Biology and Politics) of the International Political Science Association as well as the Association for Politics and the Life Sciences. Has published extensively in biopolitics.

Michael Ruse. Teaches in the Department of Philosophy at Florida State University. A prominent historian and philosopher of science, whose well-

known books include: *The Darwinian Revolution: Red in Tooth and Claw*; *Monad to Man: The Concept of Progress in Evolutionary Biology*; *Mystery of Mysteries: Is Evolution a Social Construction?*

Irwin Silverman. Received Ph.D.. in Psychology from the University of Rochester. Has taught at the State University of New York at Buffalo, the University of Florida, and York University in Toronto. Fellow of the Canadian Psychological Association. Has published widely over the past two decades in the area of evolution and human behavior.

Albert Somit. Distinguished Service Professor Emeritus at Southern Illinois University. Served as Executive Vice-President at the State University of New York at Buffalo and President of Southern Illinois University. A pioneer in biopolitics and the founder of Research Committee # 12 of the International Political Science Association. Has published extensively in biopolitics, including his book (co-authored with Steven Peterson), *Darwinism, Dominance, and Democracy*.

Eckart Voland. Studied biological anthropology and social sciences. Currently he is professor of bio-philosophy at the University of Giessen (Germany). His main research interests focus on the evolution of human social behavior (i.e. human sociobiology and behavioral ecology). His empirical research is on reproductive strategies of premodern populations, combining historical demography methods with Darwinian theory.

John C. Wahlke. Emeritus Professor of Political Science at the University of Arizona. Past-President of the American Political Science Association. Has authored research in the legislative process and biology and politics. An early pioneer in biopolitics.

Jerzy Wiatr. Professor of Political Sociology, Warsaw University. Has served as President of the Polish Political Science Association and Vice President of the International Political Science Association and President of the Central European Political Science Association. He served as a member of the Polish Parliament and was Minister of Education.

Edward O. Wilson. Pellegrino University Research Professor and Honorary Curator in Entomology at Harvard University. Born in Birmingham, Alabama, in 1929, he graduated from the University of Alabama, and in 1955 earned his Ph.D. at Harvard. His research interests have focused on the biology of ants, from which base he has ranged broadly into diverse areas of evolutionary

biology. He is best known for creation of the theory of island biogeography, as well as research on chemical communication in animals, sociobiology, and biodiversity. Among the more influential of his more than 20 books are: *The Theory of Island Biogeography* (1967, with Robert H. MacArthur), *The Insect Societies* (1971), *Sociobiology: The New Synthesis* (1975), *The Diversity of Life* (1992), *Naturalist* (1994), and *Consilience: The Unity of Knowledge* (1998).

Paul Zak. Assistant Professor in the School of Politics and Economics at Claremont Graduate University. Ph.D.. in Economics from the University of Pennsylvania. Research focus has been on co-evolution of populations, institutions, and economic performance.